Springer Series in Geomechanics and Geoengineering

Series Editor

Wei Wu, Universität für Bodenkultur, Vienna, Austria

Geomechanics deals with the application of the principle of mechanics to geomaterials including experimental, analytical and numerical investigations into the mechanical, physical, hydraulic and thermal properties of geomaterials as multiphase media. Geoengineering covers a wide range of engineering disciplines related to geomaterials from traditional to emerging areas.

The objective of the book series is to publish monographs, handbooks, workshop proceedings and textbooks. The book series is intended to cover both the state-of-the-art and the recent developments in geomechanics and geoengineering. Besides researchers, the series provides valuable references for engineering practitioners and graduate students.

** Now indexed by SCOPUS, EI and Springerlink**

More information about this series at http://www.springer.com/series/8069

Jian-Min Zhang · Limin Zhang · Rui Wang
Editors

Dam Breach Modelling and Risk Disposal

Proceedings of the First International Conference on Embankment Dams (ICED 2020)

Springer

Editors
Jian-Min Zhang
Department of Hydraulic Engineering
Tsinghua University
Beijing, China

Rui Wang
Department of Hydraulic Engineering
Tsinghua University
Beijing, China

Limin Zhang
Department of Civil
and Environmental Engineering
Hong Kong University of Science
and Technology
Kowloon, Hong Kong

ISSN 1866-8755   ISSN 1866-8763   (electronic)
Springer Series in Geomechanics and Geoengineering
ISBN 978-3-030-46350-2   ISBN 978-3-030-46351-9   (eBook)
https://doi.org/10.1007/978-3-030-46351-9

This Springer imprint is published by the registered company Springer Nature Switzerland AG
The registered company address is: Gewerbestrasse 11, 6330 Cham, Switzerland

Preface

The first International Conference on Embankment Dams (ICED 2020) was the inaugural conference of Technical Committee TC210 on Embankment Dams (Chair, Limin Zhang) of the International Society of Soil Mechanics and Geotechnical Engineering (ISSMGE). It was organised by Tsinghua University and the Hong Kong University of Science and Technology under the auspices of TC210. The conference was also supported by the Chinese Institution of Soil Mechanics and Geotechnical Engineering, Chinese National Committee on Large Dams, China Renewable Energy Engineering Institute, China Institute of Water Resources and Hydropower Research, TC304 on Engineering Practice of Risk Assessment and Management of ISSMGE, and the Risk Assessment and Management Committee of the American Society of Civil Engineers.

ISSMGE's Technical Committee 210 on Embankment Dams consists of over 40 leading experts on embankment dams from around the world. The committee aims to promote co-operation and exchange of information concerning research and developments in geotechnical issues of dams; develop guidelines and bulletins for the design, construction, and safe operation of embankment dams; assist with technical programs of international and regional conferences organised by the ISSMGE; and interact with industry and overlapping organizations working in areas related to TC210's specialist areas.

The past few years witnessed several major dam failures: The Gongo Soco tailing dam failure on 25 January 2019 in Brazil that claimed the lives of nearly 300 people; the Mount Polley open-pit copper and gold mine tailing dam failure in British Columbia, Canada on 4 August 2014, which released 10 million cubic metres of water and 4.5 million cubic metres of slurry; the failure of the Xe-Pian Xe-Namnoy dam in Laos on 23 July 2018 that affected both Laos and Cambodia and killed 39 people; the Jinsha River landslide dam failures in October and November 2018 in China, which endangered a series of cascade reservoirs along the river with a peak flow rate of 33,900 m^3/s; and the Yarlung Tsangbo Grand Canyon landslide failure in October 2018, which affected China, India, and Bangladesh. In the light of the occurrence of these major dam failure incidents, the theme of this inaugural TC210 conference was "Dam Breach Modelling and Risk

Disposal". Recent developments in research and practice in embankment dam safety and risk management are presented and discussed, stimulating fruitful scientific and technical interactions among the fields of soil mechanics, engineering geology, hydrology, fluid mechanics, structural and infrastructural engineering, and social sciences.

The proceedings of the ICED 2020 include eight keynote papers, three ISSMGE Bright Spark lecture papers, one invited paper, and 30 accepted technical papers from 12 countries and regions. Each accepted paper in the conference proceedings was subject to review by two peers. These papers cover five themes: (1) case histories of failure of embankment dams and landslide dams; (2) dam failure process modelling; (3) soil mechanics for embankment dams; (4) dam risk assessment and management; (5) monitoring, early warning, and emergency response.

One of the highlights of this conference was the ISSMGE TC210 International Workshop on Prediction of Jinsha River Dam Breaching and Flood Routing, coordinated by Prof. Zuyu Chen of China Institute of Water Resources and Hydropower Research. Five graduate student teams were invited to predict the breaching process of two recent landslide dams on the Jinsha River and the flood routing along the river and to compete for the "Best Prediction Team Award" from the ISSMGE TC210. The workshop aimed to synthesise the state of the art of dam beaching and flood routing analysis, evaluate the capabilities of various analysis tools based on a common ground of the Jinsha River dam breaching case, and suggest new topics for future studies.

The credit for the proceedings goes to the authors and reviewers. The publication of the proceedings was financially supported by the National Key R&D Program of China (Grant No. 2018YFC1508601).

February 2020   Jian-Min Zhang
                                                                    Limin Zhang
                             Co-chairmen of the Regional Organising Committee

Acknowledgements

Manuscript Reviewers

The editors are grateful to the following people who helped to review the manuscripts and hence assisted in improving the overall technical standard and presentation of the papers in these proceedings:

Berghe James Burr
Feyza Çinicioğlu
Yifei Cui
Didiek Djarwadi
Gilson Gitirana Jr.
Behrooz Ghahreman
Koen Haelterman
Jing Hu
Duruo Huang
Wenjun Lu
Luca Pagano
Meysam Safavian
Chin Kok Toh
Jean-Francois Vanden Berghe
Rui Wang
Limin Zhang

Organisation

Regional Organising Committee

Co-chairs of the Committee and Conference

Jian-Min Zhang
Limin Zhang

Secretary

Rui Wang

Members

Xiaohu Du
Jing Hu
Duruo Huang
Wenjun Lu
Fuqiang Wang
Bisheng Wu
Wenjie Xu
Jianhong Zhang
Xuedong Zhang
Zitao Zhang
Cuiying Zheng
Xingbo Zhou

International Advisory Committees

ISSMGE-TC210 Members

Limin Zhang (Chair)
Rui Wang (Secretary)
Zuyu Chen
Malcolm Eddleston
Kaare Hoeg
Jean-Pierre Tournier
Angelo Amorosi
Jean-François Vanden Berghe
James Burr
Laura Caldeira
Eduardo Oscar Capdevila
Daniele Cazzuffi
Feyza Çinicioğlu
Fernando Delgado
Didiek Djarwadi
George Dounias
Zheng-yi Feng
Jean-Jacques Fry
S. R. Gandhi
Behrooz Ghahremannejad
Gilson Gitirana Jr.
Koen Haelterman

Duruo Huang
Hendra Jitno
Nikolaos Klimis
Jörg Klompmaker
Sven Knutsson
Siavash Litkoohi
João Marcelino
Bernhard Odenwald
Luca Pagano
Miguel Pando
Krzysztof Parylak
Paolo Pitasi
Daniel Pradel
Alberto Sayão
Mahendra Singh
Zdzisław Skutnik
Chin Kok Toh
Deniz Ulgen
José María Villarroel
Nihal Vitharana
Yoshikazu Yamaguchi

Contents

ISSMGE Bright Spark Lectures and Invited Lectures

Case Histories of Failure of Embankment Dams and Landslide Dams

Dam Failure Process Modelling

Keynote Lectures

Improvement to the Analytical Method for Dam Breach Flood Evaluation

Zuyu Chen[1(✉)], Lin Wang[2], Xingbo Zhou[3], and Shujing Chen[4]

[1] Department of Geotechnical Engineering, China Institute of Water Resources
and Hydropower Research, Beijing 100048, China
chenzuyu@cashq.ac.cn
[2] College of Water Resources and Hydropower Engineering, Xi'an University of Technology,
Xi'an 710048, China
ruoshuiya@163.com
[3] China Renewable Energy Engineering Institute, Beijing 100120, China
zhou_xingbo@126.com
[4] National Academy for Mayors of China, Beijing 100029, China
chenshujing1991@163.com

Abstract. This paper describes the improvements to the existing dam breach analysis methods based on back analyses of several giant barrier lake breaches. The main improvements include a hyperbolic soil erosion model, an empirical approach to lateral enlargement modeling, and a numerical algorithm that adopts velocity increment to allow straight forward calculation for the breach flood hydrograph. It has been shown that the calculated peak flow using this improved method is less sensitive to the input parameters. The new method has been incorporated into an Excel spreadsheet DB-IWHR which is transparent, open-source, self-explanatory and downloadable on the web.

Keywords: Dam breach · Flood evaluation · Soil erosion

1 Introduction

Evaluation of the breach flood of earthen dams and levees due to overtopping is a subject of common concern that attracted a large volume of research works [1–6]. However, their physical models and numerical approaches are still unmatured to-date. Chen et al. [7] discussed the difficulties involved in establishing the physical model that couples the disciplines of soil erosion in hydraulics and slope stability in soil mechanics. Further, shortage of field monitoring hydrographs of dam breaches makes it difficult to validate the analytical approaches developed by various researchers.

China has recently experienced a large number of barrier lake break cases with several well-documented field measured breach hydrographs [8–11]. This has allowed an in-depth study and improvements to the existing dam breach analytical methods. The research outcomes were incorporated into an Excel/VBA spreadsheet that provides an easy and quick access for practicing engineers to work during an impending dam failure emergency. A comprehensive review of this work has been presented in a latest paper [7]. This paper summaries the key novel technical approaches in its work.

© Springer Nature Switzerland AG 2020
J.-M. Zhang et al. (Eds.): ICED 2020, SSGG, pp. 3–23, 2020.
https://doi.org/10.1007/978-3-030-46351-9_1

2 Soil Erosion Models

2.1 Relationship Between the Erosion Rate and Shear Stress

The relationship between the rate of soil erosion and the shear stress is a fundamental problem that has great impact on the dam breach analytical results. The relevant studies include the effort in developing better soil erosion models and apparatuses for performing the erosion tests [12].

The decrease in the channel bed elevation due to erosion per unit time, $\dot{z}$, can be related to the shear stress τ on the eroded bed surface, as follows:

$$\dot{z} = \frac{\Delta z}{\Delta t} = \Phi(\tau) \tag{1}$$

where z is the channel bed elevation and t is time. The function $\Phi(\tau)$ proposed by researchers can be divided into the following categories.

Linear Relationship. This model is expressed as

$$\dot{z} = \frac{\Delta z}{\Delta t} = a_L(\tau - \tau_c) \tag{2}$$

where τ_c is the shear stress associated with the incipient velocity V_c, and a_L is the detachment rate coefficient [6, 13–15].

Power Law Relationship. There are a number of computer programs using the relationships based on the sedimentation dynamics, such as BREACH [5], BEED [16], and MIKE11 [17]. In general, the power relationship is [18]

$$\dot{z} = \frac{\Delta z}{\Delta t} = \Phi(\tau) = a_1(\tau - \tau_c)^{b_1} \tag{3}$$

where $\dot{z}$ is the erosion rate in 10^{-3} mm/s, τ is in Pa, t is in second, and a_1 and b_1 are coefficients either regressed from the test results or based on experience. Zhou et al. [19] compares the predicted soil erosion rates from 16 models with the measured rates in Tangjiashan Barrier Lake, as shown in Fig. 1.

Hyperbolic Relationship. The preceding mentioned linear and power models have been developed via extensive testing works based on the theory of sedimentation dynamics and validated mostly in laboratory at low flow velocity for fine materials. It is difficult to verify these models in a real dam breach, in which the flow velocity can reach 5 m/s or more and the eroded soils contain large rock fragments. As shown in Fig. 1, it can be seen that at low shear stress, the predictions from these models more or less agree with the measured data. However, for the maximum flow velocity when the corresponding shear stress is near the peak, the power law relationship continues to predict high erosion rate. Further, the soil erosion rate is very sensitive to even a small change in the exponent b_1 in Eq. (3).

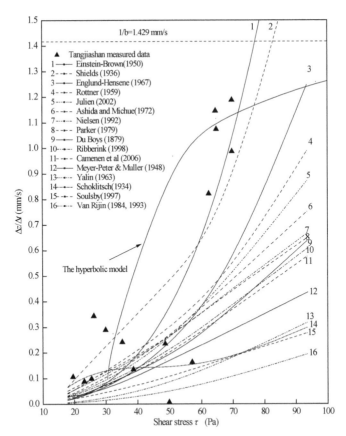

Fig. 1. Comparison of measured soil erosion rates with predicted rates by 16 models in Tangjiashan Barrier Lake. NOTE: According to the data of Tangjiashan barrier lake, the input parameters of this Figure are: water density, $\rho = 1000$ kg/m^3; $\rho_s = 2650$ kg/m^3; the medium size, $d_{50} = 5$ mm; roughness coefficient, $n = 0.025$; void ratio, $e = 0.36$; flow velocity, $V = (2.36, 2.50, 3.99, 4.61, 5.90)$ m/s and the corresponding the depth of water in channel, $h = (6.94, 7.49, 8.06, 8.19, 11.86)$ m.

In performing the back analysis for the Tangjiashan Barrier Lake, the authors proposed a hyperbolic relationship (Fig. 2) that takes the following form [1]:

$$\dot{z} = \Phi(\tau) = \frac{v}{a + bv} \tag{4}$$

where v is the shear stress with reference to its critical component τ_c

$$v = k(\tau - \tau_c) \tag{5}$$

with a unit of Pa for τ and 10^{-3} mm/s for $\dot{z}$. k is a unit conversion factor, normally takes a value of 100. The hyperbolic curve has an asymptote represented by $1/b$ as v approaches infinity, and $1/a$ represents the tangent of this curve at $v = 0$.

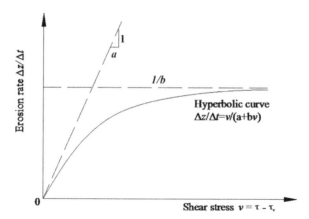

Fig. 2. Relationship between soil erosion rate and shear stress in the hyperbolic model.

The proposed model can be interpreted as a combination of the linear relationship with a 'truncation' at a certain high flow velocity. Approximately, $1/a$ equals a_L in Eq. (2). The asymptote $1/b$ represents the 'strength', which has been referred to as the 'maximum possible erosion rate' in this study. This is based on the understanding that for most structural materials, when the shear strength is sufficiently large, the erosion rate 'yields'. Use of this model can prevent the calculated erosion rate becoming too large at high flow velocity. Hence, the computation results are less sensitive to the input parameters in the model. Further, since both $1/a$ and $1/b$ have sound physical meanings, use of these parameters can be based on common sense, which can be improved with increased knowledge. The authors collected some laboratory testing results to support this model [20].

2.2 Soil Erodibility Test

A variety of testing facilities have been developed as reported by Foster et al. [21], Temple [22], Shaikh et al. [23], Hanson and Simon [24], Wan and Fell [25], Zhu et al. [26], Chang and Zhang [27], and Wu [6] conducted field jet index tests to measure the erodibility of broadly graded landslide deposits and found the soil erosion resistance coefficient of Tangjiashan materials is in a range between 0.2×10^{-3} and 1.2×10^{-3} mm/s/Pa.

A new apparatus, called Cylindrical Erosion Test Apparatus (CETA) has been developed by the authors' research group, as shown in Fig. 3 (Refer to Ma [28]). It includes a cylindrical, stainless steel container with an internal diameter of 1,040 mm. There are three glass windows to allow for visual inspection during the test. The propeller is driven by a speed adjustable motor with a maximum rotating speed of 1400 r/min. Similar to EFA, the soils under test is contained in a 160 mm long, 80 mm wide box, which can be lifted up to 150 mm manually as the erosion proceeds. Using a pitot tube and the FP111 direct-reading velocity meter, the reel rotating rate has been calibrated to relate to the flow velocity on the soil/water surface (refer to Fig. 4). Figure 5 shows the test results of a medium size sand with $d_{50} = 8$ mm. The dry density is 1.8 g/cm^3.

Fig. 3. The Cylindrical Erosion Test Apparatus (CETA).

Fig. 4. Parts of CETA: (a) Speed adjustable motor (b) propeller (c) manually operated slot (d) box containing soils fixed in the slot.

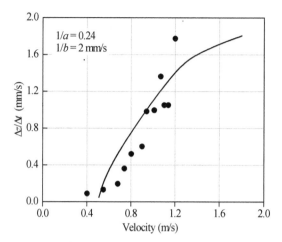

Fig. 5. Erodibility test results of sandy soil. NOTE: The solid line represents a regression curve based on the hyperbolic model. The asymptote $1/b$ was postulated due to the limitation of insufficiently large flow velocity.

3 Modeling Lateral Enlargement

3.1 Slope Stability of the Channel Banks

During the breach, the banks of the discharge channel collapse due to soil erosion that cut the toes. The wedge slide method has been commonly used in various dam breach models to model the processes in destabilizing the bank and widening the discharge channel (Singh and Scarlatos [29]). From the geotechnical point of view, the straight-line slip surface employed in this method is not representative of the actual failures observed in the field. It is better to replace the straight lines with a more generalized shape, such as a circular arc. Apart from this limitation, Wang et al. [30] also discussed the other drawbacks in the existing dam breach models, such as:

(1) A majority of the existing models fails to consider the existence of a vertical cut at the slope toe which is the key factor that destabilizes the bank.
(2) The method of determining the critical slip surface has not been clearly formulated. Wang et al. [30] discussed the proposed method by Osman and Thorne [31] who did consider the locations of the toe cut and the critical slip surface. However, they tried to find the critical toe-cutting depth by the derivatives with respect to the cohesion value of the material, which is physically unacceptable from geotechnical point of view. It is necessary to adopt the well accepted knowledge in soil mechanics that the critical slip surface is obtained by finding the minimum factor of safety associated with the geometrical coordinates that define the shape and location of the failure surface [32, 33].
(3) The pore water pressure has been invariably ignored in the analysis. The dam body prior to a breaching failure is normally saturated if the material is impervious, such as the earth core or tailings. Hence, in the analysis, the pore pressure development that causes sudden changes in the stress field need proper treatment, both in effective stress and in total stress [34–36].

3.2 Improved Approaches

In view of the drawbacks, Wang et al. [30] have made the effort to improve the model based on modern geotechnical expertise which includes: (1) stability analysis with circular slip surfaces using Bishop's simplified method, (2) assignment of a vertical cut at the toe, (3) a search technique to find the minimum factor of safety and the critical depth of toe cutting, and (4) an option for determining total stress and effective stress. Figure 6(a) shows an example that is circular slip surface of modeling lateral enlargement.

Although the lateral enlargement simulation offers better geotechnical background (Fig. 6(a)), the details of the 10 steps of circular arcs as input to a computer program are too tedious for a dam breach analysis. Wang et al. [30] found that without too much loss in accuracy, the circular arc can be simplified as a straight line as shown in Fig. 6(b).

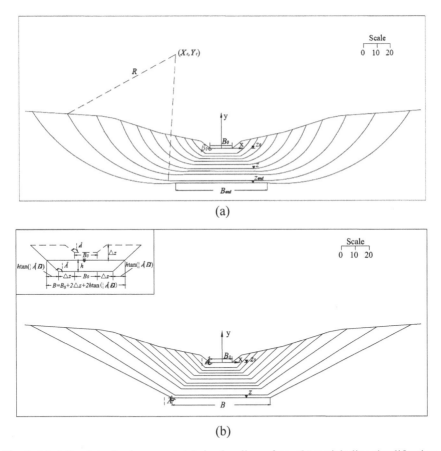

Fig. 6. Modeling lateral enlargement: (a) circular slip surfaces (b) straight line simplifications.

Based on an extensive regression work, Chen et al. [37] suggested an empirical hyperbolic relationship to determine the channel width associated with the channel bed elevation z, refer to Fig. 6(b).

$$B = B_0 + 2\Delta z + 2h \tan \left(\beta - \frac{\pi}{2}\right) \tag{6}$$

where

$$\Delta z = z - z_0 \tag{7}$$

and

$$\beta = \beta_0 + \Delta \beta = \beta_0 + \frac{\Delta z}{m_1 \Delta z + m_2} \tag{8}$$

β is the slope of the channel bed, z_0 and β_0 are initial values of bed elevation and bank inclination respectively. m_1 and m_2 are parameters of the hyperbolic relationship which are functions of cohesion c, coefficient of friction $\tan\varphi$ and the soil bulk density γ. Empirical charts for determining m_1 and m_2 have been provided by the authors.

4 Numerical Modeling of Sam Breach Flood

4.1 Governing Equations

The numerical model considers the balance of water quantity, the continuous vertical toe cutting and lateral enlargement, which can be formulated as follows.

Conservation of Energy and Mass. This condition necessitates the balance between the volumes of inflow through the breach, which is normally calculated by the broad-crested weir flow equation, and the reduction of the reservoir storage per unit time, as follows:

$$Q = CB(H - z)^{3/2} = \frac{\Delta W}{\Delta H} \frac{\Delta H}{\Delta t} + q \tag{9}$$

where C is the discharge coefficient whose theoretical value is 1.7 m$^{1/2}$/s [29]. Earlier studies have used values of C ranging from 1.3 to 1.7 [38]. H is the reservoir water elevation (Fig. 7, Chen et al. [1]), and q is the natural inflow into the reservoir.

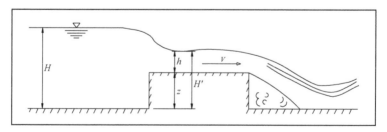

Fig. 7. Flow over a broad-crested weir.

Constitutive Model of Soil Erodibility. The formula describing the rate of channel bed erosion has been given in Eq. (4) in which the shear stress can be calculated using Manning equation, as follows:

$$\tau = \gamma R'J = \frac{\gamma n^2 V^2}{R'^{\frac{1}{3}}} \approx \frac{\gamma n^2 V^2}{h^{\frac{1}{3}}} \tag{10}$$

where γ is the density of water, n is the roughness coefficient (= 0.025 m$^{-1/3}$·s in this study), J is the slope of the channel, and R' is the hydraulic radius that can be approximated by h if the channel width B is sufficiently larger than the average flow depth h.

Determination of Water Depth Behind the Weir. The water level normally drops at the entrance of the weir if the flow is not submerged by the tailwater. Therefore, it is necessary to determine the flow depth h as shown in Fig. 7. Assuming the steady, uniform flow through the discharge channel, using Manning equation, Singh and Scarlatos [29] developed the following formula to determine h:

$$h = \left(\frac{nQ}{BJ^{0.5}}\right)^{0.6} \tag{11}$$

Use of this equation requires an input of J which is difficult to be properly defined in a dam breach problem. Experience has shown that this parameter is very sensitive to the calculated peak flow. Chen et al. [1] found that Eq. (11) can be further elaborated as

$$h = \left(\frac{nQ}{BJ^{0.5}}\right)^{0.6} = \left(\frac{nC(H-z)^{1.5}}{J^{0.5}}\right)^{0.6} = \frac{n^{0.6}C^{0.6}(H-z)^{0.9}}{J^{0.3}} \tag{12}$$

The exponent of $(H - z)$ in Eq. (12), which is 0.9, is very close to unity. Therefore, the depth can be approximately estimated by the following simplified relationship:

$$h = m(H - z) \tag{13}$$

$$m = \frac{h}{H - z} \approx \frac{n^{0.6}C^{0.6}}{J^{0.3}} \tag{14}$$

m is called the water drop ratio which has sound physical meaning. The value of m can therefore be assigned based on experience. Chen et al. [1] found that the calculated peak flow is not sensitive to the value of m, if it is between 0.5 and 0.8. The velocity can then be determined by

$$V = \frac{C(H-z)^{3/2}}{h} \tag{15}$$

4.2 Algorithm

The breach hydrograph can be obtained by solving the equations in Sect. 4.1. However, they are highly nonlinear and coupled; the authors' following treatments have greatly facilitated the numerical algorithm.

Integration Based on Velocity Increments. Equations (1), (9), and (15) allow H, z, and V to be solved at a particular time step Δt. However, these equations are highly nonlinear with concerns of numerical tractability. The new method performs the integration based on a velocity increment that completely linearizes of the governing equations and leads to a straight forwards calculation for the flow hydrograph.

At the velocity step from V_0 to $V_0 + \Delta V$, the average velocity is

$$\bar{V} = V_0 + \Delta V/2 \tag{20}$$

A variable s is defined as $(\Delta z - \Delta H)$ which can be calculated by

$$s = \Delta z - \Delta H = 2(\frac{m\bar{V}}{C})^2 - 2(H_o - z_o) \tag{21}$$

from which the average velocity can be determined by

$$\bar{V} = Cm^{-1}\sqrt{(H_o - z_o + \frac{s}{2})} \tag{22}$$

The increment in channel elevation Δz is first calculated by

$$\Delta z = \frac{s}{1 - L} \tag{23}$$

where

$$L = \frac{m\bar{V}B_o(H_o - z_o + 0.5\,s) - q}{\Phi(\bar{\tau})\dot{W}} \tag{24}$$

The other variables can be readily calculated once Δz is determined.

Transition Through the Threshold at Peak Velocity. The key technology of this new method includes a criterion that detects the threshold at which the velocity attains its maximum V_m. It has been found that both $(1 - L)$ and s in Eq. (23) approach zero at this particular point. Equation (23) is not applicable due to the limited calculation precision provided by the computer. Instead, Δz is calculated based on the Taylor series expansion for a value of V_0 less than V_m as follows:

$$V_o = V_m - \frac{dV}{dz}\Delta z + \frac{1}{2}\frac{d^2V}{dz^2}\Delta z^2 = V_m + \frac{1}{2}\frac{d^2V}{dz^2}\Delta z^2 \tag{25}$$

from which we obtain

$$\Delta z = \sqrt{\frac{2(V_o - V_m)}{d^2V/dz^2}} = \sqrt{-\frac{2\Delta V}{d^2V/dz^2}} \tag{26}$$

In DB-IWHR, Eq. (26) will be used when the spread sheet calculation finds the value L is sufficiently close to unity, in a range between 0.985 and 1.015. Figure 8 gives the flow chart of the algorithm [37].

4.3   Computer Software

The numerical algorithm has been realized by an Excel spreadsheet supported by VBA. To assist in inputting the required parameters, the spreadsheet 'Input' will guide a user to fill up the cells as described in Fig. 9. They are self-explanatory, requiring minimal tutorials. By pushing the bottom "Back to 'Calculation' sheet", the software turns to the main sheet 'Calculation'.

In sheet 'Calculation', the user will check the input and press 'Re Calculation' to start the computation (Fig. 10).

DB-IWHR and its illustrating examples are available for downloading at: https://github.com/ChenZuyuIWHR/DB-IWHR.

4.4   Suggested Values of Input for Preliminary Studies

Dam breaches often happen without early warning. This has made it difficult to perform field and laboratory tests to define the material properties. The following empirical ranges of various input parameters based on the authors' experience have been suggested for field engineers in their quick response to an impending dam failure:

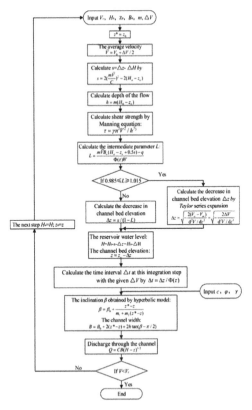

Fig. 8. Flow chart of the algorithm.

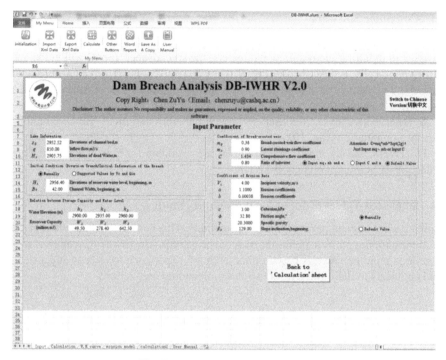

Fig. 9. Sheet 'Input' of DB-IWHR.

(1) The discharge coefficient C of a broad-crested weir is between 1.3 and 1.7 m$^{0.5}$/s [38]. For dams made of rock debris and earthen material, it may be taken to be around 1.45 and 1.50, respectively.

(2) It has been suggested that the water head drop ratio, m, should be 0.5 [39, 40]. In view of the decreasing pool water level and the rising downstream riverbed elevation Chen et al. suggested a more reasonable range for m ranged between 0.8 and 0.6.

(3) It is the responsibility of experienced geologists to suggest the dam material property parameters (γ, c, and ϕ), but conducting quick and simple field or laboratory tests would be useful, if time permits.

(4) Determining soil erosion parameters is the most difficult task. It is suggested to refer case records of similar nature and past experience. Table 1 gives some empirical suggestions for preliminary studies.

Table 1. Suggested values of a and b for preliminary studies [20].

	Erodibility	Soil materials	a	b
1	Very high	Fine sand, Non-plastic silt	1.0	0.0001–0.0003
2	High	Medium sand, Low plasticity silt	1.1	0.0003–0.0005
3	Medium	Jointed rock (spacing < 30 mm), Fine gravel, Coarse sand, High plasticity silt, Low plasticity clay, All fissured clays	1.2	0.0005–0.0007

(*continued*)

Table 1. (*continued*)

	Erodibility	Soil materials	a	b
4	Low	Jointed rock (30–150 mm spacing), cobbles, Coarse gravel, High plasticity clay	1.3	0.0007–0.001
5	Very low	Jointed rock (150–1500 mm spacing), Riprap	1.3	0.001–0.01
6	Non-erosive	Intake rock, Jointed rock (spacing > 1500 mm)	1.3	0.01–0.1

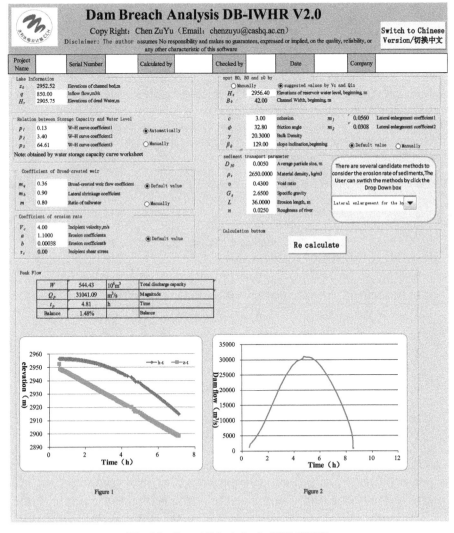

Fig. 10. Sheet 'Calculation' of DB-IWHR.

5   Case Study

5.1   Breach of Tangjiashan Barrier Lake

To illustrate the technical details described in this paper, this section uses the Tangjiashan landslide dam as a case study. The authors have in fact used this case previously [1] by investigating the breach hydrograph with the 13 m deep emergency diversion channel. This study, however, investigates the breach hydrograph without the emergency diversion channel. These comparative studies are useful when a decision has to be made.

The right part of the Tangjiashan dam is 105 m high with a crest elevation of 753 m above the sea level. The storage capacity of the reservoir is 3.2×10^9 m^3. Construction of a diversion channel enables the water in the reservoir to be drained at the lower elevation of 740 m, thereby reducing the volume of released reservoir water to 2.8×10^9 m^3. On one occasion after the reservoir water overflowed into the channel, the measured peak flow breaching the dam was 6,500 m^3/s, as compared to the peak flow of 7,160 m^3/s from the back analysis [1]. Various advanced instruments were used to measure the hydrograph of the breach, followed by a field survey and laboratory tests of geotechnical properties of the dam material [8]. This study continues the evaluation of the peak flow for the dam with no diversion channel, i.e., the reservoir water starts to drain at the elevation of 753 m.

Target Case. The inputs for the current case study are almost the same to the previous study and are summarized in Table 2.

Table 2. Input parameters for the current case study.

Item	H_0	q_{in}	C	m	V_c	a	b	c	ϕ
Parameters	753 m	80 m^3/s	1.43	0.8	2.7 m/s	1.2	0.0007	25 kPa	22°

NOTE: The relationship between the pool water level and storage can be found in Liu et al. (2010) and is approximated by $W = [p_1(H - H_r)^2 + p_2(H - H_r) + p_3] \times 10^6$ in m^3, where, p_1, p_2, p_3 are 0.61, 1.983, 44 respectively, and $H_r = 700$

Figure 6 shows the calculated results for the current case study with the inputs $\gamma_{sat} = 24$ kg/m^3, $c' = 22°$, and $\phi' = 25$ kPa. The model carried out the calculations for the 10 steps of circular failure to the elevation 727 m. After that, the width of the channel was not widened further because the rock of the right abutment was exposed as indicated by Chen et al. (2015). Figure 6(a) shows the processes of a typical failure.

For the current case study, $z_{end} = 727$ m has been used to obtain $B_{end} = 69$ m, $\beta_0 = 143°$, and $\beta_{end} = 169.5°$. These results have been applied to Eq. (8) to determine the width of the water surface at a particular channel bed elevation.

Illustration of the Computation. Table 3 shows the condensed contents of DB-IWHR calculations.

Table 3. The condensed contents of DB-IWHR calculations.

1	2	3	4	5	6	7	8	9	10	11	12	13	14	15	16	17	18	19	20	21
	H	z	V	B	Q	ΔV	h	q	Index	V	s	τ	$L = A/DE$	Δz	Δz_{taylor}	H	z	Δt	t	S_{out}
Unit	m	m	m/s	m	m³/s	m/s	m	m³/s		m/s	m	Pa	L	m	m	m	m	s	h	10⁶ m³/s
Eq.			(17)	(8)	(9)		(13)			Column 4 + Column 7	(21)	(10)	(23)	(22)	(24)					(25)
61	753.000	750.428	2.700	16.2	98.4	0.080	2.06	80	1	2.780	0.154	36.2	0.004	0.155	0.155	752.999	750.273	384	0.11	0.0
62	752.999	750.273	2.780	28.1	145.3	0.08	2.18	80	1	2.860	0.146	37.6	0.025	0.149	0.149	752.996	750.124	322	0.20	0.0
63	752.996	750.124	2.853	28.8	202.5	0.08	2.30	80	1	2.933	0.163	38.9	0.027	0.167	0.167	752.991	749.957	325	0.29	0.1
...	...	...	...	...	...	...	...		...	...	...	...	...	...	...	...	...	...	...	...
132	746.161	732.686	6.180	130.6	8590.6	0.005	10.78	80	1	6.185	0.024	106.1	0.959	0.595	0.595	745.590	732.091	502	5.03	56.2
133	745.590	732.091	6.185	134.9	8893.6	0.005	10.80	80	1	6.190	0.024	106.2	1.003	-7.659	0.788	744.827	731.303	665	5.22	60.5
134	744.827	731.303	6.191	141.0	9222.2	0.005	10.82	80	-1	6.186	-0.020	106.1	1.062	0.314	0.314	744.493	730.989	265	5.29	66.5
135	744.493	730.989	6.186	143.4	9484.2	0.005	10.80	80	-1	6.181	-0.020	106.0	1.084	0.232	0.232	744.242	730.757	196	5.35	68.9
...	...	...	...		...		...		...	...	...	...	...	...	...	...	...	...	...	...
...	...	...	...		...		...		...	...	...	...	...	...	...	...	...	...	...	...
307	694.778	691.934	2.839	177.6	1123.5	0.02	2.28	80	-1	2.819	-0.040	37.3	1.730	0.054	0.054	694.685	691.880	120	16.14	288.7
308	694.262	691.651	2.720	177.6	987.4	0.02	2.09	80	-1	2.700	-0.038	35.2	2.010	0.038	0.038	694.187	691.614	106	16.33	289.5
309	694.187	691.614	2.701	177.6	965.8	0.02	2.06	80	-1	2.681	-0.038	34.9	2.081	0.035	0.035	694.114	691.579	103	16.36	289.6

NOTE: Index '1' and '−1' represent a positive and negative ΔV respectively. Δz_{taylor} represents the value of ΔV determined by (Eq. 26).

Line 61 starts the calculation with the conditions: $H_0 = 753$ m, $V_c = 2.7$ m/s, $\Delta V = 0.08$ m/s, $z_0 = 750.42$ m and $B_0 = 16.23$ m. The calculated Δz as shown in Column 15 is 0.155 m, with $z = 750.273$ m, $H = 752.999$ m, $\Delta t = 384$ s for the next step, as shown in Columns 17, 18, and 19, respectively.

Line 62 performs the second round of integration based on the new values of H, z, and V. The calculation proceeds until Line 133 at which the velocity reaches its peak value.

Line 133 represents the transition point after which a negative ΔV has been assigned. The value of L is 1.003, which is close to unity (Column 14), and s is 0.024. Equation (23) results in an abnormal value of $\Delta z = -7.659$ m (Column 15), which is due to the insufficient precision in performing the division between two very small values by Eq. (23). Hence, the Taylors formulation (Eq. 26) was used, giving $\Delta z = 0.788$ m (Column 16).

Line 134 continues the integration with the negative input of ΔV.

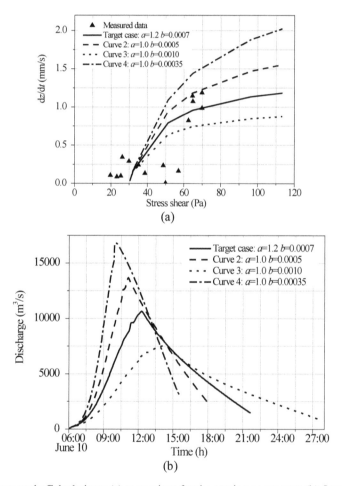

Fig. 11. Case study Calculations: (a) regressions for the erosion parameters (b) flow discharge.

Line 309 indicates the termination of the calculation at $V = 2.701$ m/s which is equal to the incipient V_c of 2.7 m/s, as specified by Eq. (26).

The Solid line in Fig. 11(b) shows the DB-IWHR calculation results of the target case.

5.2 Sensitivity Analyses

Due to the large number of uncertainties in a dam breach flood analysis, investigation on the impact of different input values is considered as part of the analytical work. In the target case with the channeled dam, Chen et al. [1] performed the sensitivity analyses on six parameters C, m, a, b and c, ϕ. This study has added one more parameter, the initial breach width B_o. Four sets of 'a and b' based on the measured data have been selected as shown in Fig. 11(a). Curve 1 (the target case), 2, 3 represent the mean, upper and lower bound of the regressions. Curve 4 is a postulated one whose value of $1/b$, the maximum possible erosion rate, is doubled to the target case. The results shown in Table 4 indicate that $1/b$ is the most sensitive parameter, compared to the others.

A comparative study has been given to a linear model as described by Eq. (2). Following the similar process, three sets of a_L have been selected representing the mean, upper and lower bound of the regressions as shown in Fig. 12(a). Figure 12(b) shows the calculated hydrographs that display quite large deviations both in terms of the shape

Table 4. Sensitivity analyses based on various characteristic parameters.

Variable	Value		Occurring time	Discharge	Deviation
			t_m (h)	Q_m (m/s)	%
	Target case	Sensitivity studies			
Initial width of the breach, B_0 (m)	16	30	12:19	10286.76	−3.6%
		50	12:23	9796.24	−8.81%
Ratio of water head drop, m	0.8	0.5	9:29	11198.60	4.8%
		0.6	11:28	11366.13	6.2%
Coefficient of broad-crested weir flow, C	1.43	$C_1 = 1.35$	12:20	10605.05	−0.5%
		$C_1 = 1.65$	14:28	10258.95	−3.9%
Coefficient of soil erosion, a and b	$a = 1.2$, $b = 0.0007$	$a = 1.0, b = 0.0005$	11:08	13,664.90	22.0%
		$a = 1.0, b = 0.0010$	14:26	7,536.02	−41.4%
		$a = 1.0, b = 0.00035$	10:02	16818.08	38.8%
Cohesion and friction angle, c (kPa) and ϕ (°)	$\phi = 22°$ $c = 25$ kPa	$c = 25$ kPa $\phi = 26°$	12:23	9705.51	−9.83%
		$c = 35$ kPa $\phi = 22°$	12:22	9546.29	−11.66%

NOTE: Q_m = peak flow, t_m = time at peak flow, Deviation = $(Q_m - Q_{m,t})/Q_m$, where $Q_{m,t}$ is 10,659 m^3/s, the peak flow of the target case.

and the magnitude of peak flow. The previous report [1] discussed the concerns on sensitivities related to the power relation model. This gives a reason of adopting the hyperbolic model (Fig. 12).

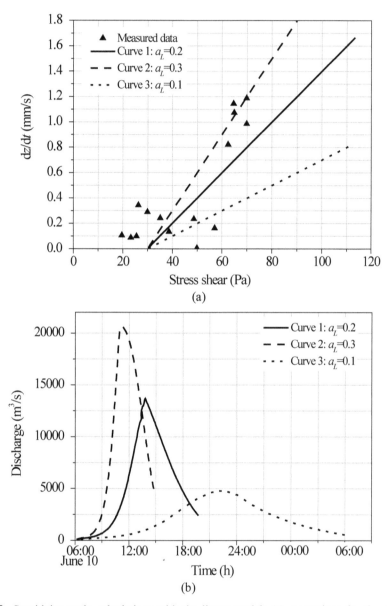

Fig. 12. Sensitivity study calculations with the linear model: (a) regressions for the erosion parameters. (b) flow discharge.

6   Conclusions

The improvements to the existing dam breach analysis methods described in this paper include a new soil erosion model, an empirical hyperbolic model for lateral enlargement modeling, and a novel numerical algorithm that allows straight forward calculation for the breach flood hydrograph. The new method has been incorporated into the Excel spreadsheets DB-IWHR, which perform the dam breach flood calculation and the lateral enlargement calculation respectively. These spreadsheets are transparent, self-explanatory and ready for group sourcing development on the web. The main highlights are:

(1)   The hyperbolic soil erosion model developed in this study includes the parameter $1/a$ which is close to the relevant coefficient in the traditional linear model, and the parameter $1/b$ which is a 'truncation' at large shear stress. The new model makes the calculated peak flow less sensitive to the input parameters. In addition, these two parameters have sound physical meaning. Therefore, with increased knowledge, their applicable values for various soils can be improved in practice. As a new model with limited experience of practical application, the authors would suggest users be cautious when using this model. It is advantageous to adopt different soil erosion models and make comparative studies. DB-IWHR allows another 5 representative models for users to select.

(2)   The procedures of modeling lateral enlargement has been updated by modern geotechnical expertise on slope stability analysis However, the solutions have been simplified to a series of trapezoidal cross sections in order to facilitate the calculation without substantial loss in accuracy.

(3)   The new method developed in this study uses velocity increment during the integration of the governing equations, thus allowing a straightforward solution. The calculation procedure has been incorporated into an Excel spreadsheet, providing easy, transparent access to practitioners.

(4)   The results of this study show that the erosion rate parameter $1/b$ is the most sensitive input that affects the calculated peak flow. Other parameters are less sensitive, which only cause 5 to10% variations in the peak flow.

References

1. Chen, Z.Y., Ma, L.Q., Yu, S., Chen, S.J., Zhou, X.B., Sun, P., Li, X.: Back analysis of the draining process of the Tangjiashan barrier lake. J. Hydraul. Eng. **141**(4), 05014011 (2015)
2. Cristofano, E.A.: Method of Computing Erosion Rate of Failure of Earth Dams. U.S. Bureau of Reclamation, Denver (1965)
3. Harris, G.W.: Outflow from breached earth dams. Doctoral Dissertation, Department of Civil Engineering, University of Utah, Salt Lake City (1967)
4. Brown, R.J., Rogers, D.C.: A simulation of the hydraulic events during and following the Teton Dam Failure. In: Proceedings of Dam-Break Flood Routing Workshop, Washington, DC, pp. 131–163 (1977)

5. Fread, D.L.: BREACH: An Erosion Model for Earthen Dam Failures (Model Description and User Manual). National Oceanic and Atmospheric Administration, National Weather Service, Silver Spring (1988)
6. Wu, W.M.: Simplified physically based model of earthen embankment breaching. J. Hydraul. Eng. **139**(8), 837–851 (2013)
7. Chen, Z., Zhang, Q., Chen, S., Wang, L., Zhou, X.: Evaluation of barrier lake breach floods: insights from recent case studies in China. Wiley Interdiscip. Rev.: Water **7**(2) (2020). https://doi.org/10.1002/wat2.1408
8. Liu, N., Chen, Z.Y., Zhang, J.X., Wei, L., Chen, W., Xu, W.J.: Draining the Tangjiashan barrier lake. J. Hydraul. Eng. **136**(11), 914–923 (2010)
9. Zhou, X.B., Chen, Z.Y., Yu, S., Wang, L., Deng, G., Sha, P.J., Li, S.Y.: Risk analysis and emergency actions for Hongshiyan barrier lake. Nat. Hazards **79**(3), 1933–1959 (2015)
10. Chen, Z.Y,, Chen, S.S., Wang L., Zhong Q., Zhang Q., Jin, S.L.: Evaluation of the breach flood of the "11.03" Baige barrier lake at the Jinsha River. Sci. China Tech. Sci. (2020). https://doi.org/10.1360/sst-2019-0297. (in Chinese)
11. Cai, Y.J., Cheng, H.Y., Wu, S.F., Yang, Q.G., Wang, L., Luan, Y.S., Chen, Z.Y.: Breaches of the Baige Barrier Lake: emergency response and dam breach flood. Sci. China Technol. Sci. (2019). https://doi.org/10.1007/s11431-019-1475-y
12. Hanson, G.J.: Channel erosion study of two compacted soils. Trans. ASAE **32**(2), 485–0490 (1989)
13. Zhang, L.M., Peng, M., Chang, D.S., Xu, Y.: Dam Failure Mechanisms and Risk Assessment. Wiley, Singapore (2016)
14. Smart, G.M.: Sediment transport formula for steep channels. J. Hydraul. Eng. **110**(3), 267–276 (1984)
15. Brown, C.B.: Sediment transportation. In: Rouse, H. (ed.) Engineering Hydraulics, pp. 711–768. Wiley, New York (1950)
16. Singh, V.P.: Dam breaching modeling technology. Kluwer Academic Publishers, Dordrecht, Netherlands (1996)
17. Engelund, F., Hansen, E.: A monograph on sediment transport in alluvial streams. Hydrotech. Constr. **33**(7), 699–703 (1967)
18. Roberts, J., Jepsen, R., Gotthard, D., Lick, W.: Effects of particle size and bulk density on erosion of quartz particles. J. Hydraul. Eng. **124**(12), 1261–1267 (1998)
19. Zhou, X.B., Chen, Z.Y., Li, S.Y., Wang, L.: Comparison of sediment transport model in dam break simulation. J Basic Sci. Eng. **23**(6), 1097–1108 (2015). (in Chinese)
20. Chen, Z.Y., Ma, L.Q., Yu, S., Chen, S.J., Zhou, X.B., Sun, P., Li, X.: Closure to "back analysis of the draining process of Tangjiashan barrier lake" by Zuyu Chen, Liqiu Ma, Shu Yu, Shujing Chen, Xingbo Zhou, Ping Sun, and Xu Li. J. Hydraul. Eng. **142**(6), 07016002 (2016)
21. Foster, G.R., Meyer, L.D., Onstad, C.A.: An erosion equation derived from basic erosion principles. Trans. ASAE **20**(4), 678–0682 (1977)
22. Temple, D.M.: Stability of grass lined channels following mowing. Trans. ASAE **28**(3), 750–0754 (1985)
23. Shaikh, A., Ruff, J.F., Abt, S.R.: Erosion rate of compacted Na-montmorillonite soils. J. Geotech. Eng. **114**(3), 296–305 (1988)
24. Hanson, G.J., Simon, A.: Erodibility of cohesive streambeds in the loess area of the midwestern USA. Hydrol. Process. **15**(1), 23–38 (2001)
25. Wan, C.F., Fell, R.: Investigation of rate of erosion of soils in embankment dams. J. Geotech. Geoenviron. **130**(4), 373–380 (2004)
26. Zhu, Y.H., Lu, J.Y., Liao, H.Z., Wang, J.S., Fan, B.L., Yao, S.M.: Research on cohesive sediment erosion by flow: an overview. Sci. China Ser. E. **51**(11), 2001–2012 (2008)

27. Chang, D.S., Zhang, L.M.: Simulation of the erosion process of landslide dams due to overtopping considering variations in soil erodibility along depth. Nat. Hazard Earth Syst. Sci. **10**(4), 933–946 (2010)
28. Ma, L.Q.: Flood analysis of landslide dam breach. Post-Doctoral Dissertation, China Institute of Water Resources and Hydropower Research, Beijing (2014). (in Chinese)
29. Singh, V.P., Scarlatos, P.D.: Analysis of gradual earth-dam failure. J. Hydraul. Eng. **114**(1), 21–42 (1988)
30. Wang, L., Chen, Z.Y., Wang, N.X., Sun, P., Yu, S., Li, S.Y., Du, X.H.: Modeling lateral enlargement in dam breaches using slope stability analysis based on circular slip mode. Eng. Geol. **209**, 70–81 (2016)
31. Osman, A.M., Thorne, C.R.: Riverbank stability analysis. I: Theory. J. Hydraul. Eng. ASCE **114**(2), 134–150 (1988)
32. Duncan, J.: State of the art: limit equilibrium and finite-element analysis of slopes. J. Geotech. Eng. **122**(7), 577–596 (1996)
33. Chen, Z.Y., Shao, C.M.: Evaluation of minimum factor of safety in slope stability analysis. Can. Geotech. J. **25**(4), 735–748 (1988)
34. Sherard, J., Woodward, R., Gzienski, S., Clevenger, W.: Failures and damages. In: Earth and Earth-Rock Dams, 1st edn, pp. 130–131. Wiley, New York (1963)
35. Lowe, J., Karafiath, L.: Stability of earth dam upon drawdown. In: First Pan-American Conference on Soil Mechanics and Foundation Engineering, Mexico City, vol. 2, pp. 537–552 (1960)
36. Johnson, J.J.: Analysis and design relating to embankments. In: Proceedings, Conference on Analysis and Design in Geotechnical Engineering, vol. 2, pp. 1–48. ASCE, New York (1974)
37. Chen, Z.Y., Ping, Z.Y., Wang, N.X., Yu, S., Chen, S.J.: An approach to quick and easy evaluation of the dam breach flood. Sci. China Tech. Sci. **62**(10), 1773–1782 (2019)
38. Jack, R.: The mechanics of embankment failure due to overtopping flow. Doctoral dissertation, University of Auckland, Auckland, New Zealand (1996)
39. Chow, V.T.: Open-Channel Hydraulics. McGraw Hill, New York (1959)
40. Doeringsfeld, H.A., Barker, C.L.: Pressure momentum theory applied to the broad crested weir. Trans. ASCE **106**, 934–969 (1941)

Failure of Saddle Dam, Xe-Pian Xe-Namnoy Project: Executive Summary

Ahmed F. Chraibi[1(✉)], Anton J. Schleiss[2], and Jean-Pierre Tournier[3]

[1] Dam Consultant, PO Box 21514, Rabat Annakhil, Rabat, Morocco
ach@damtech.ma
[2] Ecole polytechnique fédérale de Lausanne (EPFL), Lausanne, Switzerland
anton.schleiss@epfl.ch
[3] Hydro-Québec Équipement & Services partagés, 855, rue Ste-Catherine Est - 10e étage,
Montreal, QC H2L 4P5, Canada
Tournier.Jean-Pierre@hydro.qc.ca

Abstract. The Saddle Dam D at the Hydroelectric Power project Xe-Pian Xe-Namnoy in the Lao People's Democratic Republic failed on July 23, 2018. An Independent Expert Panel (IEP) was established to investigate into and report on the failure of the embankment dam. The Executive Summary of the IEP final report (Anton J. Schleiss, Jean-Pierre Tournier, and Ahmed F. Chraibi, Report of Independent Expert Panel (IEP), Xe-Pian Xe-Namnoy Project - Failure of Saddle Dam D, Final Report. 20 March 2019) is presented here.

Keywords: Dam failure · Embankment dam · Tropical residual soils

This final report on the failure of Saddle Dam D of Xe-Pian Xe-Namnoy hydropower project summarizes the findings of the Independent Expert Panel (IEP) based on the available supporting information and the observations made by the IEP during the site visits carried out in the beginning of October 2018 and end of November 2018. Furthermore, it considers the results of the recommended geotechnical investigations, made available in January 2019, and the numerical sensitivity analysis of the dam stability.

According to the available sequence of photographs and the reporting of the event, the observed movements of the sliding mass are of complex geometry. Nevertheless, the IEP is convinced that the main evidence of the incident at its beginning is a rotational sliding involving the lateritic foundation. The most important weakness in the foundation triggering deep sliding has developed along the deepest area of the saddle, respectively the highest section of the dam. Thus, the foundation of the Saddle Dam D was without doubt involved in its failure.

According to the site visit observations, the monitoring data analysis and the review of the available photographs, before, during and after the failure, as well as the results of geotechnical investigations, the IEP considers that the root cause of the incident is related to the high permeability of the foundation. The high permeability was above all favored by the presence of canaliculus interconnected path having high sensitivity to erosion. In fact, the geotechnical investigations revealed, that the foundation of the Saddle Dam D is very heterogeneous with a predominance of clayey sandy-silty soils.

© Springer Nature Switzerland AG 2020
J.-M. Zhang et al. (Eds.): ICED 2020, SSGG, pp. 24–26, 2020.
https://doi.org/10.1007/978-3-030-46351-9_2

Numerous passages rich in sand and even gravel leading to low core recoveries and higher permeability values have been observed.

The mechanism of failure of the Saddle Dam D was most probably triggered by the following successive sequences:

1. Due to the presence of high permeability horizons in the foundation, as confirmed by the investigations, groundwater level at the downstream toe was close to the surface generating resurgence in the vegetated area where topography declines rapidly. This hypothesis is supported by the observation made downstream of the very similar Saddle Dam E, where evidence of resurgence with some internal erosion was observed.
2. With continuing resurgence in the vegetated area downstream of the dam toe, regressive erosion has developed in the foundation resulting in the formation of ducts that collapsed from time to time, especially in the deepest section of the saddle where the highest seepage gradients occur. The resulting softening of the laterite triggered the speeding up of the settlement and the appearance of the first cracks on the dam crest.
3. When the erosion and softening in the foundation reached a certain extent, the static dam stability was no longer ensured and a deep rotational sliding at the highest section of the embankment developed. Simultaneously, converging embankment movements occurred from the lateral border of the sliding mass towards the middle, resulting in a bumping up of the downstream embankment face and the subsidence of the track in front of the dam toe.
4. When the remaining thin upstream edge of the embankment crest breached, the embankment was overtopped and the catastrophic uncontrolled release of water from the reservoir washed away the central section of the Saddle Dam D and its foundation.

Even if July 2018 was the wettest month over the record since 2008, with some 1350 mm falling up to the 29th of July, and the highest daily rainfall event occurring on July 22 with 438 mm, the flood event at the spillway operating with some 680 m^3/s just before Saddle Dam D failure was only in the range of a 10 to 20 years flood. Yet, the reservoir was still well below the maximum operation level at the failure incident and the embankment has to withstand safely the probably maximum flood event. Thus, the failure incident cannot be considered as "force majeure".

The failure could have been prevented by an appropriate treatment of the foundation aiming at providing the required water tightness, filtration and drainage. Furthermore, an early and correct interpretation of the monitoring data and a reinforced detailed visual inspection in the downstream toe region of the embankment, would have allowed to take actions trying to save the Saddle Dam D and/or at least trigger the warning earlier.

Since the bottom outlet has only a small capacity mainly for the release of environmental flow, there was no immediate possibility to control or to lower the level of the reservoir when the first signs of failure were observed. There remained only the possibility of removing by blasting and breaching the spillway labyrinth wall reaching almost 6 m height. Without having any control on the reservoir level during operation, at least in the most upper part by spillway gates eventually together with the powerhouse, such a concept is not acceptable according to the best international practice. In view of the catastrophic consequences in case of failure, this is particularly important for reservoirs, which volume is contained by several large embankment dams like for the Xe-Namnoy reservoir.

Saddle Dams E and F as well as the part of Xe-Pian Dam founded on lateritic soil, have similar foundation conditions compared to the failed Saddle Dam D. From the monitoring assessment, they already exhibit a comparable sudden acceleration of settlement and increase in the downstream hydrostatic pressures which are linked to the foundation quality. These dams have to be reviewed and appropriate rehabilitation measures have to be defined to ensure the safety requirements preventing any undesirable behavior.

The evolution of the groundwater level in any topographical depression present in the near downstream region of both Xe-Pian and Xe-Namnoy dam has to be monitored by piezometers or at least included in the visual inspection program.

The timely (re-)construction of new Saddle Dam D and rehabilitation of the afore-mentioned dams is of paramount importance in order to allow a safe reservoir filling during the next rainy season. The concept and design of the new saddle dam has to be robust in view of the very limited construction time, uncertainty of foundation and safety requirements. Furthermore, the new Saddle Dam D should be equipped with two high-capacity outlets which allow to control the reservoir level at least in its most upper 20 m during wet season. The reinforcement of Saddle Dams E and F as well as Xe-Pian dam requires relevant information on the depth and the quality of their lateritic foundation. Thus, prior to the detailed definition of the reinforcement works to carry out, specific and thorough investigations are necessary.

Lessons learnt from the incident comprise, among others:

- The delicate and very heterogeneous nature of Lateritic soils: they may contain highly permeable and erodible horizons, canaliculus conveying water on a long distance, they soften when saturated and may be sensitive to significant settlement. Since Laterite formations are residual soils, which may even have a potential of collapse when they are not permanently saturated. Investigations in lateritic soil should include large and deep open trenches. Positive cutoff is the most adapted seepage control arrangement in this type of foundation;
- During the first reservoir filling, highly experienced dam engineers should be mobi-lized on the site (or in permanent contact with) to carry out immediate interpretation of monitoring data and to inspect the dam and its surroundings. Experienced eyes are very important in early detection of undesirable behaviors;
- Easy access paths and vegetation-free space has to be ensured downstream of the dam in order to allow a comfortable visual inspection and early intervention in case of danger.

The IEP recommends that all large hydropower and dam projects are reviewed during the design and construction phases by an independent international panel of experts. Furthermore, a dam safety concept should be put into operation in Laos PDR by creating a dam safety supervisory authority based on a legal framework in the country.

Building Cemented Material Dam with Soft Rock Material

Jinsheng Jia[(⊠)], Lianying Ding, Yangfeng Wu, Aili Li, and Shuguang Li

State Key Laboratory of Simulation and Regulation of Water Cycle in River Basin,
China Institute of Water Resources and Hydropower Research, Beijing 100038, China
jiajsh@iwhr.com

Abstract. Cemented Material Dam (CMD) is a new type of dam first proposed in the world by the first author in order to solve overtopping failure problem of embankment dam and to decrease the cost of concrete dam. Methods have been investigated to build the Jinjigou CMD with soft rock material for the first time in the world. Mix design of materials and in-situ direct shear tests have been introduced. Meaningful results have been achieved.

Keywords: Cemented Material Dam · Soft rock material · Mix design · In-situ direct shear test

1 Introduction

Cemented Material Dam (CMD for short) is a new dam type developed in a way different from the Embankment Dam or the Concrete Dam. CMD was proposed by the first author in 2009 based on the practices of hardfill dams [1], Cemented Sand and Gravel (CSG) dams [2] and Cemented Sand, Gravel and Rock (CSGR) dams [3]. Cement, fly ash and additives are used to cement soil, sand, gravel, stone and other materials together to form new dam materials with enhanced strengths and shear resistance [4]. Its main design concepts are to optimize the functional structures of the dam by selecting proper materials to adapt different part requirement of the dam, at the same time to adjust the dam structure for better use of local materials [5]. CMD has several significant technical advantages such as high resistance against overtopping to guarantee dam safety, high adaptability to geological terrain, low requirements for materials and construction, full usage of excavation material for environmental reasons, and high construction efficiency.

CMD has been developed rapidly in dam construction practices in recent years. Up to now, more than 40 CMD projects have been built in China and other countries. Cofferdams, such as Jiemian [6], Hongkou, Gongguoqiao, Dahuaqiao, etc. have been built and proved safe for overtopping floods. The 32 m high Hongkou cofferdam withstood 8 m overtopping flood and operated safely. The Dahuaqiao cofferdam 57 m in height is the highest cofferdam up to now. It was safe also after experiencing overtopping floods. The Qianwei CMD is a successful case built on the non-rock foundation with a height of 14 m. The Shoukoubu (61 m), Shunjiangyan and Maomaohe dams are the representative CMD projects, and play an important role in practice [3]. Good results have been achieved in over 40 projects and the CMD dam type has been recognized worldwide. The Jinjigou dam was built with soft rock materials and is the first CMD in the world.

© Springer Nature Switzerland AG 2020
J.-M. Zhang et al. (Eds.): ICED 2020, SSGG, pp. 27–35, 2020.
https://doi.org/10.1007/978-3-030-46351-9_3

2   The Jinjigou CMD Constructed with Soft Rock Material

The Jinjigou CMD is located in Yingshan County, Sichuan Province, with a total storage capacity of 11.2 million m³, which is an important water supply project. In order to decrease the cost and to use local sand and gravel, the designer adopted the CMD type after comparing with a RCC dam scheme. The Jinjingou CMD has a maximum height of 33.0 m, a crest length of 72.0 m, and a crest width of 8.0 m. The upstream slope of the dam is 1:0.35 and the downstream slope is 1:0.75. A 0.6 m thick reinforced concrete layer was placed as an impervious layer and enriched cemented material was placed as downstream protection layer. The typical cross section is shown in Fig. 1.

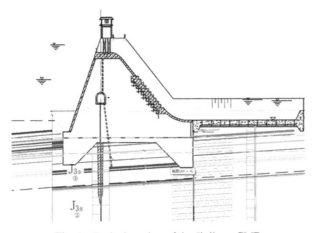

Fig. 1. Typical section of the Jinjigou CMD

2.1   Mix Design of Material

Natural sand and gravel could be used but it is located 60 km away and is limited in supply. Sandstone at the dam site is abundant but its mechanical strength is lower than the value defined by the specification for RCC dams (lower than 30 MPa). According to tests, apparent density of the sandstone is 2440~2500 kg/m³, the softening coefficient is 0.71 and the crushing value is 20.2%. To make full use of excavated sandstone, laboratory tests and in-situ tests were carried out to study the possibility of using crushed soft sandstone.

The compacted density test results of full-gradation large-size specimens show that the density of cemented material with full crushed sandstone aggregates is only 2270 kg/m³. In view of the fact that the compacted density is not large enough for the stability against sliding, the feasibility of mixed aggregate of crushed sandstone and riverbed gravel was studied. The results of the different mix proportion of lab tests are listed in Table 1.

Table 1. Mix proportion and 180-day compressive strength of wet-sieved specimen

Percentage of sandstone (%)	Sand ratio (%)	Cement (kg/m³)	Flyash (kg/m³)	Water (kg/m³)	VC(s)	Water-binder ratio	Compressive strength (MPa)
100	25	60	60	100	1.9	0.83	5.8
71.25	25	60	60	98	2.5	0.82	5.8
50	28	60	60	85	5.6	0.71	11.4
40	27	60	60	84	5.0	0.7	11.8
30	32	60	60	93	10.6	0.78	14.2

The results of the material property test show that the compressive strength of wet-sieved specimen decreases with the increase of the percentage of sandstone aggregate. Among them, the compressive strength of cemented material made with mixed aggregate of crushed sandstone and riverbed gravel (50%:50%, 40%:60%, or, 30%:70% mass ratio) was 11.4~14.2 MPa, which met the requirements of design strength (8 MPa at 180 d). Taking all the factors into consideration, the optimal scheme is that the proportion of sandstone aggregate in the mixed aggregate was 50%.

The in-situ tests (as shown in Fig 2) indicate that the material has good performance and the mix proportion is reasonable. At the same time, it also provides a reasonable construction procedure, technical parameters and quality control process for the construction.

Fig. 2. In-situ tests

2.2 In-situ Direct Shearing Test

The interlayer shear capacity of the surface is a key factor that affects the stability against sliding of CMDs [7]. There are limited experimental data on the shear strength of CMD because it is very complicated and costly. The in-situ direct shear tests of CMD were carried for different percentages of sandstone aggregate and at the same time different interlayer surfaces have been tested also, such as the surfaces between dam body interlayers, the contact surface with concrete, and the contact surface with enriched cemented material.

Test Results. On the left bank of the dam, a rock foundation pit was excavated at the Jianshaozui platform as an experimental site. It is divided into three test sections with five layers of CMD mixed with 50%, 60% and 70% sandstone. Each layer is about 50 cm thick. Due to rainfall, cold joints are formed between layers. The cold joints are brushed with a high-pressure water gun and paved with M15 cushion mortar with a thickness of 10–15 mm. The mix design of CMD is shown in Table 2.

Table 2. Mix design for in-situ direct shearing test

Percentage of sandstone	Water	Cement	Fly ash	Admixture	Sandstone aggregate	Riverbed gravel (5~40 mm)	Artificial sand
50%	85	60	60	1.2	1057	558	555
60%	90	65	65	1.35	1255	441	439
70%	90	70	70	1.4	1457	264	393
M15	280	222	222	4.44	/	/	1210
Enriched cemented	600	1184	/	11.8	/	/	/

Note: unit of material is kg/m^3.

C15 concrete and enriched cemented materials were placed in the fourth layer of the test section, and every test section is cured to an age of 90 d. Shear piers with a size of 50 × 50 × 50 cm were cut in the preset area of the fifth layer. As shown in Fig. 3, interference to the shear piers should be avoided in the production process of the shear piers. At the same time, a single-point friction test was carried out on some test pieces. The tests are shown in Table 3. Each group of direct shear test has 4 test piers, totaling 6 groups and 24 test piers.

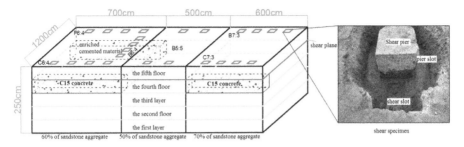

Fig. 3. Schematic of in-situ direct shearing test

Table 3. In situ direct shear tests

Percentage of sandstone	Shear plane	Serial number	Group number
50%	Dam body interlayer	B5:5	1 set (4 cells)
60%	Dam body interlayer	B6:4	1 set (4 cells)
	Contact surface with C15 concrete	C6:4	1 set (4 cells)
	Contact surface with enriched cemented material	F6:4	1 set (4 cells)
70%	Dam body interlayer	B7:3	1 set (4 cells)
	Contact surface with C15 concrete	C7:3	1 set (4 cells)

In the test, the horizontal pushing method is adopted, and the shear is carried out under the condition of artificial immersion saturation. The distance between the shear center of each specimen and the preset shear plane is controlled within the range of 3–5 mm. The maximum normal stress of the shear plane is 1.0 MPa, which is located at the center of the preset shear plane and perpendicular to the shear plane, and is applied in 3–5 stages. The shear stress direction is parallel to the predetermined shear plane and is applied in 8–10 grades according to the estimated maximum value.

Shear Characteristic Analysis. The typical sections of the shear piers are shown in Fig. 4. It can be seen that about 10~25 cm thick of the upper part is dense, while the lower part is less dense, and the coarse aggregate in some parts is concentrated.

Fig. 4. The typical sections of the shear specimens

The failure pattern of the shear plane is relatively complex. The typical cut surface is shown in Fig. 5. The shear failure modes are basically complicated. It happened along the shear plane, above the shear plane or under the shear plane.

Fig. 5. The typical failure of the shear piers

Shear Behavior Analysis. The curves of shear stress~displacement and shear stress~normal stress at typical measuring points are shown in Figs. 6, 7, 8. The direct shear test results are shown in Table 4.

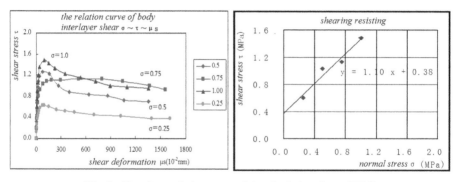

Fig. 6. Shearing resisting curve of dam body interlayer (B5:5)

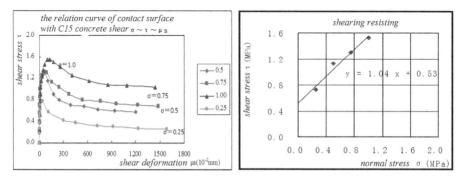

Fig. 7. Shearing resisting curve of contact surface with C15 concrete (B6:4)

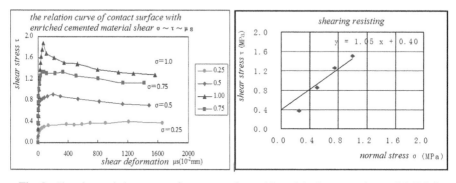

Fig. 8. Shearing resisting curve of contact surface with enriched cemented material (B6:4)

Table 4. In-situ direct shear test results

Percentage of sand-stone	Test number	Shearing resisting		Residual strength		Shear	
		f'	c'(MPa)	f_R	c_R (MPa)	f	c (MPa)
50%	B5:5	1.10	0.38	0.81	0.24	0.81	0.18
60%	B6:4	1.08	0.34	0.81	0.30	0.85	0.11
	C6:4	1.04	0.53	0.95	0.30	0.88	0.06
	F6:4	1.05	0.40	0.99	0.25	0.91	0.05
70%	B7:3	1.02	0.26	0.91	0.24	0.87	0.09
	C7:3	0.99	0.28	0.94	0.20	0.89	0.10

The cemented material with sandstone has an obvious peak strength state and a residual strength state, and the peak strength is about 10%~30% higher than the residual strength, showing a plastic~brittle failure mode. That shows a strong shear resistance at the layer or contact surface. Thus, it can also be explained that it has a strong ability to resist the risk of crash. When the proportion of sandstone aggregate is more than 60%, the phenomenon of coarse aggregate concentration is obvious in the lower part of placed layer, and the shear failure happened from the shear plane to the above. Therefore, the reasonable sandstone percentage selected is very important to ensure the stability against sliding of layers.

The curves of shear stress~displacement have an obvious linear ascending stage, indicating that the cemented material with sandstone has a certain shear capacity. From the perspective of the trend, with the increasing of the percentage of sandstone aggregate, the shear strength of the layer or contact surface gradually decreases. It can be seen from the test results that the shear friction decreases from 1.10 MPa to 0.99 MPa, and the shear cohesion decreases from 0.53 MPa to 0.26 MPa, which is consistent with the coarse aggregate concentration. The more concentrated the coarse aggregate, the lower the shearing strength.

3 Conclusion

Test results for the Jinjigou CMD built with soft rock materials indicate that the cemented material made with mixed aggregate with crushed soft sandstone and riverbed gravel has good performance and the mix design is reasonable. The stability against sliding can be guaranteed. CMD can be used in soft sandstone areas.

Acknowledgements. Financial support from the National Key R&D Plan (Grant No. 2018 YFC0406801) is gratefully acknowledged.

References

1. Batmaz, S.: Cindere dam-107 m high roller compacted hardfill dam (RCHD) in Turkey. In: CONFERENCE 2003, LNCS. Proceedings of the 4th International Symposium on Roller Compacted Concrete (RCC) Dams, AA Balkema Rotterdam, pp. 121–126 (2016)
2. Yokotsuka, T.: Application of CSG method to construction of gravity dam. In: Proceedings of 20th ICOLD, pp. 989–1007 (2000)
3. Liu, Z., Jia, J., Feng, W., et al.: Latest practice and application of cemented sand and gravel dam to mid-small scale water conservancy project. Water Resour. Hydropower Eng. **49**(5), 44–49 (2018)
4. Tian, Y., Tang, Y.: Research and utilization of dam construction technology used by elate sand gravel on the upstream cofferdam project of the Gongguo bridge. Water Resour. Plan. Des. **1**, 51–57 (2011)
5. Jia, J., Lino, M., Jin, F., et al.: The cemented material dam: a new, environmentally friendly type of dam. Engineering **2**(4), 490–497 (2016)
6. Jia, J., Ma, F., Li, X., et al.: Study on material characteristics of cement-sand-gravel dam and engineering application. J. Hydraul. Eng. **5**, 578–582 (2006)
7. Wang, H., Song, Y.: Mechanical properties of roller compacted concrete under multiracial stress state. J. Hydraul. Eng. **42**(9), 1095–1101+1109 (2011)

Cascading Mechanisms Behind the 2018 Indonesia Sulawesi Earthquake and Tsunami Disasters: Inland/Coastal Liquefaction, Landslides and Tsunami

Shinji Sassa$^{(\boxtimes)}$

Port and Airport Research Institute, National Institute of Maritime, Port and Aviation Technology, 3-1-1 Nagase, Yokosuka 239-0826, Japan
sassa@p.mpat.go.jp

Abstract. This keynote paper concisely summarizes some recent research advances on liquefied gravity flows and their consequences, with focus on coastal and submarine landslides-induced tsunamis. The review of large-scale coastal mass movements worldwide highlights the importance of such liquefied sediment flows on tsunami generation. The dynamics of liquefied gravity flows is characterized by the multi-phased physics that involves the phase change process in which the transitory fluid-like particulate sediment reestablishes a grain-supported framework during the course of flowage. The integration of fluid dynamics and soil mechanics approaches is indispensable in order to facilitate a rational prediction of the phenomena concerned. The relevant features and the concept for the analytical framework of the dynamics of liquefied gravity flows are described. The paper then presents and discusses the cascading mechanisms behind the 2018 Indonesia Sulawesi earthquake and tsunami disasters. These will facilitate a better understanding of the richness of the physics involved, and promote multidisciplinary studies for disaster prevention and mitigation.

Keywords: Costal and submarine landslide · Liquefaction · Gravity flow · Tsunami · Sulawesi earthquake · Cascading mechanisms

1 Introduction

The 2011 off the Pacific Coast of Tohoku earthquake and tsunami devastated the eastern part of Japan and resulted in more than 15,000 fatalities. The damage to infrastructures were broadly divided into the damage along the coast and at river mouths by the tsunami and the damage inland and along the coast by the seismic motion, where the latter was closely linked with the occurrence of liquefaction. Indeed, the spatial scale of the liquefaction damage caused by this earthquake was the largest recorded anywhere in the world, extending about 500 km along the eastern Japan [1]. The relevant feature of the Tohoku earthquake lies in its long duration (~200 s) that was about 10 times longer than that of the 1995 major Kobe earthquake. This fact has led to the development of a new liquefaction prediction and assessment method capable of considering the influence of the waveforms and durations of earthquakes [2], which led to upgrading of the technical standards in Japan [3]. Among the infrastructures, dams behaved generally well, suffering only minor to moderate damage, a single exception of which was the Fujinuma embankment dams that failed due to a large slide during the Tohoku earthquake [4, 5]. By contrast, the Tohoku earthquake tsunami affected a 2,000 km stretch of Japan's Pacific coast and propagated more than 5 km inland with a maximum inundation height of 19.5 m [6]. The tsunami caused significant damage and destruction of breakwaters, whose failure mechanisms were owing to the coupling processes of overflow and seepage [7].

The 2018 Indonesia Sulawesi earthquake and tsunami disasters, which resulted in more than 4,000 fatalities, were indeed cascading disasters that involved the phenomena of liquefaction, landslides and tsunami. The focal mechanism of the earthquake was strike-slip faulting, being unlikely to produce significant tsunamis, the earthquake in fact caused devastating tsunamis. This earthquake brought about extensive liquefaction and liquefaction-induced flow slides inland as well as at coastal land [8–10].

This paper presents and discusses the cascading mechanisms behind the 2018 Indonesia Sulawesi earthquake and tsunami disasters. This is based on our recent studies on liquefied gravity flows and tsunamis [8, 11–13]. Below, the paper is organized as follows. Large-scale coastal mass movements and their impacts will first be outlined, with focus on the submarine landslide-induced tsunamis. The relevant features and the concept for the analytical framework of the dynamics of liquefied gravity flows will then be presented. This is followed by the description and discussion of the cascading mechanisms of the 2018 Indonesia Sulawesi earthquake and tsunami disasters.

2   Review of Large-Scale Coastal Mass Movements and Their Impacts: Submarine Landslide-Induced Tsunamis

A plenty of large-scale coastal mass movements and their impacts have been reported to date. Some representative examples are outlined below, showing the submarine landslide-induced tsunamis.

The volcanic activity of Stromboli Island in 2002 induced submarine landslides that caused a tsunami, inflicting damage on housing and infrastructures with no fatalities [14]. The submarine landslide off Papua-New Guinea in 1998 caused a tsunami that resulted in 2,200 deaths [15]. The Great Alaskan earthquake that took place in Seward and Valdez, Alaska in 1964 induced submarine landslides and tsunamis. The submarine landslides-generated tsunamis caused the loss of more than one hundred lives and properties, such that almost 90% of the lives lost in the earthquake were attributed to tsunamis, and about 80% of those deaths were caused by the submarine landslides-generated tsunamis rather than tectonically generated tsunamis [16]. The Grand Banks earthquake in 1929 induced submarine landslides that caused a tsunami. A submarine cable broke over a distance of 1,000 km from its source, and the submarine landslide-induced tsunami devastated a small village in Newfoundland with dozens of fatalities [17, 18]. The Storegga Slides that took place offshore Norway about 6000 BC caused tsunamis that were recorded all along the coasts of the Greenland-Norwegian Sea [19]. In Japan, a major earthquake in 1596 caused an extensive submarine landslide that gave rise to the total collapse and sink of Uryu Island located in the Beppu Bay of Oita Prefecture, Japan. This submarine landslide caused a tsunami that resulted in numerous casualties with the death of over 700 people, according to historical records [20]. Notably, liquefaction-induced submarine landslide has been reported to cause the devastating tsunami. Other documented submarine landslide-induced tsunamis involve the 2009 event where the Suruga Bay earthquake induced a submarine landslide that caused a tsunami with a height more than double the tectonically induced tsunami [21]. More recently, the Haiti earthquake in 2010 induced extensive liquefaction in the river delta area that caused coastal landslides and tsunamis [22]. This earthquake was of strike-slip faulting and hence, rather than tectonically induced tsunamis, the liquefaction-induced coastal landslides caused the most severe tsunamis locally [22].

3   Dynamics of Liquefied Gravity Flows

This section describes some relevant features of the dynamics of liquefied gravity flows. With reference to Fig. 1, a gravity flow of liquefied sediment is triggered by significant coastal liquefaction, which is followed by the collapse and flow of the liquefied

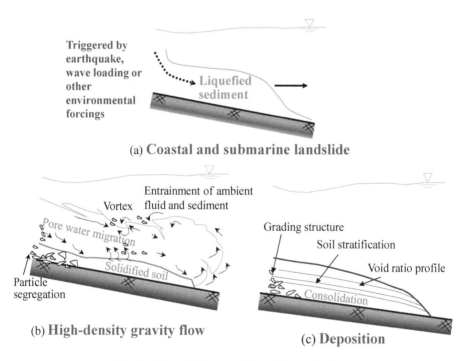

Fig. 1. Relevant features of liquefied gravity flows

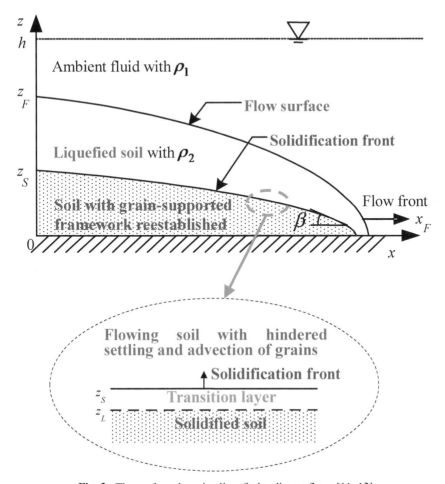

Fig. 2. Theory for submarine liquefied sediment flows [11, 12]

soil under gravity. It is categorized as a coastal/submarine landslide that transforms itself into a high-density gravity flow and subsequently flows out over a long distance, leading to re-deposition. The concurrence processes and the dynamics of liquefied gravity flow that may have a significant impact on tsunami generation described above are governed by two-phase physics [11, 12]. Indeed, pore water migration and associated solidification accompanies the process of flowage leading to redeposition, affecting the dynamics of liquefied gravity flows (Fig. 1). Namely, a solidified region develops in the course of flowage, reestablishing a grain-supported framework. Highly concentrated sediment flows above the solidification front. In contrast, the region above the flow surface is effected by turbulence/vortex and entrainment of ambient water, since this region contains only dilute sediment clouds in suspension.

The definition of the theory for the dynamics of liquefied gravity flows is shown in Fig. 2. A body of submerged granular soil undergoes liquefaction and collapses under gravity. The liquefied flow then undergoes hindered settling and advection of grains while undergoing progressive solidification (Fig. 1). Progressive solidification is a sort of phase-change process that allows transitory fluid-like particulate sediment to reestablish a grain-supported framework during the process of flowage [11, 12]. The pore water pressure p at a generic point in the sediment may be divided into two components. That is to say,

$$p = p_s + p_e \tag{1}$$

where p_s is the hydrostatic pressure and p_e represents the excess pore pressure due to contractancy (i.e., tendency for volume reduction under shearing) of the sediment. The domain of liquefied flow is formulated based on a two or three dimensional system of Navier-Stokes equations considering the effect of the excess pore pressure P_e, accompanied by the equations for hindered settling and advection of grains during the course of liquefied flow, while the domain undergoing solidification is formulated based on a two or three dimensional equation of consolidation [11, 12]. The liquefied flow domain and the solidifying domain are consistently coupled through a thin transition layer with zero effective stress yet having marginally discernable stiffness, as shown in Fig. 2. By introducing the transition layer, one can reproduce the phase change in which the transitory fluid-like particulate sediment reestablishes a grain-supported framework involving effective stress, thereby enabling the advance of the solidification front. The evolution of the flow and solidification surfaces are traced as part of solution by using a volume-of-fluid (VOF) technique. Here the soil undergoing solidification has zero velocities and acts as being an obstacle to the flowing liquefied soil. The details of the related numerical scheme are summarized in [11]. The analytical framework called LIQSED-FLOW described above has been validated against a set of two-dimensional flume tests. It consistently simulates the dynamics of liquefied gravity flows leading to redeposition. Indeed, the predicted features of flow stratification, deceleration and redeposition following the occurrence of liquefaction conform well to the experimentally observed features of the liquefied flows [11, 23]. The LIQSEDFLOW was also applied to a field event where a major earthquake on November 8, 1980 triggered extensive liquefaction and submarine flow slides that ranged 20 km along and 1 km across shelf on a very gentle 0.25° slope on the Klamath River delta, California [24]. The predicted flow dynamics and the associated morphology of the 1980 submarine slide are found to be consistent with the field evidence, and demonstrate the crucial role of the two-phase physics in the formation of scarps and terrace in submarine liquefied sediment flows [12].

4   Cascading Mechanisms of the 2018 Indonesia Sulawesi Earthquake and Tsunami Disasters

The cascading mechanisms behind the 2018 Indonesia Sulawesi earthquake and tsunami disasters are discussed below. As shown in Fig. 3, the earthquake with a moment magnitude of 7.5 was of strike-slip faulting whose fault length was over 150 km. It caused extensive liquefaction and liquefaction-induced flow slides at three major areas inland where several hundred meters to kilometer scale flow-out distances were observed. Sassa and Takagawa [8] suggested that the basic mechanism at work in the liquefaction and liquefied flows may be described as follows. First, alluvial thick loose deposits of sand with high groundwater level were subjected to a strong ground shaking of the Mw = 7.5 earthquake with its shallow focal depth of <10 km. Second, the ground contained a certain amount of silt and clay, with essential fractions being sand, and these may have enhanced the flowability of the liquefied soil. Third, the presence of confined groundwater imposed osmotic pressure, namely, seepage, on the occurrence of liquefaction and further promoted the flowability of the liquefied soil. All these factors may have helped the substantial liquefied flows occur on a very gentle inland slope of around only 1° (Fig. 3). Here, the above mechanism proves to be the case, based on the results of our recent experiments on liquefied gravity flows [13], where a submerged granular slope was subjected to a strong ground shaking in the presence and absence of percolation (osmotic pressure) and with and without involvement of fines. Figure 4 illustrates the experimental setup for earthquake-induced liquefied flows in a centrifuge [13]. The centrifuge test results on the flowability of liquefied soils under earthquake loading are

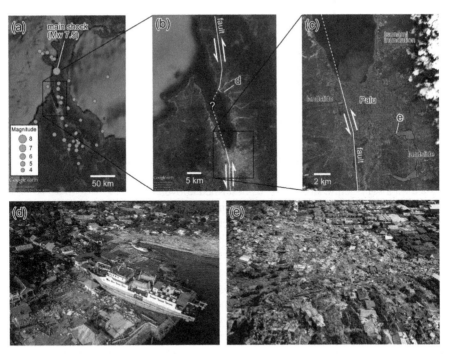

Fig. 3. (a) Spatial distribution of the epicenters of the Sulawesi earthquake and aftershocks within 3 days [25]. (b) The fault crosses the Palu Bay and Palu city in the North-South direction [26]. (c) The areas of tsunami inundation and landslides are marked (satellite image: [27]). Destructive tsunamis struck the coastal area (d) and kilometer-scale landslides (e) occurred in several places (photos: [28], both locations are shown in (b) and (c), respectively). Adapted from Fig. 1 of Sassa and Takagawa [8].

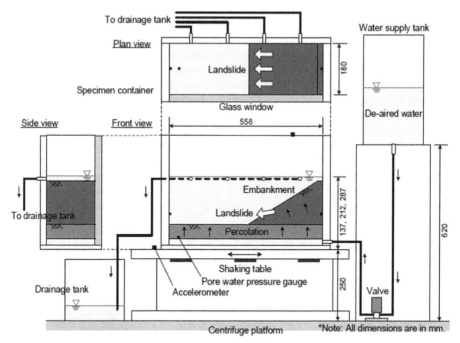

Fig. 4. Experimental setup for earthquake-induced liquefied flows in a centrifuge. Adapted from Takahashi et al. [13].

shown in Fig. 5. The flow speeds as well as the flow-out distances increased substantially depending on the presence of osmotic pressure and fines contents. Indeed, the flow speed and the flow-out distance of the liquefied sands with certain fines contents and osmotic pressure were more than ten times higher and fivefold greater than those of liquefied sands, respectively. These results could account for how such substantial liquefied flows ensued on a very gentle slope of 1° as described above.

Extensive liquefaction and liquefaction-induced flow slides took place not only inland but also along the coast. In fact, Sects. 2 and 3 of this paper have shown that significant coastal liquefaction can result in a gravity flow of liquefied soil that can cause a tsunami. A comparison with the 2010 Haiti earthquake tsunami described above indicates that essentially the same occurred at both coasts [8]. Namely, the strong strike-slip fault earthquake caused liquefaction, triggering the total collapse of the coastal land and its flow that resulted in a substantial tsunami. A notable difference between the 2010 Haiti earthquake and the 2018 Sulawesi earthquake is that such liquefaction-induced total collapse and the subsequent flow of coastal land occurred not only at one place, as in the Haiti earthquake, but at several (at least nine) places, simultaneously causing multiple substantial tsunamis throughout the bay and coastal areas [8]. The comparison

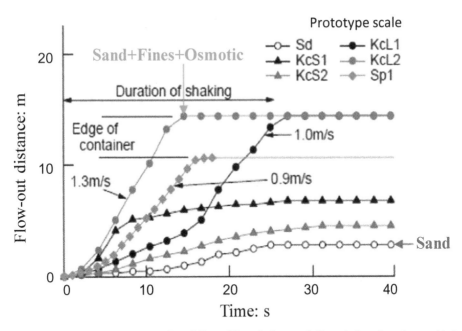

Fig. 5. Centrifuge test results on flowability of liquefied ground. Descriptions have been added to Takahashi et al. [13]. Tests were conducted at 50 g. The data shown correspond to a prototype scale. KC and SP denote kaolin clay and silica powder, and the intermediate S and L represent the fines contents, 4.8% and 16.7%, and the number 1 and 2 show without and with osmotic pressure (percolation), respectively

between the Haiti case [22] and the Sulawesi case [8] shows that the spatial area of collapse of the coastal landslides in the Sulawesi case was about ten times greater than that of the Haiti case that generated a substantial tsunami with the maximum flow depth of 3 m and a greater inundation height at the coast. Confronting the locations of the coastal retreats with those of tsunami generations [8, 29], as shown in Fig. 6, indicates that all of the locations of the multiple tsunami generations, alongshore distributions and directions conform to the locations, distributions and directions where the coastal lands collapsed and flowed due to the occurrence of liquefaction. Hence, the above results and discussion demonstrate that the strong strike-slip earthquake induced extensive inland/coastal liquefaction that caused flow slides inland in the former, and coastal and submarine landslides in the latter, as characterized by liquefied gravity flows, which resulted in multiple tsunamis. It is a subject for future study to clarify how coastal and submarine landslides-induced tsunamis could be quantitatively compared with tectonically induced tsunamis.

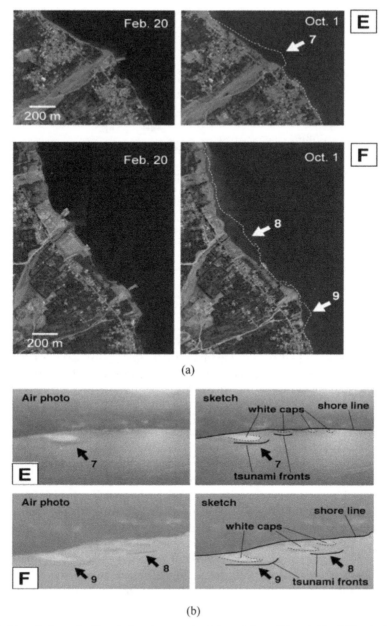

Fig. 6. Liquefied gravity flow-induced tsunamis. Adapted from (b) and (c) of Fig. 4 of Sassa and Takagawa [8]. (a) Comparisons of satellite images before and after the earthquake (E, F: [27]). Significant coastal retreat areas are indicated by white arrows. (b) Aerial photos of multiple tsunami occurrence at the areas E and F, taken just after the earthquake by Mafella [29]. Interpreted sketches are shown aside. The black arrows indicate tsunami fronts and the arrows with numbers 7–9 match those with the same numbers in the satellite images of areas E and F in (a). White caps of breaking waves are shown on the land side of the tsunami fronts.

5 Conclusions

The paper has concisely summarized some recent research advances on liquefied gravity flows and their consequences, with emphasis on the coastal and submarine landslides-induced tsunamis. Large-scale coastal mass movements worldwide have been reviewed, which highlight the importance of such liquefied sediment flows on tsunami generation. The dynamics of liquefied gravity flows is characterized by the multi-phased physics involving the phase change process in which the transitory fluid-like particulate sediment reestablishes a grain-supported framework during flowage. The integration of fluid dynamics and soil mechanics approaches is indispensable for a rational prediction of the relevant concurrent phenomena. The paper has then presented and discussed the cascading mechanisms behind the 2018 Indonesia Sulawesi earthquake and tsunami disasters. It is hoped that these may facilitate a better understanding of the richness of the physics involved and promote multidisciplinary studies for disaster prevention and mitigation.

References

1. Kazama, M., Noda, T.: Damage statistics (Summary of the 2011 off the Pacific Coast of Tohoku earthquake damage). Soils Found. **52**(5), 780–792 (2012)
2. Sassa, S., Yamazaki, H.: Simplified liquefaction prediction and assessment method considering waveforms and durations of earthquakes. J. Geotech. Geoenviron. Eng. **143**(2), 1–13 (2017). https://doi.org/10.1061/(ASCE)GT.1943-5606.0001597
3. MLIT: Ministry of Land, Infrastructure, Transport and Tourism, Japan.: Ground Liquefaction. Technical standards and commentaries for port and harbor facilities of Japan (2018)
4. JCOLD: Review of the cause of Fujinuma-Ike failure - Summary report – 25 January 2012 (2012). http://www.jcold.or.jp/e/activity/FujinumaSummary-rev.120228.pdf
5. Pradel, D., Wartman J., Tiwari, B.: Failure of the Fujinuma dams during the 2011 Tohoku earthquake. In: Geo-Congress 2013: Stability and Performance of Slopes and Embankments, vol. 231, pp. 1559–1573. Geotechnical Special Publication (2013)
6. Mori, N., Takahashi, T., Yasuda, T., Yanagisawa, H.: Survey of 2011 Tohoku earthquake tsunami inundation and run-up. Geophys. Res. Lett. **38**, L00G14 (2011). https://doi.org/10.1029/2011GL049210
7. Sassa, S., Takahashi, H., Morikawa, Y., Takano, D.: Effect of overflow and seepage coupling on tsunami-induced instability of caisson breakwaters. Coast. Eng. **117**, 157–165 (2016). https://doi.org/10.1016/j.coastaleng.2016.08.004
8. Sassa, S., Takagawa, T.: Liquefied gravity flow-induced tsunami: first evidence and comparison from the 2018 Indonesia Sulawesi earthquake and tsunami disasters. Landslides **16**(1), 195–200 (2019). https://doi.org/10.1007/s10346-018-1114-x
9. Carvajal, M., Araya-Cornejo, C., Sepulveda, I., Melnick, D., Haase, J.S.: Nearly instantaneous tsunamis following the *Mw* 7.5 2018 Palu earthquake. Geophys. Res. Lett. **46**, 5117–5126 (2019). https://doi.org/10.1029/2019GL082578
10. Polcari, M., Tolomei, C., Bignami, C., Stramondo, S.: SAR and optical data comparison for detecting co-seismic slip and induced phenomena during the 2018 Mw 7.5 Sulawesi earthquake. Sensors **19**(18), 3976 (2019). https://doi.org/10.3390/s19183976
11. Sassa, S., Sekiguchi, H.: LIQSEDFLOW: role of two-phase physics in subaqueous sediment gravity flows. Soils Found. **50**(4), 495–504 (2010)
12. Sassa, S., Sekiguchi, H.: Dynamics of submarine liquefied sediment flows: theory, experiments and analysis of field behavior. In: Advances in Natural and Technological Hazards Research, vol. 31, pp. 405–416. Springer (2012)

13. Takahashi, H., Fujii, N., Sassa, S.: Centrifuge model tests of earthquake-induced submarine landslides. Int. J. Phys. Model. Geotech. (2019). https://doi.org/10.1680/jphmg.18.00048

14. Tinti, S., Manucci, A., Pagnoni, G., Armigliato, A., Zaniboni, F.: The 30 December 2002 landslide-induced tsunamis in stromboli: sequence of the events reconstructed from the eyewitness accounts. Nat. Hazards Earth Syst. Sci. 5, 763–775 (2005)

15. Tappin, D.R., Watts, P., McMurtry, G.M., Lafoy, Y., Matsumoto, T.: The Sissano, Papua New Guinea tsunami of July 1998: offshore evidence on the source mechanism. Mar. Geol. 175, 1–23 (2001)

16. Haeussler, P., et al.: Submarine landslides and tsunamis at Seward and Valdez triggered by the 1964 magnitude 9.2 Alaska earthquake. Alaska Geol. 39(2), 1–2 (2008)

17. Heezen, B.C., Ewing, M.: Turbidity currents and submarine slumps and the 1929 grand banks earthquake. Am. J. Sci. 250, 849–873 (1952)

18. Whelan, M.: The night the sea smashed Lord's Cove. Can. Geograph. 114, 70–73 (1994)

19. Solheim, A., Bryn, P.B., Sejrup, H.P., Mienert, J., Berg, K.: Ormen Lange-an integrated study for safe development of a deep-water gas field within the Storegga slide complex, NE Atlantic continental margin; executive summary. Mar. Pet. Geol. 22, 1–9 (2005)

20. Kato, K.: Collapse of Uryu island: Unraveling Japan's Atlantis legend. Newton, August issue, pp. 88–97 (1991). (in Japanese)

21. Shimbun, Y.: Submarine landslide induced amplification in tsunami: six-fold in the Suruga Bay earthquake, YOMIURI ONLINE (2010)

22. Hornbach, M.J., et al.: High tsunami frequency as a result of combined strike-slip faulting and coastal landslides. Nat. Geosci. 3, 783–788 (2010)

23. Amiruddin, Sekiguchi, H., Sassa, S.: Subaqueous sediment gravity flows undergoing progressive solidification. Norwegian J. Geol. 86, 285–293 (2006)

24. Field, M.E., Gardner, J.V., Jennings, A.E., Edwards, E.D.: Earthquake-induced sediment failures on a 0.25° slope, Klamath river delta, California. Geology 10, 542–546 (1982)

25. USGS: Earthquake Hazards Program. https://earthquake.usgs.gov/earthquakes/search/. Accessed 30 Oct 2018

26. Valkaniotis, S.: Displacement from #Sentinel2 @CopernicusEU image frames for the whole length of the #Palu #earthquake sequence. [twitter post]. https://twitter.com/SotisValkan/status/1047515941570007042. Accessed 30 Oct 2018

27. Digital Globe Open data Program: Satellite Images of Palu and Donggala, Sulawesi, Indonesia. https://www.digitalglobe.com/opendata/indonesia-earthquake-tsunami/. Accessed 30 Oct 2018

28. Reuters: Indonesia's quake-hit Sulawesi island from above. https://www.reuters.com/news/picture/indonesias-quake-hit-sulawesi-island-fro-idUSRTS23SN1. Accessed 30 Oct 2018

29. Mafella, R.: Batik Air ID 6231 scheduled to depart at 17.55, door closed at 17.52 then pushed back. 18.02 after cleared for take off, tower building collapsed. [Instagram Post]. https://www.instagram.com/p/BoRttnsn5po/?taken-by=icoze_ricochet. Accessed 30 Oct 2018

Update on ICOLD Embankment Dam Technical Committee Works

Jean-Pierre Tournier[✉]

Chair of ICOLD Embankment Dam Technical Committee Hydro-Quebec, Montreal, Canada
Tournier.Jean-Pierre@hydro.qc.ca

Abstract. The ICOLD Embankment Dams Technical Committee is one of the most active among the ICOLD TC. The paper presents a short resume of the last eight (8) Bulletins published, on the ongoing proposed Bulletins and an update of the activities of the working groups.

1 Part A – A Brief Review of the Last Already Published Bulletins

1.1 Embankment Dams on Permafrost A Review of the Russian Experience (Bulletin 133)

Zones where permafrost is found (cryolitic zones) are characterized by a rigorous climate and complicated topographic, geological and hydrogeological conditions. During the construction of hydraulic structures in these zones, we are faced with a number of specific problems, not existing outside these zones. Therefore, it is reasonable to give an insight into the structural features and the engineering characterizing the operation and repair of dams built in these zones; particularly on the experience gained in exploration, design, construction and operation.

The operation of hydropower plants built in the regions of the Polar North proved highly efficient. The possibility to build highly efficient hydropower stations is provided by deep-water rivers, relatively favorable engineering-geological conditions and a low agricultural value of the flooded areas. The specific natural conditions and the lack of developed transport connections with industrial regions give preference to dam construction using local materials. Embankment dams are built with soils using a homogeneous or non-homogeneous type of design and various materials as the impervious part of the dam. The main type of embankment dams, especially high dams, is earth and rockfill with a central core and sometimes an inclined core, and an effective drainage of the downstream shell to improve structural stability and prevent frost penetration and icing inside the dam. Rockfill dams with impervious elements of steel, asphalt concrete and polymer materials are also promising. It is possible to make an impervious core by grouting a coarse gravel soil with a clay-cement or a clay binder.

The regions with permafrost occupy over 25% of the total land area on the earth and 65% of the territory of Russia. The current paper presents the analysis of natural conditions in cryolitic zones of Russia, the special features of the layout, type and structure of fill dams, the ice and water discharged through the spilling structures and the special parameters of fill dam technology. Typical examples are given of dam deformation, repair and reconstruction, as well as operation in a severe climate.

© Springer Nature Switzerland AG 2020
J.-M. Zhang et al. (Eds.): ICED 2020, SSGG, pp. 47–55, 2020.
https://doi.org/10.1007/978-3-030-46351-9_5

1.2   Weak Rocks and Shales in Dams (Bulletin 134)

Shales together with calcareous rock predominate among the sedimentary rock in the earth's crust. It is estimated that shale constitute about 50% of the rocks exposed at or just below the earth's surface. There are therefore many potential dam sites on shale or other weak rocks. In the past such sites have often been avoided because of the uncertainties and reports of the difficulties of constructing dams on weak foundations and of weak rock fills. This bulletin aims to assist dam engineers Io develop such sites effectively in future. It advise on the means of using these materials by developing an understanding of their properties and by learning from the experiences of others in their successful use. It provide information for the international dam community on the use of weak rock and shales as a dam building material. It is predominantly about weak rock and shales used as fill in dam but because foundations on shales often present challenge some reference is made to shales in the foundations of dams.

However, the bulletin is not a completely comprehensive manual giving guidance on the use of weak rocks in dam fill and many aspects will need to be investigated further following the guidance given in other ICOLD Bulletin and the many references listed at the end.

1.3   Concrete Face Rockfill Dams – Concepts for Design and Construction (Bulletin 141)

The concrete face rockfill dam, CFRD, had its origin in the mining region of the Sierra Nevada in California in the 1850s. Experience up to 1960 using dumped rockfill, demonstrated the CFRD to be a safe and economical type of dam, but subject to concrete face damage and leakage caused by the high compressibility of the segregated dumped rockfill. As a result, the CFRD became unpopular, although rockfill had been demonstrated to be a high strength and economical dam building material. Partly in response to these problems, the earth core rockfill dam, with compressible dumped rockfill, was developed. The dumped rockfill was found to be compatible with the earth core and its filters. With the advent of vibratory roller compacted rockfill in the 1950s, the development of the CFRD resumed. During the 1965–2000 development period, many CFRDs were adopted to replace a previously selected arch, gravity or earth-core-rockfill dam type. Reason s for the change included the late discovery of adverse foundation conditions for a concrete dam, cost, or lack of appropriate core material for an earth-core-rockfill dam. Today, the CFRD is an established major dam type to be included in initial project feasibility studies.

The bulletin 70 "Rockfill Dams with Concrete Facing" was published in 1989. Several symposium were held on CFRD during the decade of the 1990s and the Barry Cooke Volume was published in 2000. The update bulletin 141 contains eleven chapters devoted to design concepts, analysis, foundation treatment, instrumentation, construction and performance. The CFRD dam has many attractive features in design, construction and schedule. However in the beginning of the 2000s, the leakage performance of the CFRD has been criticized in the literature and several causes were suggested for the high leakage rates. Particular design details and/or construction defects lead to larger absolute and differential deformations of the face slab, which can then lead to face slab cracking. Avoiding these causes by means of appropriate selection of filter and rockfill materials, upstream and downstream shell placement in thinner layers along with generous use of water during compaction, elimination of rock protrusions downstream of the perimeter joint, and avoiding inappropriate shell construction sequences, will lead to smaller deformations and a reduction in face slab cracking. One of the purposes of this bulletin is to emphasize that careful selection of design and construction details is extremely important to avoid face cracking and embarrassing leakage rates.

1.4 Cutoffs for Dams (Bulletin 150)

Dams are constructed to retain or store water. To minimize the flow of water through the dam/foundation system special impervious zones or elements must be designed and constructed. Dams constructed of concrete can be considered practically impervious, except for possible leaky joints. Embankment or fill dams require a zone of low permeability soil, asphalt, or concrete, which can be placed either in the interior of the dam or with the latter two materials also on the upstream face. Flow of water through the foundation below the dam is controlled by the prevailing geological conditions. Seepage through pervious strata (alluvial deposits, residual soils, etc.) can be controlled by a barrier or cutoff consisting of a sequence of impervious elements (piles, panels) reaching down to a stratum of much lower permeability, usually rock. Seepage through rock is usually controlled by single or multiple row grout curtains. Rock, however, can also be highly pervious, for example in the case of karst, which may reach to great depth, or when rock is intensely broken or crushed in regions of high tectonic stresses. For such cases cutoffs may be more appropriate than grout curtains.

This Bulletin is limited to foundation treatment methods using cutoff-type barriers. Due to recent experiences, high emphasis is given to alluvial deposits throughout this document; however, different materials, such as pervious residual soil, pervious laterites and saprolites, highly fractured and weathered rock, and karst may require cutoff. The construction of cutoffs has made significant advances during the last two decades, mainly through the development of more powerful machinery for drilling and excavation, but also through the introduction of new concepts and techniques, such as jet grouting and deep soil mixing. The concept of diaphragm wall techniques using a cutter to provide continuous excavation originated in Japan in 1980. Rapid developments in Europe followed and cutter wheels with rock-roller bits were introduced around 1990. Since then, cutoff depths exceeding 100 m with vertical deviation of less than one percent have been accomplished. In addition, there has been development of the materials used for the sealing elements, such as plastic concrete. These less rigid materials provide better compatibility with the in-situ ground conditions surrounding the cutoff wall elements.

The following types of cutoffs are presented in this bulletin:

- Diaphragm walls
- Vib walls
- Pile walls
- Superimposed concreted galleries
- Jet grouting
- Deep mixing

These methods are first briefly described and then more explicitly in different chapters. In addition, the practical application of each method is illustrated by selected case histories. These case histories also demonstrate how certain difficulties specific to a particular dam site have been dealt with.

1.5   Tropical Residual Soils as Dam Foundation and Fill Material (Bulletin 151)

Dam construction across the world has recently acquired an accelerated pace as needs for water supply and renewable energy sources have increased in many countries. Many of these countries are located in areas where tropical residual soils are abundant. Unlike transported soils, residual soils originate from the in situ weathering and decomposition of the parent rock, which has not been transported from its original setting. The conditions present in humid and tropical climates provide the adequate moisture and temperature conditions to transform, through weathering processes, the underlying rocks into residual soils faster than they can be removed by erosion. The main difficulty in dealing with these soils for engineering purposes is that their characteristics are very different from those of transported soils. Furthermore, concepts and methodologies typically used in geotechnical engineering were developed in regions dominated by temperate climates. As a consequence, many of the concepts of soil and rock behavior and properties, and geological conditions have been conditioned by earth materials found there.

Despite these inconveniences, the widespread presence of residual soils does not allow us to avoid them, but challenges us, to deepen our understanding and knowledge about them to be able to accept and employ them to our advantage in future projects. Success on this challenge is determined by a proper understanding and appreciation of the engineering properties of these soils and weathered rock encountered. The purpose of this bulletin is to illustrate how these materials have been accepted and used in dam projects without imposing selection of better known materials that could jeopardize the economic viability of a project. The intention is not to provide detailed geological descriptions or characterization methods. The geologic literature offers a plethora of articles on these issues. The bulletin rather focuses on the dam engineering implications of dealing with residual soils. To understand the properties of tropical soils requires rigorous study of their composition and structure, both of which are unique consequences of the climate conditions that prevail during their formation. Therefore, the formation, composition, and the micro and macro structures of tropical residual soils are reviewed first follow by a description of the unique aspects of soil profiles, sampling and testing, engineering properties, design criteria, construction techniques and registered behavior. Finally, an extensive gathering of case histories around the world of dams built on or with residual soils are presented to illustrate these topics in a pragmatic matter.

1.6   Internal Erosion in Existing Dams, Levees and Dikes and Their Foundations

Internal erosion occurs when soil particles within an embankment dam or its foundation are carried downstream by seepage flow. It starts when the erosive forces imposed by the hydraulic loads exceed the resistance of the materials in the dam to erosion. The erosive forces are directly related to reservoir water level and the highest hydraulic loads occur when the water level in the reservoir is high during floods.

Internal erosion and piping in embankments and their foundations is the main cause of failures and accidents at embankment dams. For new dams, the potential for internal erosion can be controlled by good design and construction of the core of the dam and provision of filters to intercept seepage through the embankment and the foundations. However many existing dams, dikes and levees were inadequately zoned and not provided with filters or have filters or transitions which do not satisfy modern filter design criteria, and are susceptible to internal erosion failure, with a risk of internal erosion increasing with ageing. Accordingly, the reassessment of the safety of these dams is an important issue.

This has been recognized by the European Club of ICOLD which devoted a Working Group to the problem and much research has been carried out in recent years to better understand the physical processes and mechanics of internal erosion. This has built upon the knowledge gained from many years of earlier research and successful design of new dams. As the state of the art of internal erosion of dams, dikes and their foundations is an evolving science, readers should look in the literature for new developments following the production of this bulletin which is presented in two volumes.

1.6.1 Volume 1 (Bulletin 164)

The volume 1 of the bulletin deals predominantly with internal erosion processes and the engineering assessment of the vulnerability of a dam to failure or damage by internal erosion, with a brief oversight of monitoring for and detection of internal erosion and remediation to protect dams against internal erosion. It includes a comprehensive listing of the Terminology used in internal erosion. Many references are also given, including ICOLD publications related to internal erosion.

This volume 1 gives a statement of the problem, explaining why internal erosion is a threat to existing dams and the importance of assessing the vulnerability of individual dams to it. Then it goes through the overall process of erosion from initiation, through continuation (or arrest) of erosion, through progression, and on to breach, unless erosion is detected early enough by appropriate monitoring systems, to allow timely intervention to halt or slow the development of a breach and failure. Volume 1 provides methods to estimate the water level at which internal erosion will initiate and lead to failure in the four initiating mechanisms: concentrated leaks, suffusion, backward erosion and contact erosion.

1.6.2 Volume 2: Cases Histories, Investigations, Testing, Remediation and Surveillance (Bulletin 164a)

The volume 2 of the bulletin presents case histories of internal erosion failures and incidents. It advises and gives more details on the investigations, the appropriate sampling and testing that can be used to provide the data needed to carry out and support the analyses and decisions that engineers must make to determine and improve, if necessary, the vulnerability of dams to internal erosion. It also advises on detection and remediation, if analyses have demonstrated that it is necessary, and on surveillance and monitoring systems to check and confirm the continuing ability of the dam to resist internal erosion in the long-term.

This volume 2 provides, also, details of laboratory tests which measure properties of erosion and filter capability of soils, monitoring methods including several newly developed techniques and information on case studies relevant to the subject. It concludes with a comprehensive list of internal erosion references adding to those in volume 1.

1.7   Asphalt Concrete Cores for Embankment Dams (Bulletin 179)

Approximately 5000 years ago, a small dam was built in the Indus river valley, using asphalt as a mortar between stones. That was one of the first utilization of asphalt as an impervious material. However, the first "modern" asphalt concrete core dam, mechanically placed and compacted, was built in Germany in 1962. This method expanded first in Europe, particularly in Germany and Austria. Since the last fifteen years, this type of dam is gaining in popularity. Comprehensive research in the technology and development of asphalt concrete core machinery has been done. The Committee has decided to update the Bulletin which was published in 1992. That ambitious task was taken by specialists from Austrian, German, Norwegian, Chinese and Canadian national committee representatives.

As of today, nearly 200 asphalt concrete core embankment dams (ACED) have been built worldwide with excellent field performance or are under construction in different countries around the world. Among them, the 170 m high Quxue Dam in China is the highest; however even higher ACEDs are under design. There are now more than 50 years of successful experience for this dam type. The gain of popularity, particularly the last fifteen years, in the water retaining, hydropower and mining industries, follows the excellent recorded field performance and behavior for this type of dam, and also comprehensive researches in the technology and development of asphalt concrete core machinery. This bulletin covers the state-of-the-art of current practice after the important development in design and construction during the last 25 years. It addresses all aspects of the design, construction, performance and operation. Characteristics of asphalt concrete cores, requirements for the mix design, laboratory testing and quality control are discussed. Technical specifications are also presented and proposed. Finally, several typical case histories with characteristics and performance are given in Appendices.

2   Part B – A Brief Review of the Works in Progress

2.1   Geotextiles in Dams

A brief recognition of the history of dam engineering over the past 4 000 years shows that filter technology is a relatively young science of less than a 100 years old with the formalization of the Terzaghi criteria for non-cohesive granular filters in the early 20th century. The implementation of this filter technology contributed to a substantial improvement in dam safety from before 1950 to the period that followed. The advancement of granular filter technology for fine grained cohesive soils and geotextile filter technology has been commensurate since the 1980s. While both granular and geotextile filters have controversial issues, there are also many similarities to be found in case histories which analyses their inadequate performance. Nevertheless the change in water resource quality with time is also to be recognized. The further advancement in technology of geotextile filters and polymer science has led to a significant improvement in understanding the beneficial use of geotextiles as filters as well as the limitations thereof.

So, it was decided to update the 1986 Bulletin 55 "Geotextiles as Filters and Transitions in Fill Dams". Since 1986 significant developments in the materials of geotextiles occurred and various applications of geotextiles have been developed. The South African National Committee took the lead of that revision with the participation along with countries contributing case histories. A first draft version was proposed for comments to all the national committees. This whole process of revision and comment is not yet completed. Hence it may be concluded that geotextiles may be used as primary filters in non-critical situations and could be used as adjuncts to granular filters in some critical situations in dams. This Bulletin will contain the latest characteristics of geotextiles as well as natural filters and includes applications in dams for these materials with several case histories.

2.2 Cofferdams

The existing Bulletin 48 "River control during dam construction" was published in 1986 and focused mainly on construction for new dam. It was proposed to update and expand the topic on "Water management for dam construction and rehabilitation" with the introduction of the concept of risk evaluation and assessment. Various type of cofferdams should be presented. It was also decided to work jointly with the technical committee on "Operation, Maintenance and Rehabilitation of Dams" for that bulletin. A working group is being formed and the US National Committee will lead and develop the bulletin from USSD guidance. Members are asked to send also guidance from their country. The principle of owner design will be preferred.

2.3 Materials for Cutoffs

At the Vienna ICOLD Congress in 2018, a workshop was organized by the committee with the European Federation of Foundation Contractor (EFFC), the Deep Foundation Institute (DFI) and well-known specialists such as Trevi and Bauer to present the latest new material and materials used for the treatment of foundations particularly for the cutoffs. Trevi has presented also their special 250 m deep test excavation which was carried out as trial for possible solution for Mosul dam. A lot of advancement has been done in the technology of support and drilling fluids. Since several years, more and more companies (contractors and suppliers) use polymers in foundation works as replacement of bentonite but the knowledge is not shared and if the drilling fluids are not managed and balanced properly it could have an impact on the foundation quality. It was proposed that a working group, led by Dave Paul (USA), with the support of specialized organizations and companies makes an update and complement of the Bulletin 150 on Cutoffs, particularly on the materials used for cutoffs. It will cover all types, except soil mixing. Members are asked to send case histories and particularly records of performance of installed cutoffs.

2.4 Deformations and Cracking in Embankment Dams

The recent published Bulletin 164 on internal erosion outlines the current state of practice for assessment of the various mechanisms of internal erosion and piping, and provides guidelines for the estimation of width and depth of cracks. These are a guide only. It was decided that the development of these guidelines will be very helpful in providing tools to reduce the uncertainty in the estimation of cracks and the potential for hydraulic fracture and also improve outcomes for assessment of internal erosion and piping by concentrated leak erosion. Focus would be mainly on deformation associated with formation of potential transverse crack and hydraulic fracture mechanisms but settlement and deformation in general during and post-construction will be treated. Methods will be described for interpretation of surveillance records to assist in the crack assessment process, including evaluation of strains and likelihood of strain concentration and potential for formation of low stress zones susceptible to cracking and hydraulic fracture. These methods (in conjunction with judgement based methods) include numerical ones for assisting with evaluation of the stress and strain conditions within embankment dams. A working group, led by G. Hunter (Australia), is in charge of that proposed bulletin which will be a complementary to the Bulletin 164 and will include also a summary of case histories where cracks, particularly transverse ones, within embankments have been described.

2.5 Updating the CFRD Bulletin

According to the development of some new technologies and the construction of very high (above 200 m) CFRD, particularly in China, it will be important to have an update document on CFRD. This document will include the new developments regarding the compression failure of vertical joints in the upstream face slab, the guidance on analysis and practice to avoid cracking on concrete face particularly on high dams in narrow valleys. A working group is being formed with A. Marulanda (Colombia), J.J. Fry (France) and CHINCOLD to work on that document.

2.6 Compaction of Earth Fill in Embankment Dams

At the 2019 meeting in Ottawa presentations have done regarding compaction and showing how different functions of fills in embankment dams led to different compaction processes comparing those normally used in earthworks to limit settlement in road or railway embankments for example. In embankment dams, recommendations are generally made to limit differential settlements between shells, transitions, filters and core. The meeting suggested and agreed that the proposed bulletin must be in two volumes or separate bulletins, one covering the earth fill (core material) and one for the other material (sand, gravel, rockfill). The French committee will lead the volume on core material.

References

ICOLD Bulletin 133: Embankment dams on permafrost-a review of the Russian experience. International Commission on Large Dams, Paris (2008)

ICOLD Bulletin 134: Weak rocks and shales in dams. International Commission on Large Dams, Paris (2008)

ICOLD Bulletin 141: Concrete face rockfill dams-Concept for design and construction. International Commission on Large Dams, Paris (2011)

ICOLD Bulletin 150: Cutoffs for dams. International Commission on Large Dams, Paris (2018)

ICOLD Bulletin 151: Tropical residual soils as dam foundation and fill material. International Commission on Large Dams, Paris (2017)

ICOLD Bulletin 164: Internal erosion of existing dams, levees and dikes, and their foundations, vol. 1. International Commission on Large Dams, Paris (2017)

ICOLD Bulletin 164: Cases histories, investigations, testing, remediation and surveillance, vol. 2. International Commission on Large Dams, Paris. (Preprint)

ICOLD Bulletin 179: Asphalt concrete cores for embankment dams. International Commission on Large Dams, Paris. (Preprint)

Particle Mobilization and Piping Erosion of Granular Soil Under Various Fluid Characteristics and Flow Conditions

Ming Xiao$^{(\boxtimes)}$, Benjamin Adams, Asghar Gholizadeh-Vayghan, and Yuetan Ma

Pennsylvania State University, University Park, PA 16802, USA
mzx102@psu.edu

Abstract. This paper presents the soil's internal erosion in earthen embankments due to seepage with various physicochemical fluid characteristics and flow conditions. The paper includes two components: (1) experimental study on the relative and interactive effects of fluid's viscosity, pH, and ionic strength on the incipient motion of a single granular particle under laminar flow condition, and (2) bench-scale piping erosion progression of sand considering the same seepage's physicochemical characteristics under turbulent flow. An innovative experimental setup was designed and constructed that can simultaneously adjust fluid's viscosity, pH, and ionic strength and provide repeatable test results on particle's incipient motion and soil erosion rate index. This research showed the relative and interactive effects of three physicochemical characteristics (viscosity, pH and ionic strength) of permeating fluid on particle mobilization and on piping erosion of a sandy soil. This paper suggests in the field evaluation of piping in earthen embankments, if the subsurface seepage is known to possess different physicochemical characteristics from those of tap water or distilled water, the fluid's properties should be considered in the laboratory tests of piping.

Keywords: Particle mobilization · Piping erosion · Viscosity · pH · Ionic strength

1 Introduction

Permeating fluids through earthen dams or levees may have various physicochemical characteristics such as viscosity (due to temperature variation and particulate concentration), ionic strength (due to seawater invasion), and pH. Various combinations of these characteristics can be significantly different from those of tap water, which is used in most laboratory erosion testing to understand erosion behavior of soils. The past experimental research has demonstrated that permeating fluids consisting of de-ionized water and various concentrations of sodium chloride (thus different ionic strength) can induce different erosion behaviors [1–3]. The studies conducted by Hubbe [4, 5], Sharma et al. [6], and McDowell-Boyer [7] attempted to understand the hydrodynamic forces that are required to dislodge particles from flat surfaces. These studies pointed out that the hydrodynamic forces vary with flow rate, particle size, particle elasticity, ionic strength, pH,

© Springer Nature Switzerland AG 2020
J.-M. Zhang et al. (Eds.): ICED 2020, SSGG, pp. 56–74, 2020.
https://doi.org/10.1007/978-3-030-46351-9_6

and the London-van der Waal forces and electrical double layer forces between colloidal particles and the surface of the solid matrix. They concluded that larger particle size, lower pH, and higher ionic strength increase the critical hydrodynamic force required for particle dislodging, resulting in less erosion. Previous researchers also conducted erosion tests on soil columns using permeating fluids of varying ionic strength. Sherard et al. [8], Arulanandan et al. [1], and Reddi et al. [2] reported that higher content of dissolved salts in the water (i.e., higher ionic strength) resulted in lower erosion of clayey soil.

Particle mobilization is a fundamental aspect in the process of soil erosion and involves the balance of forces acting on individual grains within a soil matrix. Some of the potential forces involved in the mobilization process include self-weight, buoyancy, hydrostatic fluid pressure, inter-particle contact and electrical forces, rotational resistance, and shear forces [9–11]. The condition for incipient motion of particles is known as the critical or threshold condition and is shown to be affected by fluid's physicochemical characteristics. After soil particles mobilization, the subsequent erosion behavior can still be influenced by seepage's physicochemical properties. Knowledge of the near-threshold flow conditions is important for a variety of applications ranging from civil and environmental engineering to stream ecology [12]. Threshold conditions for particle mobilization (entrainment) also depend on flow conditions. In laminar flow, horizontal hydraulic drag force (or hydraulic shear stress) is known as the driving force to overcome resistance between the particle and the bed (smooth or rough) beneath. In turbulent flow, particle mobilization occurs when turbulent lift force overcomes gravity [13, 14].

The past research preliminarily revealed the individual effects of a fluid's viscosity, pH, and ionic strength on the mobilization of colloidal particles and the erosion of clayey soils. Whether and how these physicochemical characteristics affect the incipient motion of granular particles (individual grain diameters on the order of millimeters) and subsequent erosion process such as piping erosion is unknown. The effect of these physicochemical characteristics may also be interactive in that the magnitude of their individual effects may be dependent on the levels of the other characteristics. The relative and interactive effects of fluid physicochemical characteristics on the incipient motion of granular particles under laminar flow condition and piping erosion progression under turbulent flow condition are presented in this paper.

2 Experimental Methodology on Particle Mobilization

2.1 Experimental Setup

The experimental setup, as shown in Fig. 1, includes a flow cell that houses a single spherical test particle, constant-head upstream and downstream reservoirs that create a hydraulic head difference to drive the flow through the flow cell, a microscopic video camera and image capturing system to monitor the test particle mobilization. The test particles were polished soda-lime glass spheres specially designed and manufactured for applications requiring maximum uniformity between particles and minimum variation in sphericity. The particles had densities of 2.5 g/cm^3 ± 0.1 g/cm^3 and diameters of 0.69 ± 0.02 mm. The high uniformity of these particles was expected to help reduce the variance introduced when different experimental runs were performed with different test particles or with the same test particle placed in a different orientation with respect to the particle bed and the approaching flow. The flow cell includes two components: a rectangular, open-topped box and a removable top plate. The box had inner dimensions of 2.50 cm (high) × 7.62 cm (wide) × 21.50 cm (long). The upstream side of the flow cell was packed with glass spheres ranging between 0.6 and 0.9 cm (nominal) in diameter to distribute the incoming jet-motion flow before it reached the test particle. The top plate included a valve to facilitate easy removal of air during the initial saturation process. During testing, the entire flow cell was filled with water and there was no free water surface.

Fluid was provided to the flow cell by an upstream constant-head reservoir while a second constant-head reservoir received the flow cell's effluent. Fluid was re-circulated from the downstream reservoir back to the upstream reservoir with a submersible pump. The hydraulic head difference was controlled by raising or lowering a mechanical jack that supported the upstream reservoir, and additional flow control was gained through the use of a high-precision needle valve positioned upstream of the flow cell. In order to provide a uniform flow field in the flow cell, a "honeycomb" insert was designed and 3-D printed and incorporated into the flow cell by bonding it atop the existing trapezoidal block. The "honeycomb" flow straightener has a row of 29 tubes spanning the entire width of the flow cell. The flow cell's inside dimensions are 7.62 cm wide, 0.26 cm high, and 18.30 long. The flow straightener's height is the same as that of the flow cell, so that the flow was not allowed to pass below or above the flow straightener. The flow and particle Reynolds numbers showed the flow is within laminar regime. In order to provide a consistent and repeatable initial resting position of the test particle in each test, a particle-supporting "pocket" was manufactured, as shown in Fig. 2. The pocket is located approximately 9 cm downstream from the outlet of the flow straightener along the centerline of the flow direction. This particle-supporting pocket was comprised of three domes protruding from the originally flat, smooth surface of the flow bed. The domes were intended to provide three consistent points of support for the test particle by elevating it just above the surface of the underlying flat plane.

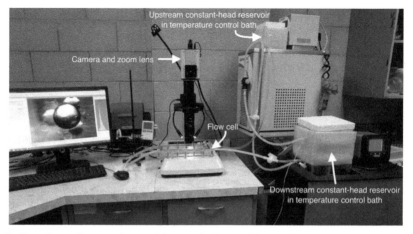

Fig. 1. Experimental setup for studying a single particle mobilization under various fluid characteristics [15]

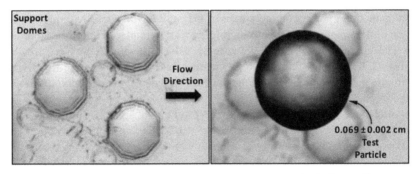

Fig. 2. Single spherical glass bead on support dome in the flow cell

2.2 Design of Experiments

Three physicochemical characteristics of fluid were studied: pH, ionic strength, and viscosity. Distilled water was used as the base of each test fluid. The pH was lowered to a targeted value by adding an appropriate quantity of hydrochloric acid (HCl), and raised to a targeted value by adding an appropriate quantity of sodium hydroxide (NaOH). HCl and NaOH are examples of a strong acid and base, respectively, which are known to dissolve completely in water and therefore yield reliable concentrations of hydrogen ions (and thus pH) at equilibrium [16]. Adjustment of pH considered the temperature effect. A similar incremental approach was used to adjust the solution to the desired ionic strength. In this research, sodium chloride (NaCl) was used because it effectively alters the ionic strength with little or no effect on pH. However, adding acids or bases to a solution necessarily alters the ionic strength. The magnitude of this effect and the required adjustment to reach the targeted ionic strength can be determined by Eq. (1) [16]:

$$I = \frac{1}{2} \sum_{all\,ions} c_i \cdot z_i^2 \tag{1}$$

where I = ionic strength of the solution, c_i = concentration of the ith ionic species, and z_i = charge on the ith ionic species.

The target viscosities were achieved by controlling the temperature of the fluid using the upstream and downstream temperature control baths. The targeted temperatures were selected based on the assumption that each fluid would follow the well-established relationship between the temperature of distilled water and its corresponding viscosity, as described in Eq. (2) [17].

$$\mu = 2.414 \times 10^{-4} \times 10^{247.8/(T-140)} \tag{2}$$

where μ is dynamic viscosity (Poise, or g/(cm $\cdot$ s)), and T is temperature (Kelvin). Equation (2) applies to temperature from 0 °C to 370 °C. The targeted high level of viscosity is 0.01317 Poise and corresponds to a low fluid temperature of 9.76 °C; the targeted low level of viscosity is 0.00547 Poise and corresponds to a high fluid temperature of 50.00 °C. This technique of using temperature to adjust viscosity allowed simultaneously adjusting the three characteristics to their target values according to the experimental design methodology. To accurately control the fluid temperature (and thus, viscosity) both the upstream and downstream fluid reservoirs were placed in temperature control baths. The temperature difference between the up- and down-stream reservoirs was less than 1.0 °C. The actual temperature of the test fluid in each test used the average of temperature measurements taken in the upstream and downstream reservoirs. The effect of salt concentration (up to 0.15 mol/L in this research) on viscosity is negligible. Zhang and Han [18] measured the viscosity of water with NaCl solution at 25 °C, the data showed the dynamic viscosity increased by 0.46% when the ionic strength increased from 0 to 0.15 mol/L. The density of water is 999.7 kg/m^3 at 9.79 °C and 988.1 kg/m^3 at 50.00 °C, which are the two extreme temperatures achieved in this research. Salt was added to water, the highest salt mass concentration is 100.8 g per 11.5 L, or 8.8 kg/m^3. Acid or base solutions were added to water, the highest mass concentration is 0.36 g per 11.5 L or 3.1 × 10^{-8} kg/m^3. The fluid densities considering the effects of temperature and added salt and acid or base solutions are listed in Table 2. The density ranges from 992.480 to 1006.170 kg/m^3 and the variation range is 1.7%. In this study, the effects of temperature, added salt and acid/base mass on fluid density were not considered. The experiments were designed using the concepts of the response surface methodology (RSM), which is known to be an effective method for the simultaneous estimation of the linear and quadratic effects as well as the interactions of the test variables (i.e., viscosity: μ, pH: P and ionic strength: I) on the response parameters of interest (i.e., the mean critical velocity and its standard deviation). The advantage of the RSM is that it allows for the study of five distinct levels of each test variable and the extraction of important statistical information from a relatively small number of experimental runs. There are two popular design approaches in RSM: the Box-Behken design and the central composite design (CCD). The latter approach was adopted in designing the experiments for the present study. CCD showed that total of 20 experiments are necessary for estimating the linear and quadratic effects of the three fluid characteristics (three test variables) as well as their two-way interactions. The three test variables and their five coded and natural levels are shown in Table 1.

Table 1. Coded and natural levels of the main test variables

	Low: L	Intermediate-Low: IL	Intermediate: I	Intermediate-High: IH	High: H
Viscosity: μ (g/cm · s = Poise)	0.00547	0.00699	0.00914	0.01150	0.01317
pH: P	3.5	4.92	7	9.08	10.50
Ionic strength: I (mol/L)	0.002	0.032	0.076	0.120	0.150

The fluids are labeled by assigning the corresponding factor levels of viscosity, pH, and ionic strength as the lowercase suffixes to μ, P and I, respectively. For instance, $\mu_{IL}P_{IL}I_{IH}$ denotes a fluid having intermediate-low levels of viscosity and pH (i.e., 0.00699 g/cm · s and 4.92, respectively), and an intermediate-high level of ionic strength (i.e. 0.12 mol/L). The 20 tests of various variable combinations as outcomes of CCD are listed in Table 2. Among the 20 tests, six duplicate tests were obtained from the RSM and were used to verify the tests' reproducibility. The 20 tests were then conducted based on a randomized run order to eliminate any systematic error that might occur when following a specific order of preparing the fluids. After the fluid preparation, the ionic strength and pH were assessed using pH and ion meter; and the effects of ionic strength and pH on viscosity were neglected, and the viscosity was based on Eq. (2).

Table 2. Recipes for preparing the experimental fluids

Run order	Fluid ID	T (°C)	Acid/base to add	NaCl for 11.5 L of fluid (g)	Fluid density (kg/m³)
1	$\mu_I P_I I_I$	23.89	HCl	51.077	1001.768
2	$\mu_I P_I I_I$	23.89	HCl	51.077	1001.768
3	$\mu_I P_H I_I$	23.89	NaOH	50.879	1001.751
4	$\mu_I P_L I_I$	23.89	HCl	50.864	1001.75
5	$\mu_{IH} P_{IL} I_{IH}$	14.63	HCl	80.639	1006.169
6	$\mu_I P_I I_I$	23.89	HCl	51.077	1001.768
7	$\mu_I P_I I_H$	23.89	HCl	100.809	1006.093
8	$\mu_I P_I I_I$	23.89	HCl	51.077	1001.768
9	$\mu_{IL} P_{IH} I_{IL}$	36.46	NaOH	21.487	995.392
10	$\mu_I P_I I_I$	23.89	HCl	51.077	1001.768
11	$\mu_L P_I I_I$	50.00	NaOH	51.076	992.480
12	$\mu_{IH} P_{IH} I_{IL}$	14.63	NaOH	21.502	1001.027
13	$\mu_{IH} P_{IH} I_{IH}$	14.63	NaOH	80.644	1006.170
14	$\mu_{IL} P_{IL} I_{IL}$	36.46	HCl	21.497	995.393
15	$\mu_I P_I I_L$	23.89	HCl	1.344	997.444
16	$\mu_{IH} P_{IL} I_{IL}$	14.63	HCl	21.497	1001.026
17	$\mu_{IL} P_{IL} I_{IH}$	36.46	HCl	80.639	1000.536
18	$\mu_H P_I I_I$	9.76	HCl	51.076	1004.164
19	$\mu_{IL} P_{IH} I_{IH}$	36.46	NaOH	80.629	1000.535
20	$\mu_I P_I I_I$	23.89	HCl	51.077	1001.768

2.3 Experimental Procedure and Analysis Approach

For each of the 20 fluids, seven repeat tests were conducted after a specified fluid was prepared with the target parameter (test variable) levels. In each test, the test particle was positioned on the particle-support pocket, the top cap was securely fastened to the flow cell to form an enclosed flow chamber, which was carefully saturated to remove all air. Continuous flow circulation through the system at a low flow rate ensures the temperature of the fluid reached and maintained at the target value before the start of each test. To reach the critical flow condition, the high-precision flow control valve was slowly opened to allow increasing flow rates (and thus flow velocities) to reach the test particle. Manual observation of a test particle's mobilization was made in real-time (with the aid of the camera, zoom lens, and real time display on the monitor). Further opening of the flow control valve stopped at the moment of particle mobilization. The flow rate at particle incipient motion was determined manually by measuring the time required to collect a given volume of effluent that exited the flow cell. To increase measurement precision, a high-precision balance was first used to measure the mass of the effluent, then the mass was converted to volume based on the density as determined by the measured temperature and added salt concentration. The suitability of this method was confirmed by demonstrating its repeatability in 8 different trial cases using distilled water. It was concluded from the relatively low standard deviations of flow rate measurements in all cases that the method showed good reproducibility and could be expected to contribute insignificantly to the overall variance in measuring the critical velocity. After each repeat test for the same fluid, the flow cell cap was removed, and the test particle was re-positioned on the particle-support pocket to start a new repeat test. A regression analysis of the experimental data obtained from the RSM was conducted to provide an approximate polynomial relationship between the response parameter of interest (critical velocity) and the test variables (viscosity, pH and ionic strength). The regression function was then used to determine the relative and interactive effects of the three fluid properties on particle's mobilization.

3 Experimental Methodology on Piping Erosion Progression

3.1 Experimental Setup

Hole erosion test (HET) was employed to evaluate the individual and interactive effects of the fluid's physicochemical characteristics on piping erosion progression of sandy soil. The soil was an engineered fill and sieved through U.S. #20 sieve (0.84 mm), the portion that passed #20 sieve was then added with 6% (by mass) of Kaolin clay. The soil has $D_{50} = 0.24$ mm, liquid limit = 10.6, and plasticity index = 4. It is classified as SW (well-graded sand) according to the Unified Soil Classification system. The soil was reconstituted in a specially designed Plexiglas flow cell. The diameter of the soil specimen after compaction was 7.0 cm (2.75 in.) and its length was 7.6 cm (3 in.). Figure 3 shows the experimental setup. Fluid was provided to the flow cell by an upstream constant-head reservoir while a downstream constant-head reservoir received the flow cell's effluent. The fluid was circulated from the downstream reservoir back to the upstream reservoir using a submersible pump. The hydraulic head difference was controlled by raising or lowering a mechanical jack that supported the upstream reservoir. A flowmeter was used to constantly record the flow rate variations during the test. In order to measure the hydraulic gradient across the soil specimen, two tubes (manometers) were connected with the flow cell and were used to measure the head difference across the soil specimen during the erosion testing.

3.2 Experimental Design and Fluid Preparation

The same physicochemical characteristics (factors) of fluid were studied: pH, ionic strength, and viscosity. Distilled water was used as the base of each test fluid. Each fluid parameter has a target low and high levels as shown in Table 3. Using full-factorial experimental design, the number of fluid types is: (number of levels)$^{\text{number of factors}}$ = $2^3 = 8$. The three-factor, two-level, full factorial design and preparation of the experimental fluids are listed in Table 4. The same fluid preparation approaches in the particle mobilization experiments were used.

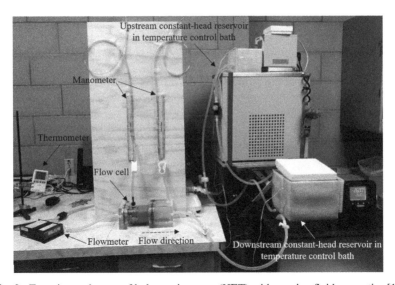

Fig. 3. Experimental setup of hole erosion tests (HET) with varying fluid properties [19]

Table 3. Two levels of fluid's physicochemical properties

Fluid properties	Low level	High level
Viscosity: μ (g/(cm · s) = Poise)	0.0054 (at 50.00 °C)	0.0142 (at 7.00 °C)
pH: P	3.5 (add HCl)	10.50 (add NaOH)
Ionic strength: I (mol/L)	0.05 (add NaCl)	0.5 (add NaCl)

Table 4. Three-factor, two-level, full factorial design of the experimental fluids in HET

Fluid number	Fluid descriptions	Fluid temperature (°C)
1	Low μ, low P, low I	50.00
2	Low μ, low P, high I	50.00
3	Low μ, high P, low I	50.00
4	Low μ, high P, high I	50.00
5	High μ, low P, low I	7.00
6	High μ, low P, high I	7.00
7	High μ, high P, low I	7.00
8	High μ, high P, high I	7.00

3.3 Experimental Procedure of HET and Results Analysis Approach

For each of the 8 fluids, two repeat tests were conducted after each specific fluid was prepared in the upstream and downstream reservoirs at target temperature (thus viscosity), pH, and ionic strength. The soil was compacted in three uniform layers at 95% of the maximum dry density ($\rho_{dmax} = 2.10$ g/cm^3) at the optimum water content of 11% (based on the standard proctor test). A hole was formed in the middle of the specimen using a drill; the initial hole diameter was 2.5 mm. Filter papers were attached to both surface of the soil specimen, a hole was cut in the filter paper to allow eroded particles to exit the soil specimen without impedance. Gravels were used to fill the upstream space of the flow cell. The upstream and downstream reservoirs were initially connected without the flow cell to circulate the fluid to achieve uniform temperature in the system. Then the soil specimen was saturated using the fluid by positioning the soil specimen in vertical orientation and slightly pushing the fluid upward in the specimen to expel the air. After soil saturation, the flow cell was positioned in horizontal orientation, and a control valve was fully opened. During the HET, the head difference and flow rate were measured every 1 min during the maximum of 90-min duration. After each test, the flow cell was disconnected and allowed to reach the room temperature. Then the soil specimen was carefully extruded and cut into halves. The final diameter of the piping channel was measured at three different locations using a caliper to obtain the average hole diameter.

The erosion rate indexes (I_e) of the soil, defined in Eq. (3), under various permeating fluids were obtained to quantify the relative and interactive effects of fluid's physicochemical properties on the fluid's erosive capacity.

$$I_e = -\log C_e \tag{3}$$

where C_e = coefficient of soil erosion (s/m), which is expressed in the assumed linear relationship between the rate of erosion and the applied hydraulic shear stress, which are obtained in HET. C_e can be obtained from HET and is usually in the order of 10^{-1} to 10^{-6} (s/m) [18]. Since $\log C_e$ is easier to use to reflect the result of erosion rate, past researchers often use Eq. (3) to define the erosion rate index. Therefore, I_e is in the range of 0 to 6. A lower value of I_e indicates a higher erosion rate. The HET data processing approach to derive the erosion index followed that of Wan and Fell [20]. A regression model was used to analyze the results in terms of erosion rate indexes of the three-factor, two-level full factorial experiments, following the same approach as in the particle mobilization experiments.

4 Results and Analysis of Particle Mobilization Experiments

The observed critical velocities at particle incipient motion (V_{cr}) for the 20 fluids were analyzed with respect to the three fluid characteristics (i.e., test variables) (μ, P, I), and their individual and interactive influences on V_{cr} were evaluated. Table 5 shows the measured critical velocities of the 20 experimental runs with various combinations of the three test variables. Since the regression analysis is based on the assumption of normally-distributed and homoscedastic residuals, the Box-Cox transformation [21] was first performed to transform the critical velocity values. The fluid characteristics (μ, P, I) and their interactions were analyzed with respect to the transformed response. Table 6 shows the analysis of variance (ANOVA) for the transformed response. The significance of the effects of viscosity, pH, and ionic strength and their interactive effects on the critical velocity for a particle's incipient motion can be judged by the p-values in Table 6. This analysis suggests that at a level of confidence of 95%, viscosity has a strongly significant effect on critical velocity (p-value < 0.0005) and pH has a marginally significant effect (p-value $= 0.059$). The ionic strength was not found to have a statistically significant effect (p-value $= 0.137$). It should be noted that this is not a positive indication of the irrelevance of ionic strength to variations in critical velocity. However, no significant effect of ionic strength was captured in this study at the designated significance level. Moreover, the viscosity and pH were found to have a strong antagonistic interaction (p-value $= 0.076$).

Table 5. Measured mean critical velocity variations with the fluid properties

Run order	Viscosity (Poise)	pH	Ionic strength (mol/L)	Mean critical velocity: V_{cr} (cm/sec)
1	0.0091	7.00	0.076	4.92
2	0.0091	7.00	0.076	4.87
3	0.0091	10.50	0.076	4.90
4	0.0091	3.50	0.076	5.80
5	0.0115	4.92	0.120	4.57
6	0.0091	7.00	0.076	5.39
7	0.0091	7.00	0.150	5.51
8	0.0091	7.00	0.076	5.26
9	0.0070	9.08	0.032	5.51
10	0.0091	7.00	0.076	4.93

(continued)

Table 5. *(continued)*

Run order	Viscosity (Poise)	pH	Ionic strength (mol/L)	Mean critical velocity: V_{cr} (cm/sec)
11	0.0055	7.00	0.076	5.76
12	0.0115	9.08	0.032	4.16
13	0.0115	9.08	0.120	4.34
14	0.0070	4.92	0.032	5.53
15	0.0091	7.00	0.002	4.74
16	0.0115	4.92	0.032	4.63
17	0.0070	4.92	0.120	5.25
18	0.0132	7.00	0.076	4.37
19	0.0070	9.08	0.120	5.78
20	0.0091	7.00	0.076	5.09

Figure 4 shows the effect on the mean critical velocity caused by the interaction between viscosity and pH. While increasing viscosity always promotes particle mobilization (i.e., reduces the critical velocity), its reducing effect is more pronounced at a higher level of pH (e.g., 10.5) than at a lower level of pH (e.g., 3.5). This is indicated by the greater slope of the line for the higher pH value. Figure 5 shows the contour plot of the critical velocity responses to viscosity and pH at a constant ionic strength $I = 0.076$ mol/L. The projection of the spherical design of experiment's space is also shown on the graph. The grayed area falls outside the design of experiments (DOE)

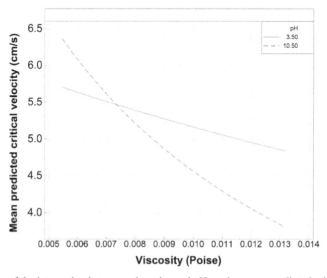

Fig. 4. Effect of the interaction between viscosity and pH on the mean predicted critical velocity.

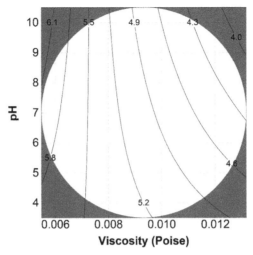

Fig. 5. Contour plot of the critical velocity response to viscosity and pH in particle mobilization experiments

space and the predictions in that region shall be used and interpreted with caution. The plot reveals the same observation as in Fig. 5: at a higher pH, the effect of viscosity on critical velocity was more pronounced than at a lower pH. The plot also reveals the same conclusion that was derived from Table 6: viscosity affected the critical velocity more than pH. In addition, at a viscosity of approximately 0.0074 g/cm · s, the critical velocity remained relatively unchanged as pH increased from 3.5 to 10.5. At a viscosity greater than 0.0074 g/cm · s, the critical velocity decreased with increasing pH. At a viscosity less than 0.0074 g/cm · s, the effect of pH on the critical velocity reversed: the critical velocity increased with an increase in pH. At any pH value critical velocity decreases with the increase of viscosity.

Table 6. The analysis of variance (ANOVA) for the critical velocity versus the test variables (viscosity: μ, pH: P and ionic strength: I)

Source		Coefficient	Contribution (%)	p-Value
Regression:		–	82.51	0.000
	Constant	−0.195	–	0.001
	μ	0.750	70.52	0.000
	P	0.010	4.88	0.059
	I	0.092	2.87	0.137
	$\mu \times P$	−1.376	4.24	0.076
Error:		–	17.49	–
	Lack-of-Fit	–	13.38	0.307
	Pure Error	–	4.10	–
Total		–	100.00	–

5   Results and Analysis of Piping Erosion Progression Experiments

Table 7 lists the determined average erosion rate indexes of the 8 fluids. Two duplicate tests were conducted for each fluid, and the duplicate tests show good repeatability of the HETs.

Table 7. Erosion rate indexes of the 8 fluids

Fluid	Average erosion rate index
Low μ, low pH, low I	3.47
Low μ, low pH, high I	3.56
Low μ, high pH, low I	3.11
Low μ, high pH, high I	3.16
High μ, low pH, low I	4.10
High μ, low pH, high I	4.50
High μ, high pH, low I	4.04
High μ, high pH, high I	4.44

ANOVA was conducted on the average erosion rate indexes of the 8 fluids. In the analysis, the erosion rate index is known as a response. 95% confidence interval was used for testing the significance of various components. The significance of the effect of viscosity (μ), pH and ionic strength (I) and their interactive effects on the erosion rate index can be judged by the p-values in Table 8. It can be concluded that viscosity, pH and ionic strength all have significant effect on erosion rate (i.e., p-value < 0.0005). Moreover, combined with viscosity, both pH and ionic strength are found to have a strong interaction (i.e., p-value < 0.0005), while the interaction of pH and ionic strength has less significant effect (with p-value = 0.072 > 0.0005). No significant 3-way interaction among viscosity, pH and ionic strength was captured in this study (with p-value = 0.576). In Table 8, the coefficient describes the significance (as indicated by the absolute value) and direction (as indicated by the sign) of the relationship between a factor (i.e., viscosity, pH, ionic strength) and the response variable (i.e., erosion rate index). Comparisons of the coefficients in Table 8 show that viscosity plays the most important role and ionic strength has slightly more effect than pH on soil erosion. In terms of the interactive effect, when combined with viscosity, both pH and ionic strength have the same interactive effect while pH and ionic strength have the least interactive effect on erosion.

Figure 6 quantitatively shows the relative effects of viscosity, pH and ionic strength on the mean response of erosion rate index. When viscosity increased from 0.0054 to 0.142 g/(cm · s) or Poise, the erosion rate index approximately increased from 3.3 (that represents relatively higher erosion) to 4.2 (that represents relatively lower erosion).

Table 8. Analysis of variance of the erosion rate indexes

Linear model	Coefficient value	P-value
Constant	3.79650	–
μ	−0.47275	0.000
pH	0.10900	0.000
I	−0.11575	0.000
2-way interaction		
$\mu \times$pH	0.008250	0.000
$\mu \times I$	0.008250	0.000
$pH \times I$	−0.00350	0.072
3-way interaction		
$\mu \times pH \times I$	−0.00350	0.576

Such trend suggests lower viscosity results in higher erosion and conflicts with the understanding as implied by Eq. (4) proposed by Kakuturu and Reddi [22, 23] and the findings in this study on individual particle mobilization under laminar flow conditions.

$$\tau(t) = \frac{4Q\eta}{\pi r_{cc}^3} \tag{4}$$

In the above equation, $\tau(t) =$ hydraulic shear stress acting on a piping channel, $Q =$ flow rate, $\eta =$ dynamic viscosity of the permeating fluid, $r_{cc} =$ radius of the idealized cylindrical piping channel. This equation suggests under laminar flow, hydraulic shear stress increases with viscosity. Such conflict may be explained in the following two aspects. Based on the Shields diagram [24], Annandale [25] suggested that the commonly held belief that soil erosion is caused by hydraulic shear stress is only valid in laminar flow, while in turbulent flow normal lift force dominates. Turbulent flow occurred in the HETs in this study. Meanwhile, the soil's resistance to erosion may be affected by temperature, which complicates the quantitative understanding of the relative erosive capacity of the eight fluids. Figure 7 shows the Mohr-Coulomb failure criteria lines of the soil under two temperatures of 8 °C and 40 °C that were obtained from temperature-controlled direct shear tests. At 8 °C, the cohesion of soil is 5.53 kPa, and the friction angle is 26.06°; at 40 °C, the cohesion of soil is 1.2 kPa, and the friction angle is 31.87°. On the surface of the piping channel that is exposed to the pipe flow, the normal stress can be assumed to be zero, the shear strength is controlled by cohesion. Higher temperature resulted in lower cohesion, thus more erosion. It should be noted that the cohesion indicated by the Mohr-Coulomb failure envelope is the apparent cohesion that includes interlocking of granular particles and true cohesion due to the added 6% kaolinite. Both interlocking and true cohesion contribute to the resistance to particle detachment.

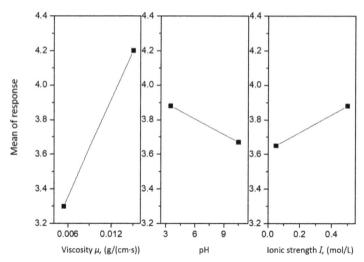

Fig. 6. Variation of erosion rate index with viscosity, pH, and ionic strength. (The mean of response is the mean erosion rate index).

Figure 6 also shows that the fluid's erosive capacity has a positive relationship with pH, i.e., higher pH causes higher erosive capacity (that is represented by lower erosion rate index); this trend is consistent with previous studies by Hubbe [4, 5], Sharma et al. [6], McDowell-Boyer [7], and the particle mobilization results in this study. Further, Fig. 6 suggests that fluid's erosive capacity has a negative relationship with ionic strength; this trend is consistent with previous studies by Sherard et al. [8], Arulanandan et al. [1], Reddi et al. [2], as well as the individual particle mobilization results in this study.

Figure 8 shows the interactive effect of each two parameters on the mean response of the erosion rate index. As for the interaction between viscosity and ionic strength, ionic strength does not affect the erosion at low viscosity ($I \approx 3.3$) since the erosion rate indexes are the same; but when the viscosity is higher (or the fluid temperature is colder), higher ionic strength causes more significant increase in erosion rate index, which means less erosion. As for the interaction between pH and ionic strength, the two lines of the low and high ionic strength are nearly parallel, which means that there is almost no interactive effect between pH and ionic strength.

It should be noted that this research focused on the relative and interactive effects of fluid's physicochemical characteristics, specifically viscosity, pH, and ionic strength, on the piping erosion of a sandy soil. Only the fluids were varied and tested on one soil type. Those effects that were revealed in this study may vary with different soils. The effects of fluid's parameters on soil's piping erosion are closely related to the fluid's temperature and soil's temperature. This research did not isolate the effects of the fluid's parameters with soil's temperature.

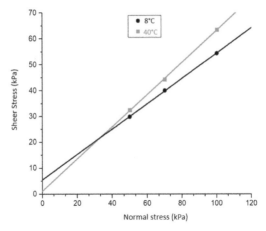

Fig. 7. Mohr-Coulomb failure envelopes of the sandy soil under two temperatures

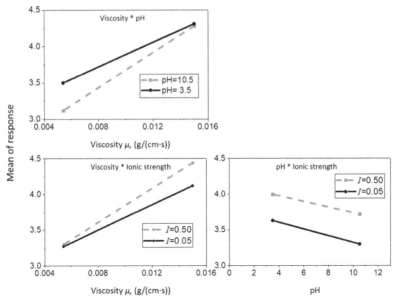

Fig. 8. Interactive effects between fluids viscosity, pH, and ionic strength on soil erosion rate index. (The mean of response is the mean erosion rate index).

6   Summary and Conclusions

This research employed a statistical experimental design approach to investigate the relative and interactive effects of three physicochemical characteristics of fluid on a single granular particle's incipient motion and on bench-scale piping erosion progression of a sandy soil. The incipient motion is characterized by the critical fluid velocity. An innovative experimental setup was designed and constructed that can simultaneously adjust fluid's viscosity, pH, and ionic strength and provide repeatable test results on particle's incipient motion. The knowledge gained from this research is summarized as follows.

- The response surface methodology proved to be an effective and efficient method to quantify the relative and interactive effects of the three physicochemical fluid characteristics on particle's incipient motion and piping progression of a sandy soil.
- For particle mobilization, viscosity, pH, and their two-way interaction were determined to be the most influencing factors on the critical velocity. The ionic strength did not prove to be statistically significant at $\alpha = 0.1$ significance level. Viscosity and pH were both shown to have an inverse relationship with critical velocity. This is in agreement with the findings of previous studies. Viscosity was shown to be of greater influence on critical velocity than pH.
- The research revealed an interactive effect between viscosity and pH on the critical velocity for particle mobilization. When pH is 10.5 the viscosity had considerably higher influence on the critical velocity than when the pH 3.5.
- The statistical analysis shows that the effects of pH on critical velocity reversed trends (from positively correlated to negatively correlated) as viscosity increased. At pH values from 3.5 to 10.5, critical velocity always decreased with an increase in viscosity.
- For piping progression of a sandy soil, viscosity, pH and ionic strength were all determined to be significant factors on the rate of erosion. The two-way interactions between viscosity and pH, and between viscosity and ionic strength were also determined to be significant interaction factors, while the interaction between pH and ionic strength did not prove to be statistically significant.
- While the statistical experimental design and methodology may be used to study the relative and interactive effects of fluids with different physicochemical characteristics, the fluid's temperature also affected soil's strength and consequently erosion resistance. This complicated the quantitative understanding of the relative erosive capacity of the eight fluids.
- Fluid's erosive capacity has a positive relationship with pH, i.e., higher pH causes higher erosive capacity. Fluid's erosive capacity has a negative relationship with ionic strength, i.e., higher ionic strength causes lower erosive capacity of fluid.

This research showed the relative and interactive effects of three physicochemical characteristics (viscosity, pH and ionic strength) of permeating fluid on piping of a sandy soil. Further research is need to provided fundamental explanations for the above phenomena. This paper suggests that in the field evaluation of piping in earthen embankments, if the subsurface seepage is known to possess different physicochemical characteristics from those of tap water or distilled water, the fluid's properties should be considered in the laboratory tests of piping.

Acknowledgement. This research was funded by the National Science Foundation CMMI Geotechnical Engineering and Materials Program under award number CMMI 1200081 and The Pennsylvania State University.

References

1. Arulanandan, K., Krone, R.B., Longanathan, P.: Pore and eroding fluid influences on surface erosion on soil. J. Geotech. Eng. Div. **101**(1), 51–66 (1975)
2. Reddi, L., Lee, I.-M., Bonala, M.: Comparison of internal and surface erosion using flow pump tests on a sand-kaolinite mixture. Geotech. Test. J. **23**(1), 116–122 (2000)
3. Yong, R.N.Y., Jorgensen, M.A., Ludwig, H.P., Sethi, A.J.: Interparticle action and rheology of dispersive clays. J. Geotech. Eng. Div. **105**(10), 1193–1209 (1979)
4. Hubbe, M.A.: Detachment of colloidal hydrous oxide spheres from flat solids exposed to flow. 1: experimental system. Colloids Surf. **16**, 227–248 (1985)
5. Hubbe, M.A.: Detachment of colloidal hydrous oxide spheres from flat solids exposed to flow. 3: forces of adhesion. Colloids Surf. **25**, 311–324 (1987)
6. Sharma, M.M., Chamoun, H., Sita Rama Sarma, D.S.H., Schechter, R.S.: Factors controlling the hydrodynamic detachment of particles from surfaces. J Colloid Interface Sci. **149**(1), 121–134 (1992)
7. McDowell-Boyer, L.M.: Chemical mobilization of micron-sized particles in saturated porous media under steady flow conditions. Environ. Sci. Technol. **26**(3), 586–593 (1992)
8. Sherard, J.L., Decker, R.S., Ryker, N.L.: Piping in earth dams of dispersive clay. In: Proceedings, Special Conference on Performance of Earth and Earth-Supported Structures, vol. 1, Part 1, pp. 589–626. ASCE, Reston (1972)
9. Briaud, J.L., Chen, H.C., Govindasamy, A.V., Storesund, R.: Levee erosion by overtopping in New Orleans during the Katrina Hurricane. J. Geotech. Geoenviron. Eng. **134**(5), 618–632 (2008)
10. Fournier, Z., Geromichalos, D., Herminghaus, S., Kohonen, M.M., Mugele, F., Scheel, M., Schulz, M., Schulz, B., Schier, C., Seemann, R., Skudelny, A.: Mechanical properties of wet granular materials. J. Phys. Condens. Matter **17**, 477–502 (2005)
11. Santamarina, J.C.: Soil behavior at the microscale: particle forces. Geotechnical Special Publication 119, pp. 25–56. ASCE, Cambridge, Massachusetts (2001)
12. Diplas, P., Celik, A.O., Dancey, C.L., Valyrakis, M.: Nonintrusive method for detecting particle movement characteristics near threshold flow conditions. J. Irrig. Drain. Eng. **136**(11), 774–780 (2010)
13. Niño, Y., Garcia, M.H.: Experiments on particle-turbulence interactions in the near-wall region of an open channel flows: implications for sediment transport. J. Fluid Mech. **326**, 285–319 (1996)
14. Niño, Y., Lopez, F., Garcia, M.: Threshold for particle entrainment into suspension. Sedimentology **50**, 247–263 (2003)
15. Xiao, M., Gholizadeh-Vayghan, A., Adams, B.T., Rajabipour, F.: Experimental investigation of the relative and interactive effects of physicochemical fluid characteristics on the incipient motion of granular particles under laminar flow conditions. ASCE J. Hydraul. Eng. **144**(5), 04018013-1–04018013-14 (2018)
16. Benjamin, M.M.: Water Chemistry. Waveland Press Inc., Long Grove (2010)
17. Al-Shemmeri, T.: Engineering Fluid Mechnanics, pp. 17–18. Ventus Publishing ApS, Frederiksberg (2012). ISBN 978-87-403-0114-4
18. Zhang, H.-L., Han, S.-J.: Viscosity and density of water + sodium chloride + potassium chloride solutions at 298.15 K. J. Chem. Eng. Data **41**, 516–520 (1996)
19. Ma, Y., Xiao, M., Kermani, B.: Experimental investigation of the effects of fluid's physicochemical characteristics on piping erosion of a sandy soil under turbulent flow. ASTM Geotech. Test. J. **43**, 436–451 (2020)
20. Wan, C.F., Fell, R.: Laboratory tests on the rate of piping erosion of soils in embankment dams. Geotech. Test. J. **27**(3), 295–303 (2004)

21. Box, G.E., Cox, D.R.: An analysis of transformations. J. Roy. Stat. Soc. Ser. B (Methodol.) **26**, 211–252 (1964)
22. Kakuturu, S., Reddi, L.N.: Evaluation of the parameters influencing self-healing in earth dams. ASCE J. Geotech. Geoenviron. Eng. **132**(7), 879–889 (2006)
23. Kakuturu, S., Reddi, L.N.: Mechanistic model for self-healing of core cracks in earth dams. ASCE J. Geotech. Geoenviron. Eng. **132**(7), 890–901 (2006)
24. Shields, A.: Application of similarity principles and turbulence research to bed-load movement. Hydrodynamics Laboratory, California Institute of Technology, Pasadena, CA, USA (1936). (W. P. Ott and J. C. van Uchelen, trans.)
25. Annandale, G.W.: Scour Technology - Mechanics and Engineering Practice. McGraw-Hill Companies Inc., New York (2006)

Key Technology in the Remediation Project of Hongshiyan Landslide-Dammed Lake on Niulan River Caused by "8•03" Earthquake in Ludian, Yunnan

Zongliang Zhang[1,2(✉)], Kai Cheng[1,2], Zaihong Yang[1,2], and Fuping Peng[1,2]

[1] PowerChina Kunming Engineering Corporation Limited, Kunming 650051, China
zhang_zl@powerchina.cn
[2] Sub-Center of High Earthfill/Rockfill Dam, National Energy and Hydropower Engineering Technology R&D Center, Kunming 650051, China

Abstract. The Hongshiyan landslide dammed lake on the Niulan River is a large-sized one caused by the Ludian 8•03 earthquake in 2014. Based on the uniqueness of the Hongshiyan dammed lake, after the completion of the emergency rescue, the treatment concept of "*Eliminating Hazard, Utilizing Resources and Turning Waste into Treasure*" was innovatively proposed and this dammed lake was then rebuilt into a large-integrated water conservancy complex with the multiple functions of flood control, water supply, irrigation, and power generation. The project is the first in the world to be developed and utilized immediately after the formation of the barrier dam. This paper systematically introduces the formation of the Hongshiyan landslide dam, the emergency rescue as well as the remediation in the later period. The reservoir has been impounded in 2019, the landslide dam has already retained to a high water level and operated safely. Through the practice of the project, this paper summaries the key technologies such as emergency treatment technology and remediation plan under the lack of information conditions, the comprehensive treatment of the 750 m high slope affected by the intense earthquake and the comprehensive treatment of the 130 m grade landslide dam composed of materials with discontinuous wide gradation, which can provide reference for the emergency rescue, development and utilization of similar landslide-dammed lakes.

Keywords: Landslide-dammed lake · Water conservancy complex · Emergency treatment under lack of information · Remediation scheme · High slope caused by strong earthquake · Landslide dam

1 Background

China is a country with frequent geological disasters. The landslide material from the mountains can block rivers and form dams, causing not only massive submerge loss and endangering enormous life and property along the river at downstream in case of dam break. In 2008 Wenchuan earthquake, 34 dam lakes were induced by the earthquake, and the Tangjiashan dam lake was one of them. It threatened 1.3 million people

© Springer Nature Switzerland AG 2020
J.-M. Zhang et al. (Eds.): ICED 2020, SSGG, pp. 75–95, 2020.
https://doi.org/10.1007/978-3-030-46351-9_7

in the downstream area, nearly 280,000 people were evacuated immediately. Hence, emergency action and comprehensive treatment of landslide dams are the national key requirements for the natural disaster prevention and control. Restricted by environmental risks, harsh conditions, tight time, inconvenient transportation, and unknown geological and hydrological conditions of the landslide dam, the type of dam break is uncertain, and emergency rescue and disposal are extremely difficult. Comprehensive treatment of landslide dams is a major measure to eradicate the risks of landslide dams, reduce disasters and make profits. Currently there is no precedent case to treat landslide dams in China and around the world. It is hard to solve the problems of investigation, design and construction of landslide dams with present technologies.

2   Overview of Hongshiyan Landslide Dammed Lake

At 16:30 on August 3, 2014, a magnitude 6.5 earthquake occurred in Ludian County, Yunnan Province, China, on the main stream of the Niulan River at the junction of Lijiashan Village, Huodehong Township, Ludian County, and Hongshiyan Village, Baogunao Township, Qiaojia County. The earthquake triggered mountain collapses and river blockage, resulting in a 260 million m^3 dammed lake. The dammed block was located between the dam and the powerhouse of the existing Hongshiyan Hydropower Project (HPP) and was a rapid toppling landslide. The materials were mainly from the higher elevations of the right bank. There were also slide and collapse materials from the left bank, which was mainly composed of fragmented stones and blocks. The maximum crest elevation is about 1,222 m and the maximum height of the dammed block is about 103 m. The width vertical to the flow direction is about 307 m, while the width along the flow direction is about 911 m. The overall upstream slope is about 1:2.5 (horizontal: vertical), the overall downstream slope is 1:5.5 and the total volume is about 10 million m^3, as shown in Fig. 1.

After the Hongshiyan landslide dammed lake formed, its water level continued to rise. The highest water level reached up to 1,182 m, the backwater was 25 km long, and more than 5,000 μ of land and residential houses along the Niulan River in the upper reaches of Huize and Ludian were submerged. Nearly 13,000 villagers were evacuated and resettled. At the same time, the risk of collapse of the dam will seriously threaten the downstream towns along the river, including 10 townships, over 30,000 people, 33,000 μ of arable land in Ludian, Qiaojia, Zhaoyang counties (districts), and threaten the safety of the downstream hydropower projects (HPPs) on Niulan River such as Tianhuaban and Huangjiaoshu. According to the SL 450–2009 Standard for Classification of Risk Degree of the Landslide Dammed Lake, the Hongshiyan landslide dammed lake belongs to a large-sized lake with a very high risk level and serious damage severity. The risk level of the landslide dammed lake is classified as Grade I (the highest level) according to the risk level and the severity of the damage.

(a) Post-earthquake photo of landslide dam

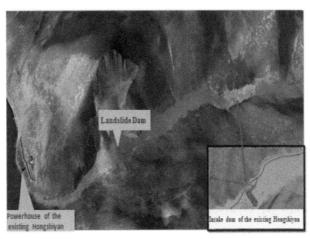

(b) Landslide dam and power station in 3D BIM model

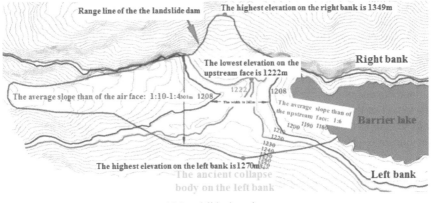

(c) Landslide dam shape

Fig. 1. Overview of the Hongshiyan landslide dam

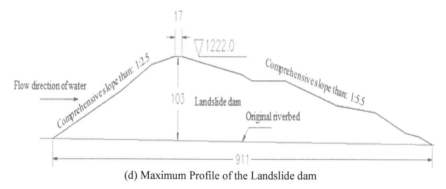

(d) Maximum Profile of the Landslide dam

Fig. 1. (*continued*)

The remediation of Hongshiyan landslide dammed lake is mainly divided into three stages:

(1) Emergency treatment stage: From August 4 to August 12, 2014, five reports including the Report on Emergency Disposal Plan were completed using the Lack of Information Technology, which provided important support for the decision-making of the hazard removal and disposal of the landslide dammed lake, and reduced the danger as early as possible for the downstream people.
(2) Subsequent disposal stage: From August 13 to October 3, 2014, the new emergency flood release tunnel emptied the lake and eased the risk of dam collapse.
(3) Remediation stage in late period: Since August 19, 2014, the feasibility study of reconstruction of the power station has been completed, and the comprehensive management concept of "Eliminating Hazard, Utilizing Resources and Turning Waste into Treasure" has been innovatively proposed, and the lake was rebuilt into a large integrated water conservancy complex providing the functions of flood control, water supply, irrigation, and power generation.

3 Emergency Treatment and Subsequent Disposal Under Lack of Information

3.1 Uniqueness of the Location of the Dammed Block

The headrace tunnel of the existing Hongshiyan HPP is located on the right bank of the dammed block. The intake gate of the power station was open during earthquake, and an open downstream surge tank was connected to the landslide dammed lake. After the water level of the lake rose, on 4 August 2014 the lake water level was higher than the surge tank top El.1171.80 m. Water spilled freely from the top of the surge tank, forming a unique emergency flood release channel.

3.2 Definition of Lack of Information Condition

Lack of Information (LoI) Condition refers to the working background of basic data and information such as hydrology, meteorology, topography, geology, social impact, transportation, building materials and cost, which should be obtained only by actual measurement or comprehensive survey, such as the emergency cases of remote areas both at home and abroad or emergency rescue condition. Lack of Information Investigation and Design Technology refers to a method and technique using the basic data obtained by the Internet, satellite, etc. in case of the lack of information, and using specific software and other means to process and transform the basic data, and complete the preliminary investigation and design quickly, efficiently at low costs. The development of spatial geographic information and next-generation information technology has expanded the sources and methods of data collection. Basic geographic information data can be obtained without reaching the project site. The development of network, next-generation information technology and big data technology provides a means for data collection.

3.3   Establishment of Terrain and Hydrology Model Under LoI Condition

Considering the hydrology and meteorology, topography and geology, social-environmental uncertainty, by way of space-air-land panoramic image under lack of information to quickly collect and process data, the hydrological model was configured of the landslide dam lake and its adjacent areas. The three-dimensional laser scanning operation of the Hongshiyan Dammed Block and the low-altitude unmanned aerial vehicle (UAV) photography system was used to acquire image data and high-resolution video of the landslide dam location, geometric parameters and volume of the accumulated mass. The three-dimensional topography model was established. The results of the survey and the terrain model are shown in Fig. 2.

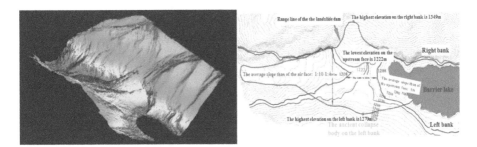

Fig. 2. Terrain model and topography of the dammed block

3.4   Calculation of Dam-Break Flood and Its Impact Analysis on Upstream and Downstream Reaches

Based on the results of the shape of the dammed block after data processing, the reservoir volume and the upstream and downstream topography, etc., a dam break flood analysis software DB-IEHR, proposed by the China Institute of Water Resources and Hydropower Research, is used to analyze the flood routing (Hongshiyan landslide dam-Tianhuaban HPP-Huangjiaoshu HPP) and study the impact of the lake on the upstream and downstream reaches.

Four alternatives are proposed to study: bottom elevations of discharge chute are El.1208 m (bottom widths 20 m and 5 m) and El.1214 m (bottom widths 20 m and 5 m). Finally, combined with the site construction conditions, the bottom elevation of the discharge chute is selected as 1,208 m and the bottom width as 5 m. The calculated peak flow of the dam break is 6,345 m^3/s, the peak time is 6.47 h, the total flood release capacity is 203 million m^3, and the total flood release time is 16.4 h.

After the formation of the flood discharging chute at the top of the dam, the high water level of the flood control in the upstream reservoir area will be reduced, and the boundary of the evacuation zone and the warning zone will be reduced accordingly. The flood control standard of the Xiaoyantou powerhouse at the head of the reservoir can be improved accordingly. The dam safety of the two downstream hydropower projects (Tianhuaban and Huangjiaoshu HPPs) will not be affected even if a dam break occurs. The peak flood of the Tianhuaban HPP is large (Qm = 6,345 m^3/s, exceeding the check flood of the power station) and the high-water level in front of the dam will have a certain impact on the flooding of the reservoir area. The maximum discharge of the flood release structure of the Huangjiaoshu HPP is 6,781 m^3/s, and the inflow peak flood to the reservoir of the Huangjiaoshu HPP is only 4,868 m^3/s as the flood released from the landslide dammed lake is regulated by the Tianhuaban reservoir. The excavated discharge chute can also improve the flood control capacity of the downstream residents and power stations.

3.5 Emergency Treatment Complete Technology

The emergency complete treatment technology includes non-engineering measures and engineering measures. A number of non-engineering measures were taken during the disposal stage of the landslide dammed lake: Transferring and resettlement work of the local people were well done. The upstream Deze Reservoir was utilized for flood detention, the storage capacity of the downstream Tianhuaban and Huangjiaoshu reservoirs were vacated, the site monitoring was intensified and the plan for the disposal of the landslide dammed lake was organized, etc. At the same time, there were a number of engineering measures: at noon on August 11th, the blasting demolition of the maintenance gate for the construction plug at surge tank bottom. After the demolition of the 1.8 m maintenance manhole, the released flow is about 60 m^3/s. On August 12, when the bottom width of the discharge chute reached 5 m and the bottom elevation reached 1,214 m, it can withstand P = 20% (5-yr) flood. The scheme met the flood control requirements during the emergency disposal period, and the emergency treatment was completed. On August 28, the discharge chute was excavated down to El.1,208 m with a bottom width of about 5 m. The new 280 m-long emergency flood release tunnel on the right bank connected the headrace tunnel of the existing Hongshiyan HPP was broken through on October 3, 2014. On the evening of October 4, the water of the landslide dammed lake has been basically released and the subsequent disposal was completed.

The above-mentioned measures enabled the dammed block to safely pass the 2014 flood season, temporarily relieved the risk of collapse and created conditions for the later permanent remediation.

4   Implementation Plan for Remediation Project

4.1   Urgency and Necessity of Remediation Project

The Danger of the Lake Still Exists and the Flood-Passing Situation is Severe. After the emergency disposal and subsequent disposal of the landslide dammed lake, it can only meet the requirements of the annual flood control standard. The catchment area of the lake is 12,087 km^2, which is nearly four times that of the Tangjiashan landslide dammed lake formed by the Wenchuan 5·12 earthquake in Sichuan. The peak flood is large. If it is not timely treated, in case of dam-break flood resulted from a larger flood event in the following 2015, it will be devastating and difficult to estimate the damage to the people's lives and property on the upper and lower reaches of the river and the existing power stations and it is easy to trigger a disaster chain, so the permanent remediation of the lake is imminent.

The Geological Hazards of the Lake Caused by the Earthquake are Serious. The right bank collapsed to form a nearly 750 m high slope. The slope was steep and the cracks are intensely developed. The dangerous rock mass still has collapsed and slide, which seriously endangered the safety of the lower portions. On the left bank, there was an ancient landslide with a volume of 56.7 million m^3. The earthquake has also produced dangerous rock mass at the trailing edge of the ancient landslide. There were 12 landslides, 7 collapsed deposits, 14 unstable slopes and 4 debris flow gullies in the reservoir area. The above unfavorable geological bodies posed a great threat to the safety of dammed block and the residents living around the lake, and must be treated as soon as possible.

Affected People's Production and Life are in Urgent Need of Resettlement. After the formation of the Hongshiyan dammed lake, most of the affected people were resettled in the villages near the dammed lake or on the roadside where the living conditions were extremely harsh and the safety risks were high. The impact of the earthquake and the landslide dammed lake broke the people's original living and production environment. It is necessary to use the storage capacity formed by the lake according to the local conditions. It can irrigate 66,200 μ of farmland in the dry and hot valley, provide source of drinking water for 80,800 people, increase the capacity of production and living environment and alleviate the resettlement pressure of the people whose land resources were damaged. Therefore, the resettlement work of the affected people should be started and implemented immediately.

4.2   Concept of "Eliminating Hazard and Utilizing Resources"

As the landslide dam is massive body with a gentle slope and favorable material compo-nents, the overall stability meets the requirements. The right bank slope is stable, a usable water head is available, and there is water release channel and construction time during the construction period after the completion of subsequent disposal. Therefore, the con-cept of "Eliminating Hazard, Utilizing Resources and Turning Waste into Treasure" is proposed.

4.2.1   Stability of the Dammed Block

The maximum cross-section of the dammed block is obtained from the longitudinal profile of the site topography along the riverbed. The thickness of the alluvium of the riverbed is considered to be 15 m. It is calculated that when the dammed block retains the water at the highest water level of 1,222 m and below, the stability of seepage, deformation and the dam slopes meet the requirement. Therefore the dammed block is safe.

4.2.2   Preliminary Analysis of Slope Stability

The slope of the right bank is nearly 750 m high before the collapse, and the slope is $70°$–$85°$. From the analysis of geological data surveyed on site, the upper part of the collapse is thick-bedded and giant thick bedded limestone, dolomite, dolomitic limestone, and the lower part is medium-thick and thin sandstone and shale; the slope is a rocky slope, hard in the upper elevations and soft in the lower elevations. The rock formation on the right bank dips towards the mountain and downstream with an occurrence of $N20°-60°E$, $NW\angle 10°-30°$. The occurrence of rock formation is favorable for the anti-sliding stability of the abutment. Three types of joints are developed in the slope rock mass, including the joints with steep dip angle across the river, the joints with steep dip angle along the river and the bedding joints. The rock mass of the slope is hard alternated with soft; the soft rock of the slope under the action of the weight of the upper rock mass will enable the upper brittle rock mass to crack and disintegrate, forming a tensile deformation, and to relax along the river under the action of earthquake. The cracks produce a wide range of collapse under the combined action of other structural planes (such as bedding joints) and slide downstream along the F_5 fault. The length of the collapse along the river is about 890 m, and the height of the trailing edge is about 500 m (the maximum height of the nearly vertical cliff is about 350 m), which is an extremely large collapsed deposit.

After the slope collapse, the slope stress is adjusted, the relaxation fissures of the slope surface develop and form dangerous blocks with the unfavorable structural planes. Due to the influence of aftershock, rainfall and adjusted slope stress, the unstable block will collapse and fall off intermittently. According to the preliminary stability analysis, the slope is generally stable, and the rock mass at the top of the slope is in a critical stable state, for which treatment is required.

4.2.3   Available Hydropower Resources

The Luojiaping HPP is under planning between the Xiaoyantou HPP and the existing Hongshiyan HPP. With the presence of the landslide dam, the water level of the landslide-dammed lake can be directly connected with the upstream Xiaoyantou HPP due to its dam height. In order to make full use of water resources, according to the actual conditions of the lake, the Hongshiyan HPP and Luojiaping HPP will be merged as one cascade, which can form a seasonally regulated reservoir to improve the efficiency of the project itself and compensation benefits of the downstream cascades.

4.2.4 Water Release Passage and Construction Time During Construction Period

After the completion of the follow-up treatment, the existing surge tank and maintenance gate of the construction adit and emergency flood release tunnels can meet the 10-year flood control standards in the construction period. Therefore, it is necessary to immediately use the 2014 dry season to excavate a flood release tunnel so as to meet the requirements of subsequent flood control and create conditions for the remediation project.

4.3 Comparison of Remediation Schemes

According to the practice of remediation of dammed lakes at home and abroad, the remediation of the landslide dammed lake is divided into two categories: permanent remediation and demolition remediation. For this reason, a comprehensive comparison of the two remediation schemes has been carried out.

To remove the huge landslide body at Hongshiyan 103 m in height and 10 million m^3 in total volume is too difficult to implement with high cost. If the removal scheme is adopted, the excavation material volume will be up to 9.91 million m^3, it is difficult to find a suitable disposal area for the storage of the landslide dam. Further, the removal difficulty by explosion is huge, in addition, the stability of the collapse deposit on the left bank will be reduced, the strengthening measures is required and hard to implement, Most likely the secondly disater may occur.

From the perspectives of compensation subsidy policy, quality of living standards in the resettlement areas, basic public service level and infrastructure support, alleviation of ecological pressures in the hazard areas, improvement of environmental quality in hazard areas and acceleration of resettlement progress, the permanent remediation scheme is recommended.

In the permanent remediation scheme, a large-sized water conservancy project will be rebuilt by landslide dam. A new 201 MW power station with an annual power generation of 800 GWh can be developed with the features of a reservoir total capacity of 185 million m^3, a 2000-year flood standard, water supply for 80,800 persons, and an irrigation area of 66,200 μ.

From the perspectives of comprehensive benefits, difficulty in the implementation, and balance of funds, the permanent remediation scheme is recommended.

4.4 Remediation Project and Key Technologies

The project complex of the Hongshiyan-dammed lake consists of the remediation of the landslide dam, high slope treatment, a flood release tunnel on the right bank, a flood release/sand flushing/empty tunnel on the right bank, waterway & power generation structures on the right bank and downstream water supply and irrigation structures, as shown in Fig. 3.

The key technology of the Hongshiyan landslide-dammed lake remediation project includes: key technologies for the comprehensive treatment of 750 m-high slope resulting from intense earthquake and comprehensive treatment of 130 m-grade landslide dam composed of materials with a discontinuous wide gradation.

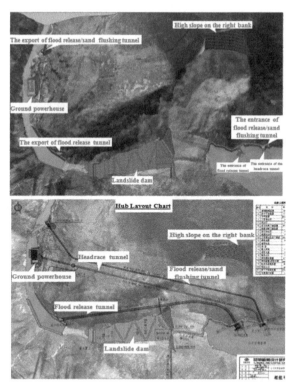

Fig. 3. Three-dimensional BIM layout of the project complex of Hongshiyan landslide-dammed lake remediation project (Right figure is Perspective View)

5   Key Technology for Comprehensive Treatment of 750 m-High Slope Resulting from Intense Earthquake

5.1   Survey and Geological Analysis

Stage 1: UAV mapping and geological survey, which provides the basis for the analysis and design of the slope

The rock slope formed by the collapse of the right abutment is up to 750 m high; the slope stress is adjusted; and the relaxation fissures of the slope are developed in large quantities and form dangerous blocks with the contribution of the unfavorable structural planes. There are many inverted suspensions on the steep cliff. There are many deposits of scum in the interface of the steep and gentle areas and the collapsed trough; the cracks are distributed in the about 60 m area at the top of the steep cliff. According to the development degree of the cracks at the top of the slope and the degree of slope risk, the stability of the slope can be divided into three zones. (See Fig. 4):

Zone I is the strip-like ground and unstable slope from 10 m–30 m from the edge of the cliff, intermittently extended. In addition to the boundary crack, the cracks in the zone are crossed, with a width of 1 cm–40 cm; the rock mass is obviously loosened and may collapse at any time;

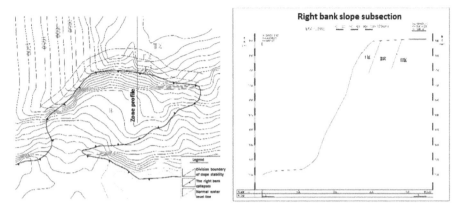

Fig. 4. Schematic zoning of stability of slope collapse deposit on right bank

Zone II is located outside Zone 1, 40 m–60 m away from the edge of cliff with poor slope stability. Cracks develop in this area, but the density is relatively small; the width is generally less than 15 cm; dislocated platforms are not obviously seen, but there is still the possibility of instability during the deformation adjustment of the lower slope.

Zone III is basically stable, and is located outside Zone II. No obvious cracks are found in the area, and the slope is basically stable. In the natural state, the slope reconstruction will not extend to such zone.

Stage 2: Tilt photography technique for identifying in detail the dangerous rock masses

The high and steep slopes and the dangerous rock masses are widely distributed. The surveyors cannot reach the steep cliffs for geological mapping. In order to accurately find out the distribution of cracks, the tilt photography technique is used to establish a high-resolution slope model of the right bank with a precision of 2 cm (Fig. 5).

Fig. 5. High-slope tilt photography on the right bank

5.2 Mechanism of High-Slope Collapse and Formation of Landslide Dam and Stability Analysis

5.2.1 Formation Mechanism

The formation model of the landslide on the right bank of Hongshiyan was determined according to the on-site investigation and analysis. The formation mechanism of landslide was determined by numerical simulation of the process of slope collapse.

The area of the landslide dam is mainly in the medium and high mountain narrow valley area where the tectonic erosion is dominant. The valleys are deep and steep, and the bedrock is exposed. Due to the development of adverse physical geology and the combination of unfavorable structural planes and strong rock mass weathering and relaxation, the natural slope is mostly at a critical stable state. The overall failure of the slope is controlled by the steep dip relaxation fissures and the weak interbeds located in the lower part. When the 6.5-magnitude earthquake occurred in Ludian, due to the horizontal and vertical ground motions, the weak interbeds in the middle and lower part of the slope was extruded and slid. The middle and upper rock mass lost support and was affected by the cracks with steep dip angles. The relaxation cracks along the river direction separated completely from the trailing edge under the action of seismic force, resulting in a wide range of collapse and failure. It moved to the riverbed and downstream under the action of self-weight, toppled and collapsed at a high speed, and quickly deposited in the riverbed to form a landslide dam.

5.2.2 Inversion of Physical and Mechanical Parameters of Rock Mass

Based on the analysis of slope instability pattern, the strength reduction method is used to invert the mechanical parameters of rock masses in weak interbeds, and the impact of the parameters of other rock formations on the inversion results is analyzed. Combined with the test results and engineering practice, the physical and mechanical parameters of the slope rock masses are determined.

5.2.3 Stability Analysis

After the occurrence of collapse, the overall stability safety factor of the slope meets the requirements of the specification under all working conditions, but the safety factor of the local upper overturned rock masses in Zone A is relatively low under accidental conditions (earthquake), and there is possibility of toppling failure. After the excavation and removal of the upper overturned rock mass, the slope stability coefficient meets the requirements of the specification.

5.3 Comprehensive Measures

Slope cutting treatment was taken to excavate the slope rock mass developed in Zone I and Zone II, remove the unstable rock mass and slow down the slope. Positive support measures were taken, such as prestressed anchor cables were used locally to increase the stability of the slope. Protection was carried out for the slope and drainage measures were strengthened. The soft argillaceous layer distributed on the slope is the main cause of the collapse, as a result, the P_1l Liangshan Formation in the middle and lower part of the slope was required to be enclosed; system monitoring was required for the slope (Fig. 6).

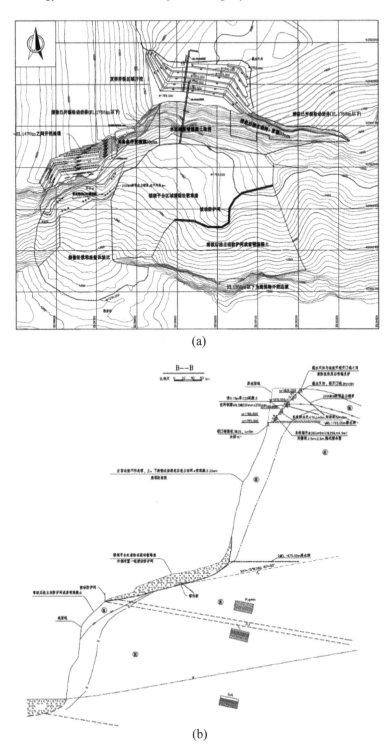

(a)

(b)

Fig. 6. Remediation plan for the high-slope on right bank

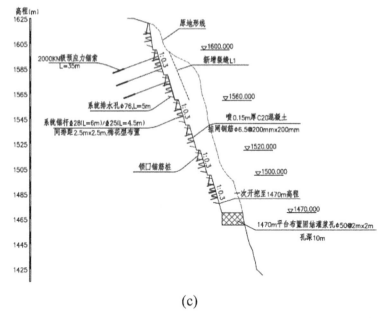

(c)

Fig. 6. (*continued*)

5.4 Analysis of Monitoring Data

At present, the slope treatment is nearly completed. The horizontal displacement of the surface deformation monitoring points of the collapsed slope on the right bank and its upstream side slope is between 2.7 mm and 17.5 mm, and the vertical displacement is between −28.5 mm and 18.7 mm. The vertical displacement shows that the settlement occurs. The displacement of each monitoring point fluctuates, and there is no obvious trend of displacement increase. The deformation of the deep cracking area at the top of the collapsed slope on the right bank is between 0.7 mm and 0.73 mm, indicating that the deformation value is small. The monitoring data show that the high-slope after treatment on the right bank has good stability and is in good consistence with the analysis results.

6 Key Technology for Remediating the 130 m Landslide Dam Composed of Discontinuous Wide Gradation Materials

6.1 Comprehensive Geophysical and Heavy-Duty Investigation Technologies for Identifying the Components of Landslide Dam and Physical-Mechanical Parameters

6.1.1 Method of Investigation

Considering the large undulations, many individual rocks, big gaps, poor stratification, and thick accumulation of landslide dam, conventional geophysical methods cannot be used. Instead, the passive source surface wave method, transient electromagnetic method, comprehensive detecting well are combined in the investigation shaft to detect the dam materials, together with comprehensive interpretation. Inhomogeneous, relative density, scale and space distribution of landslide dam materials are fully and clearly investigated.

At top of landslide dam, three large diameter investigation shafts with diameter of 1.5 m are arranged. The maximum depth of shaft is up to 97 m, which reach the bedrock, clearly identified the landslide dam.

6.1.2 Field Test Results of the Landslide Dam and the Ancient Landslide on the Left Bank

A large number of tests were carried out on the landslide dam and the left-bank ancient landslide. The results are shown in Tables 1 and 2 and Fig. 7.

Table 1. Test results of the landslide dam

Test item	Density test by water-filling method			Relative density		Grain composition (%)			
	Dry density (g/cm^3)	Specific weight	Porosity (%)	Maximum dry density (g/cm^3)	Minimum dry density (g/cm^3)	>200 (mm)	200–60 (mm)	60–2 (mm)	<2 (mm)
Range value	1.83–2.28	2.71–2.83	19.4–33.9	/	/	0–18.3	1.8–41.1	31.1–85.3	5.5–37.4
Average value	2.01	2.78	26.2	2.11	1.55	6.82	19.9	49.8	26.8
No. of Tests	7	15	3	1	1	15	15	15	15

6.1.3 Analysis of the Test Results

According to the analysis of the field test results, the specific weight of the landslide dam is between 2.71 g/cm^3 and 2.83 g/cm^3, the maximum dry density of the seven test points is 2.28 g/cm^3, the minimum dry density is 1.66 g/cm^3, and the average dry density is 1.98 g/cm^3. The corresponding minimum porosity is 19.4%, and the maximum porosity is 40.7%. Because the grain portion with >60 mm was eliminated in the laboratory relative density test, the results of relative density (maximum dry density 2.11 g/cm^3, minimum dry density 1.55 g/cm^3) in test group 1 are only for reference.

According to the gradation curve of the materials excavated from the pits, the grading between 100 and 5 mm belongs to a relatively steep descending curve, which occupies a relatively large portion. Most of the soil samples smaller than 2 mm are relatively few. According to the results of the grading curve of the boreholes, the maximum diameter of the core sample from ZK107 is 400 mm. The portion above 2 mm is relatively large, and the content of fine grains below 0.075 mm is very small. Due to the crushing effect of the bore core during the cutting process of the drill bit, the grain size of the actual landslide deposit is larger than that of the sample of the drill core.

Table 2. Field test results of the ancient landslide on the left bank

Test statistics	Test item									
	Density test by water-filling method			Field permeability test K_{20}(cm/s)	Relative density		Grain composition (%)			
	Dry density g/cm³	Specific weight	Porosity (%)		Maximum dry density (g/cm³)	Minimum dry density (g/cm³)	>200 (mm)	200–60 (mm)	60–2 (mm)	<2 (mm)
Range value	2.20–2.35	2.69–2.89	17.5–23.1	2.13–12.4×10^{-2}	/	/	0–19.6	1.4–46.7	7.2–68.8	7.2–51.4
Average value	2.30	2.83	19.5	/	2.34	1.77	4.4	19.2	53.6	22.8
No. of Tests	6	34	6	4	1	1	22	22	22	22

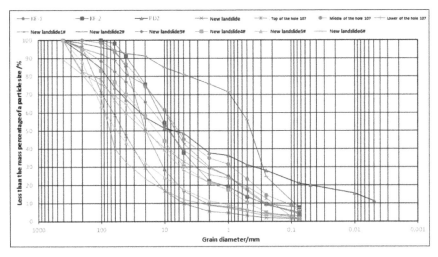

Fig. 7. Grading curve of landslide block

6.1.4 Analysis of the Results of the Ancient Landslide on the Left Bank

The in-situ density test of 6 points was completed in the adit of the ancient landslide. The dry density range was 2.20–2.35 g/cm^3, the average value was 2.30 g/cm^3, the average specific weight was 2.83, and the porosity of the 6 points was 17.5%–23.1% with an average of 19.5%. One group of soil samples was taken for laboratory relative density test. The grain portion with >60 mm was eliminated before the test. The maximum dry density was 2.34 g/cm^3 and the minimum dry density was 1.77 g/cm^3.

The analysis of the grading test for 22 groups of grains showed that the proportion of grain size was mainly concentrated in the gravel section of 60–2 mm, with an average value of 53.6%, followed by the portion below <2 mm, accounting for 22.8%, and the average value of giant grains larger than 60 mm was 23.6%.

The on-site permeability test was carried out in an adit and a total of 4-point tests were completed. The test result showed that the minimum permeability coefficient was 2.13×10^{-2} cm/s and the maximum value was 1.24×10^{-1} cm/s, indicating that the permeability was high.

6.2 Anti-seepage Proposal

As the water head retained by the landslide block is more than 100 m, several proposals including a anti-seepage proposal for the upstream slope, a concrete cut-off wall proposal and a self-compacting concrete reinforcement proposal were compared, of which a cut-off wall for the landslide dam and a grouting curtain for the ancient landslide on left bank combination proposal were selected.

The anti-seepage system of "137 m-in-depth cut-off wall on main river channel plus 125 m-in-depth grouting curtain of ancient landslide on left bank" was adopted, the treated landslide dam is formed.

The axis of the cut-off wall is arranged along the crest of the dam. The total length is 267 m, the wall is 1.2 m thick, with a depth of 1 m into the rock mass, and the maximum depth is about 130 m. The left bank ancient landslide is provided with grouting tunnel and double-row curtain grouting for seepage control. The maximum depth of the curtain in the area of the deposit is about 92 m, and the single-row grouting is used for seepage control in the bedrock, with a grouting spacing of 1.5 m. The bottom boundary of the anti-seepage is controlled by the 5 Lu line, and the anti-seepage profile is shown in Fig. 8.

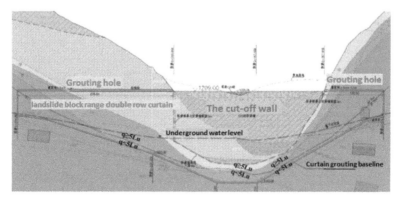

Fig. 8. Profile along the axis of the anti-seepage alignment

6.3 Analysis of Deformation, Seepage and Stability of the Landslide Dam

A finite element grid for stress and deformation analysis is established considering the ancient landslide and the landslide dam. The computation grid includes the paleo-bed al-1, the paleo-bed al-2, the ancient landslide del-1, the ancient landslide del-2, the landslide block col-1, the landslide block col-2, the anti-seepage curtain and the concrete cut-off wall (Fig. 9).

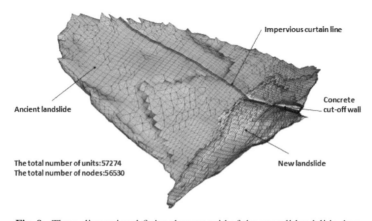

Fig. 9. Three-dimensional finite element grid of the overall landslide dam

According to the calculation and analysis, the maximum settlement of the dam is 31.4 cm, which occurred near the surface of the dam on the left bank; the maximum displacement along the river occurred in the middle of the cut-off wall with a maximum value of 9.78 cm; the maximum settlement is 8.44 cm. The maximum principal stress of the cut-off wall is about 12.6 MPa, and the tensile stress in the tensile stress zone of the cut-off wall is generally less than -2 MPa. When the full supply level is at El.1,200 m, the leakage is 45.6 L/s. See Table 3 for details.

Table 3. Calculation results of maximum stress and deformation of Hongshiyan landslide dam and cut-off wall

Calculation plan	Settlement (cm)	Displacement along river(cm)		Displacement along dam axis(cm)		Principal stress (MPa)	Minimum stress (MPa)
		Towards upstream	Towards downstream	Towards left bank	Towards right bank		
Landslide block during formation period	101.5	10.6	14.5	26.2	11.0	3.27	1.03
Landslide block during deposit period	31.4	4.14	3.65	5.92	11.0	3.37	1.00
Landslide block during impoundment period	8.44	1.12	9.78	3.61	2.92	3.67	1.22
Cut-off wall during impoundment period	2.42	0.00	9.78	2.80	0.28	12.6	6.90

The upstream and downstream slopes of the Hongshiyan landslide dam are gentler than those of conventional earth-rock dams. The safety factor of anti-sliding stability is greater than the allowable value of the specification, and there is a certain safety margin. The locations of most dangerous slip arcs of the upstream and downstream slopes under each working condition along the maximum cross section are shown in Fig. 10.

6.4 Construction Technology of Cut-off Wall and Grouting Technology of Landslide Dam

With the use of pre-blasting and pre-grouting and other processes, the maximum depth of the left bank anti-seepage wall implemented is about 131 m.

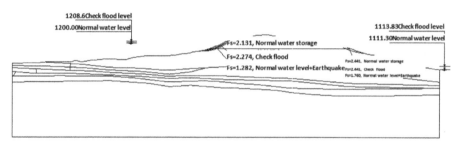

Fig. 10. Locations of the most dangerous slip arcs of the upstream and downstream slopes under each working condition along the maximum cross section

6.4.1 Analysis of Key and Difficult Points in the Construction

The maximum depth of the concrete cut-off wall of the project is more than 137 m. The wall passes through the rockfill deposit and the sand/pebble/gravel layer. The grain size of the rockfill deposit is extremely uneven, the trench is easy to be inclined, the work efficiency is low, and the slurry leakage and collapse, giant blocks resulted in difficulties during the formation process. The trench entering the rock of the steep slope on the right bank is the difficult and key point of the construction of the project; the cut-off wall of the project is a permanent building, the thickness and depth of the wall are large, and the joint connection of the wall is the key of the project.

The grouting tunnel of left ancient landslide is quite small with limited space. The efficiency of common drilling equipment is quite lower. There are some large-porosity rock layers under landslide dam foundation, the cement mortar of grouting is easy to loss and disperse. Grouting of cement mortar in the foundation of the landslide dam with concentrated fines tends to be difficult.

6.4.2 Construction Procedure and Method

According to the characteristics of the project, the drilling and grasping method is adopted, which is based on the impact drilling rig and supplemented by the mechanical grab. The joint construction adopts "joint piping method" and "drilling method":

(1) Before the construction of the concrete cut-off wall, a row of consolidation grouting boreholes are arranged on the upper and lower sides of the axis of the cut-off wall. The grouting borehole is 1.5 m away from the axis of the cut-off wall, with a borehole spacing of 1.5 m, and the paste slurry is used for grouting. The large leakage channel is blocked to reduce the leakage during the process of the cut-off wall to ensure the smooth implementation of the construction.
(2) In order to prevent the trench from collapsing, the construction platform is rolled and compacted. The guide trench is made of reinforced concrete structure and compacted with clay on the back side of the guide wall.
(3) The mud for fixing the wall is made of high quality MMH positive electric glue mud, and the tackifier or plugging agent is added according to the actual situation.
(4) For large blocks, pre-blasting and grouting are used in the drilling before construction. In the formation process of trench, it is equipped with a 12t heavy hammer. If

the effect is not ideal or a large boulder is encountered, the in-hole energy blasting or small-diameter drilling blasting is adopted.

(5) The accuracy detection of trench or borehole is performed using KD-400 or KM684 ultrasonic logging tool.

(6) In order to ensure the success rate of tube drawing, a layer of foam is wrapped in the joint tube, and then wrapped with geotextile to reduce the resistance of the drawing, and the tube is drawn by a vibration hydraulic tube puller.

(7) During the borehole-formation process, in case of voided formation, in addition to the use of conventional clay filling, sawdust and other materials, expansion powder and high-efficiency plugging agent can be used for plugging.

The full hydraulic low clearance driller is studied and manufactured special for the Hongshiyan landslide dam with small dimension and flexible operation to solve the quick drilling for grouting curtain in overburden withing limited space. New grouting material, i.e. silica sol with high adhesive plastic, large yield strength and anti-scouring is studied and made. It suits the grouting condition with large porosity rock layer and under flowing water, and can avoid or reduce the grouting mortar to be dispersed and washout. The grouting material of environment-friendly nano-degree ludox is studied and made to solve the technical difficulties of filling grouting material of wide gradation landslide dam materials.

7 Summary

The remediation project of the Hongshiyan landslide-dammed lake was started following the emergency rescue and subsequent disposal of the lake in August 2014. The reservoir has been impounded in 2019, the landslide dam has retained to a high water level with safe operation. This paper summarizes the key technologies such as emergency treatment technology under the lack of information conditions, the comprehensive treatment of the 750 m-high slope resulting from intense earthquake and comprehensive treatment of the 130 m-grade landslide dam composed of materials with discontinuous wide gradation. The paper provides reference and experiences for the development and utilization of similar landslide dammed lakes.

Emergency Response and Thinking of Risk in Baige Barrier Lake in the Jinsha River

Sheng'an Zheng$^{(\boxtimes)}$, Chao Liu, and Fuqiang Wang

China Renewable Energy Engineering Institute, No.2, Beixiaojie Street, Liupukang, Xicheng District, Beijing 100120, China
zhengsa@creei.cn

Abstract. Landslides occurred twice in Baige Village, Jiangda County, Changdu, Tibet on October 10 and November 3, 2018, blocking the main stream of the Jinsha River and forming a barrier lake. After the occurrence of the significant event, entrusted by the National Energy Administration, China Renewable Energy Engineering Institute quickly established a Technical Emergency Response Team and scientifically studied and judged the risks of the barrier lake and accurately predicted the flood caused by break of the barrier lake with the support and cooperation of all relevant organizations. In the shortest time, it studied, formulated and implemented a series of emergency response schemes, such as the manual excavation of the barrier, the removal of the cofferdam of Suwalong Hydropower Station, and the emergency water discharge of Liyuan Hydropower Station, which effectively alleviates the disaster and risk, and ensures the safety of hydropower projects under operation and construction in the lower reaches. Combined with the experience of this risk treatment, this paper considers and predicts the risk assessment, emergency safety management, joint dispatching of cascade reservoir, etc. of the river basin water at cascade level.

Keywords: Barrier lake · Dam break · River basin cascade · Emergency safety · Hydropower project

1 Hydropower Planning for the Upper Reaches of Jinsha River

The National Development and Reform Commission approved the Hydropower Planning for the Upper Reaches of the Jinsha River (FGBNY [2012] No. 2008) in 2012. From the upper reaches to the lower reaches, Xirong, Shala, Guotong, Gamtog, Yanbi, Bolo, Yebatan, Lawa, Batang, Suwalong, Changbo, Xulong and Benzilan are successively planned, and the hydropower layout of "one reservoir and thirteen cascades" is shown in Fig. 1.

Currently, hydropower stations at Yebatan, Lawa, Batang and Suwalong have been approved and are under construction; Changbo Hydropower Station has passed the review of feasibility study report, and four cascade hydropower stations, Gamtog, Boluo, Xulong and Benzilan, are in the stage of feasibility study.

J.-M. Zhang et al. (Eds.): ICED 2020, SSGG, pp. 96–104, 2020.
https://doi.org/10.1007/978-3-030-46351-9_8

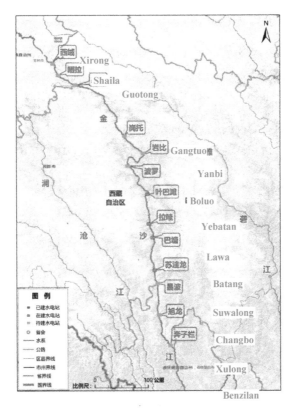

(a) Planning

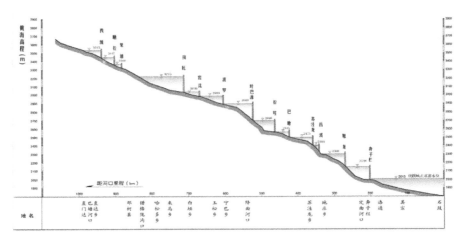

(b) Longitudinal Section

Fig. 1. Hydropower planning for the upper reaches of Jinsha River.

2   Formation and Break Process of Baige Barrier Lake

2.1   The First Barrier Lake Formed on "10.11"

On the evening of October 10, 2018, a landslide occurred at the junction of Baiyu County, Ganzi Prefecture, Sichuan Province and Jiangda County, Changdu City, Tibet Autonomous Region. The landslide blocked the main stream of the Jinsha River to form a barrier lake. The accumulated length of the barrier along the river is nearly a 1,000 m, the width of the cross river is about 300 m, and the total volume is about 12.5 million m^3.

At 17:00 on the 12th, the barrier lake began to overflow naturally, and the peak discharge of the burst flood reached 11,000 m^3/s around 3:00 on the 13th. From Yebatan Hydropower Station to Liyuan Hydropower Station, the lower reaches of the barrier lake is affected by the dam break flood. Among them, two unfinished diversion tunnels on the left and right banks of Yebatan Hydropower Station under construction were forced to be under overflow, some construction roads and tunnels with low elevations were flooded, and the roads, construction sites and facilities along the river in the dam site area were seriously damaged; the cofferdam and diversion tunnel of Suwalong Hydropower Station under construction were tested by over-limit flood. At 10:00 on the 15th, the flood peak of the barrier break reached the built Liyuan Hydropower Station, which had no effect on the lower reaches after being regulated by cascade hydropower stations such as Liyuan and Ahai. The upstream power stations also organized and carried out hidden danger investigation in projects and production recovery in time after the barrier break flood.

2.2   The Second Barrier Lake Formed on "11.03"

At about 17:00 on November 3, the original main chute of Baige Landslide collapsed again to form a secondary river blocking. The newly added volume of the barrier is 2 million m^3, with a total scale of 10 million m^3. The top elevation is about 2,966 m, and the capacity of the barrier lake formed is about 775 million m^3, which is 4–5 times of the capacity of the "10.11" barrier lake. Once the barrier breaks, it will cause disastrous consequences to the lower reaches.

After the formation of the high-risk barrier lake, the Expert Group of China Renewable Energy Engineering Institute (CREEI) considered after analysis that the barrier lake break flood would overflow the earth rock cofferdam of Suwalong Hydropower Station under construction in the lower reaches, resulting in the barrier burst, and the combination of the barrier burst flood and the barrier lake flood would aggravate the impact on the lower reaches. Hence, the Expert Group put forward several suggestions of emergency plan to the National Energy Administration on the early manual intervention of the barrier lake, including the excavation of the discharge channel, the demolition of the cofferdam of Suwalong Hydropower Station, and the early and deep water discharge of the Liyuan hydropower station.

At 5:00 on the 12th, the water level of the barrier lake rose to 2,952.52 m, an elevation from the floor of manual excavation of discharge chute, and the water gradually entered the manual excavation chute; at 10:50, the manual excavation chute officially and comprehensively overflowed. At 8:00 on the 13th, the flow of the barrier break increased to about 70 m³/s and entered the break stage; at 18:20, the maximum peak flow of the barrier break was estimated to be about 33,900 m³/s. After the break of the barrier, the flood peak formed rapidly advanced to the lower reaches, passing through the locations of Yebatan, Lawa, Batang, Suwalong, Benzilan, Shigu and Liyuan in turn. Figure 2 shows the flow process of each station on the upper and middle reaches of Jinsha River.

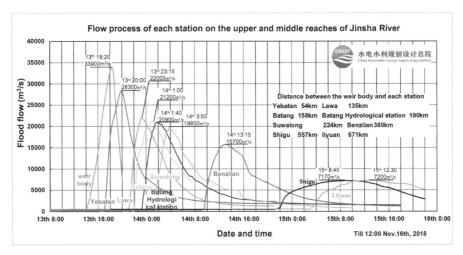

Fig. 2. Flood flow process of stations in the upper and middle reaches of the Jinsha River.

The direct economic loss caused by the breach of the barrier lake on "11.03" to Sichuan, Tibet and Yunnan Province was more than RMB 13 billion yuan. The direct economic loss caused by the second risk of barrier lake on "11.03" to Yebatan, Batang, Suwalong and other hydropower projects under construction in the upper reaches of Jinsha River was about 1.188 billion yuan, and the loss to Liyuan and other hydropower projects under construction was 64 million yuan. Through manual intervention, the top elevation of the barrier was reduced by about 13.5 m, and the storage capacity of the barrier lake was reduced by 216 million m³, which effectively reduces the risk of the lake and the downstream loss.

3   Emergency Response to the Risk of Barrier Lake

After the occurrence of Baige Barrier Lake, CREEI immediately dispatched more than 30 technical experts to set up a Technical Emergency Response Teamunder the unified deployment and guidance of the National Energy Administration, established a contact channel with the construction management and operation organizations of each cascade power station, timely and accurately obtained on-site information, and issued a risk report as soon as possible. Especially in the second emergency response of barrier lake on "11.03", CREEI has organized seven working meetings of the TechnicalEmergency Response Team to discuss and judge the risk, arrange relevant work, put forward a series of emergency response plans, and carry out dynamic tracking analysis on the real-time situation of barrier lake and barrier break flood.

3.1 Major Emergency Response Plans

On November 4, CREEI sent experts to the Baige Barrier Lake for the first time, cooperated with the National Energy Administration, the Ministry of Emergency Management of the People's Republic of China and other national ministries to carry out the site work, and collected the first-hand information about the barrier. The Technical Emergency Response Team used the preliminary information fed back from the site and the data of basin topography to estimate the capacity of the barrier lake, and immediately carried out the preliminary calculation and analysis of the barrier break flood and flood evolution, so as to comprehensively study and judge the risk and its impact on the existing power station under construction.

According to the analysis of the barrier break flood, the peak flow of the barrier break flood was expected to exceed 37,000 m^3/s without any manual intervention. If the barrier was manually pre-excavated for a discharge chute to a depth of 10 m, the pass elevation of discharge chute of the barrier could be reduced to 2,956 m, and the peak flow of the barrier flood was expected to be about 31,000 m^3/s, and the manual excavation of the barrier would effectively reduce the break flood flow.

According to the analysis of flood evolution, under the condition of manual pre-excavation of discharge channel, the peak flood flow barrier break flood to the upstream cofferdam of Suwalong Hydropower Station would reach 20,000 m^3/s, which was far greater than the discharge capacity of diversion structure of Suwalong Hydropower Station, 6,480 m^3/s. In view of that the storage capacity of "11.03" Baige Barrier Lake was about 775 million m^3, which was far more than that of the first barrier lake on "10.11", even though the diversion structure of Suwalong Hydropower Station was provided with a certain over-discharge capacity, the highest water level at the cofferdam where the break flood reached would still reach 2,448 m, which is higher than the cofferdam top elevation of 2,432 m. The possibility of overflow and break of the earth rock cofferdam top was great, which would cause superimposed effect of flood on the lower reaches. Therefore, it was recommended to carry out measures such as partial removal of cofferdam, filling water in foundation pit or overflow protection of cofferdam in advance (Fig. 3), to avoid secondary disasters caused by cofferdam collapse, and initiate emergency plans to ensure the safety of life and property in the construction area.

According to analysis of flood regulation of Liyuan Reservoir, when the barrier break flood evolved to Liyuan Reservoir, the peak flood flow may reach 22,000 m^3/s in extreme cases. If the water level of Liyuan Reservoir was lowered to 1,598 m of the crest elevation of the spillway in advance, the highest water level in front of the dam can be controlled at 1,625.88 m, 2.67 m higher than the check water level and 0.12 m lower than the crest elevation through the discharge of spillway and flood and sediment releasing channel. Therefore, in order to ensure the operation safety of Liyuan Hydropower Station, it was necessary to take the emergent measure of water discharge to lower the water level in advance. In view of the analysis of reservoir discharge duration and the prediction of the full storage time of the barrier lake, it was considered that Liyuan Hydropower Station was provided with enough time to discharge the reservoir, and the water discharge scheme mainly based on power generation can meet the requirements. First, the full discharge of the unit of 2,544 m^3/s was used to discharge the reservoir to the dead water level of 1,605 m; then, the reservoir water level was further lowered to the crest elevation of

Fig. 3. Destruction and removal of upstream and downstream cofferdams of Suwalong Hydropower Station

1,598 m of the spillway by the flood discharge facilities, or even 1,571 m if necessary according to the actual implementation of measures such as manual intervention of the barrier.

Based on the above analysis results, CREEI quickly put forward three suggestions to the National Energy Administration: (1) carrying out manual excavation to the barrier as early as possible; (2) breaking and removing the upstream and downstream cofferdams of Suwalong Hydropower Station; (3) carrying out emergent water discharge for Liyuan Hydropower Station. Combined with the results of calculation and analysis, CREEI assisted the Joint Working Group on site to complete the analysis report on the downstream water level along the river after the dam break, providing the evaluation basis for the local government on inundation impact. After discussion, the above suggestions were adopted, which provided important reference for emergency response.

3.2 Dynamic Tracking Analysis of Flood Situation of Varrier Lake and Barrier Break Flood

According to the actual situation of the final "11.03" Baige Barrier Lake, the peak flood flow at the barrier break was about $33,900 \, m^3/s$, and that at Suwalong Hydropower Station was about $19,800 \, m^3/s$, which was slightly different from the predicted peakflood flow of $31,000 \, m^3/s$ at break and $20,000 \, m^3/s$ in Suwalong Hydropower Station. The Technical Emergency Response Team of CREEI accurately predicted the key information, such as the break time of the barrier and the maximum peak flood flow.

In order to grasp the situation of Baige Barrier Lake in time and accurately, the Technical Emergency Response Team of CREEI organized all kinds of data through the channels of experts appointed to the site and comprehensive monitoring platform, analyzed and judged the possible risks and hazards continuously for 24 h, issued and predicted the real-time situation of water level of barrier lake and barrier break flood evolution, and even updated and issued information every 15 min during the peak period.

At the same time, the Technical Emergency Response Team successively carried out the analysis of barrier break flood evolution and flood regulation calculation of 10 schemes according to actual situation including the progress of manual intervention of the barrier and barrier break flood, and constantly adjusted the schemes of water discharge for Liyuan Reservoir, which provided necessary technical support for reasonable decision-making.

4   Inspiration of Risk Response and Work Prospect

4.1   Cascade Risk Assessment of Upstream Jinsha River

The upper reaches of the Jinsha River is rich in hydropower resources, and is an important hydropower base in China. However, the topographical and geological conditions in this area are complex, especially the barrier lake formed by occurrence of two continuous landslides in Baige Village, Jiangda County, Changdu City, Tibet Autonomous Region has a great impact on the construction of hydropower projects under construction in the basin. In order to deal with the risks in the hydropower development and construction of the Jinsha River, and to investigate and sort out the potential adverse geological bodies and geological disasters in the construction of the power station, CREEI organized 12 organizations to carry out the risk assessment of the upstream hydropower cascade of the Jinsha River for 8 months according to the requirements of the National Energy Administration.

Firstly, the potential unfavorable geological bodies in the upper reaches of the Jinsha River and the geological disasters in the construction of the hydropower station are systematically investigated in the evaluation. According to the investigation, the research on the shape and volume of the barriers formed by landslides, the analysis of the barrier break flood, and the calculation of the flood evolution were carried out for the unfavorable geological bodies with high risks of river blocking and affecting the safety of the river basin cascades. On this basis, the risk and countermeasures of 4 cascade hydropower stations under construction in the basin were analyzed and studied year by year and by period. Finally, systematic solutions for unfavorable geological bodies were put forward from the aspects of planning and implementation, construction sequence, design standards, engineering response measures, construction operation management, emergency plan, etc. of cascade hydropower stations in the basin, so as to ensure the construction and operation of the project and the safety of the basin.

4.2   Reasonable Arrangement of Construction Sequence of Cascade Power Stations

It can be found from the two failures of Baige Barrier Lake that the most effective way to deal with the risk of barrier break flood is to use the built cascade reservoirs in the lower reaches to retain the flood. Unified dispatching of controlled reservoirs and basin cascades is important to reduce barrier break flood of barrier lakes. Therefore, it is necessary to analyze the reasonable construction sequence of cascade hydropower stations, give full play to the role of flood prevention and mitigation of basin control projects such as the leading reservoir, and improve the overall response capacity of the basin to risks.

4.3 Demonstration and Revision on the Relevant Technical Standards for Hydropower Projects

How to consider the risks of large landslides and barrier lakes to reservoirs, and how to determine the design flood standard, safety free board and other safety standards in hydropower projects are studied and demonstrated in depth. At present, CREEI is organizing and carrying out the research and demonstration on the relevant technical standards.

4.4 Improvement of the Emergency Safety Management Level of Hydropower Projects

By the end of 2019, the total installed capacity of hydropower in China has reached about 320 million kilowatts. In the Jinsha River, Yalong River and other basins, 65%–80% of the installed hydropower capacity has been built or under construction, while the hydropower development degree has exceeded 80%–90% in Wujiang River, Nanpan River - Hongshui River, Dadu River and the upper reaches of the Yangtze River. The key development of hydropower basin in China has been gradually changed from project construction to operation management. At present, the foundation of basin risk management is relatively weak.

In the emergency response of Baige Barrier Lake, the information of river basin power stations was prepared and analyzed comprehensively in a short time. There are also some deficiencies in the technical emergency support system. Different organizations have great differences in the prediction of barrier break flood. In the calculation of flood evolution, the influence of wide valley in Shigu Section on flood evolution was not taken into account, and the loss caused by the downstream flood was not predicted accurately, which restricted the efficiency, pertinence and accuracy of emergency response to a certain extent.

Currently, relevant management measures and standards have been issued by the State Flood Control and Drought Relief Headquarters, the Ministry of Water Resources of the People's Republic of China, the National Energy Administration and other national ministries and commissions, which put forward requirements for the emergency plan for reservoir flood prevention and rescue, the risk and impact of dam break flood, the emergency management system for dam safety, etc. In recent years, the Spillway Accident of Oroville Hydropower Station in the United States and the secondary dam break accident of Xepian-Xenamnoy in Laos have brought us enlightenment. In order to correctly, quickly and properly handle the emergency accidents such as dam break, barrier lake, earthquake and flood exceeding the limit, it is necessary to further improve the emergency safety management level of hydropower projects, improve relevant technical standards and carry out risk management research on full life cycle of hydropower projects.

In addition, it is necessary to establish a safety management platform for hydropower basin as soon as possible. We need to resolve or slow down various risks such as over-limit floods, earthquakes, barrier lakes, reservoir bank landslides and debris flows, and minimize the losses through the joint operation and water regime prediction of cascade reservoirs in the basin while striving for the maximum economic benefits.

5 Conclusions

After two emergencies of Baige Barrier Lake, China Renewable Energy Engineering Institute has made efforts to scientifically study and judge the risk of barrier lakes, accurately predict the barrier break flood under the guidance of the National Energy Administration and the support and cooperation of all relevant organizations, and studied, formulated and implemented a series of emergency response plans in the shortest time, which effectively alleviated the disaster and risk, and ensured the safety of downstream hydropower projects in operation and under construction.

Through a series of events at home and abroad in recent years, such as the emergency response of Baige Barrier Lake, we have realized that it is necessary to build an emergency command platform and system for the power industry, strengthen the comprehensive monitoring of safety emergency in the basin, carry out safety risk assessment and hidden danger investigation in the basin, orderly promote the construction of cascade power stations, and improve the level of disaster prevention and mitigation in river basins.

ISSMGE Bright Spark Lectures and Invited Lectures

Stability and Failure Mechanisms of Riprap on Steep Slopes Exposed to Overtopping

Priska Helene Hiller[1](✉) ⓘ and Ganesh Hiriyanna Rao Ravindra[2] ⓘ

[1] Norwegian Water Resources and Energy Directorate (NVE), Landslides, Flood and River Management, Abels gate 9, 7030 Trondheim, Norway
phh@nve.no
[2] Department of Civil and Environmental Engineering, Norwegian University of Science and Technology (NTNU) Trondheim, S. P. Andersens veg 5, 7031 Trondheim, Norway
ganesh.h.r.ravindra@ntnu.no

Abstract. Riprap is widely used as erosion protection consisting of either dumped or, in an interlocking pattern, placed stones. Data about the stability and failure mechanism of riprap on steep slopes is scarce and hence subject to research. Dumped and placed ripraps constructed on a slope of 1:1.5 (vertical: horizontal) were exposed to overtopping scenarios in small scale model tests and in the field with large-scale riprap stones. Detected stone displacements in the riprap depict a two-dimensional deformation of the placed riprap structure. The displacements along the flow direction led to gap formation at the upstream section of the riprap where the adjacent stones lost their interlocking and became prone to erosion. In combination with a lift in the middle of the riprap, buckling is described as the failure mechanism for placed riprap with a fixed toe with analogy to Euler's Buckling theory. Dumped riprap failed by sliding down the filter layer as soon as the top of the stones were overtopped by the flow.

Keywords: Riprap · Failure mechanism · Rockfill dam · Physical modelling

1 Background

Embankment dams are the most common dam type worldwide, 78% according to the World Register on Dams (International Commission on Large Dams, 2020). The most frequent reason for embankment dam failure is overtopping (International Commission on Large Dams, 1995). A protective riprap layer on the crest and the downstream slope (Fig. 1) can delay or prevent erosion in case of accidental leakage, overtopping or sabotage of the dam (e.g. Orendorff et al. 2013 and Toledo et al. 2015).

The current construction practice in Norway is to protect the downstream slopes of embankment dams with single-layered placed riprap structures. In comparison with ripraps comprising of randomly dumped stones, placed riprap consists of sufficiently large stones which are placed in an interlocking pattern with their longest axes towards the dam (Ministry of Petroleum and Energy 2009, [*Dam safety regulations*]). The riprap stones sizing is usually of the order of 0.3 to 0.7 m in diameter. Most of the large embankment dams in Norway, exceeding 15 m in height, are rockfill dams and were constructed before 1990 (Norwegian Water Resources and Energy Directorate 2016).

© Springer Nature Switzerland AG 2020
J.-M. Zhang et al. (Eds.): ICED 2020, SSGG, pp. 107–116, 2020.
https://doi.org/10.1007/978-3-030-46351-9_9

Fig. 1. Depiction of placed riprap constructed on the downstream slope of the 143 m high dam Oddatjørn, highest rockfill dam in Norway. (Photo: NTNU)

The support fill of rockfill dams consists of rockfill for more than 50% of the dam volume (Kjærnsli et al. 1992). The downstream slopes of rockfill dams are usually inclined with 1:1.5 (vertical: horizontal) corresponding to a slope of $S = 0.67$. An ongoing research program focuses on investigating stability aspects of rockfill dams subjected to extreme loading scenarios. Part of it is to describe the stability aspects and failure mechanisms of placed riprap on steep slopes exposed to overtopping flows.

Data of riprap stability, either dumped or placed, on steep slopes up to $S = 0.67$ are scarce. A few studies are available from Germany, where small earth dams (<10 m in height) are used for flood retention reservoirs and the spillways in some cases are formed as an overtoppable dam section with erosion protection (Siebel 2007 and Dornack 2001). A single study (Dornack 2001) with placed ripraps on slopes as steep as $S = 0.67$ was found. Siebel (2007) investigated placed riprap stability on slopes of $0.067 < S < 0.33$ and described the failure mechanisms: "erosion of single stones", "sliding of the protection layer" and "disruption of the protection layer".

Further, available information regarding practical aspects of placed riprap construction such as adopted construction practices and existing state of placed ripraps constructed over the past several decades is rare. With an objective of addressing this concern, a field survey of placed ripraps on 33 different rockfill dams was conducted by Hiller (2016). A subsequent field survey of 9 rockfill dams was conducted by Ravindra et al. (2019). These studies document details describing placed riprap construction such as stone sizing, orientation, packing density and existing state of toe conditions for placed ripraps constructed on several Norwegian rockfill dams.

The research presented in this paper aims at adding to available literature on riprap design on the downstream slopes of rockfill dams. Physical model test and field test with large-scale riprap stones were executed to increase the knowledge and technical expertise on performance of ripraps on steep slopes ($S = 0.67$) under overtopping conditions. This paper summarizes the so far conducted studies. The findings are discussed with focus on riprap stability and failure mechanisms.

2 Methodology

Model ripraps constructed with stones of median size $d_{50} = 0.057$ m were tested in the hydraulic laboratory at the Norwegian University of Science and Technology (NTNU) Trondheim. Furthermore, two series of tests were run on large-scale ripraps constructed with stones of sized $d_{50} = 0.54$ m at a temporary field site in 2013 and with $d_{50} = 0.37$ m in 2015. Most tests were conducted with placed ripraps, and few tests were also carried out with dumped ripraps for comparison with literature.

Essential parameters to describe riprap stability are the packing factor and the critical stone-related Froude number. The packing factor P_c in Eq. (1) relates the number of stones per m^2 surface area, N, to the area of an average stone represented with the median stone size, d_{50}^2. It was introduced by Linford & Saunders (1967) and Olivier (1967).

$$P_c = \frac{1}{N \cdot d_{50}^2} \tag{1}$$

The critical stone-related Froude number $F_{s,c}$ in Eq. (2) was used to quantify riprap stability with the critical unit discharge at riprap failure q_c, the gravitational acceleration g and the nominal stone size d.

$$F_{s,c} = \frac{q_c}{\sqrt{g d^3}} \tag{2}$$

2.1 Model Tests

The model test setup was situated in a hydraulic flume (25 m long, 1 m wide, 2 m high) and consisted of a horizontal crest and an inclined 1.8 m long chute with $S = 0.67$ (Fig. 2). The tests P05, P06, P07 and D02 were run with reduced chute length to achieve scaling of 1:6.5 to the field tests run in 2015, using Froude similarity. The single layered ripraps were constructed by hand and the model setup was limited to the filter layer and the riprap. Overtopping discharge magnitude was increased stepwise for specific time intervals. Between each step, the discharge was stopped, and the riprap inspected carefully. The position of selected stones was determined employing a 3D laser traverse system. The procedure was repeated until ultimate riprap collapse was achieved. Discharge, water levels and video footages were recorded during the tests. Observers followed the tests carefully and noted removal of stones. The details for the model studies are available in Hiller et al. (2018) and Ravindra et al. (2020).

2.2 Field Tests

The field tests with large-scale riprap stones were situated within the outlet channel of dam Svartevatn as shown in Fig. 3. The 3 m high and 12 m wide test dams were built employing large construction machinery. Discharge was supplied through a middle outlet from the reservoir. The field tests are documented in detail within Hiller et al. (2019) and Hiller & Lia (2015).

Fig. 2. Sideview of the model setup in the laboratory with placed riprap. Horizontal crest and inclined chute with $S = 0.67$. (Photo: NTNU)

3 Results

The results from ten model tests and five field tests are summarized in Table 1 as well as in Hiller (2017). The results of the model tests are in detail described and discussed in Hiller et al. (2018). Data sets obtained from Hiller et al. (2018) were subjected to further statistical analysis to describe 2D riprap stone displacements. Results from the study are presented in Ravindra et al. (2020). The model and field tests are systematically analysed and compared in Hiller et al. (2019). Videos footages are available online for test F13P1 (Hiller 2015) and test F15P2 (Hiller et al. 2019).

Placed ripraps were found to be more stable than dumped ripraps as demonstrated by the higher critical stone-related Froude numbers in Table 1. The stability for placed riprap, ending with riprap failure, was in the range between $F_{s,c} = 1.6$ for F13P2 and $F_{s,c} = 10.6$–11.3 for F15P2. There were three tests P05–P07 with placed riprap which withstood the maximum available discharge of $q = 0.49$ m^{2}s^{-1} in the test flume. They are hence indicated with $F_{s,c} > 11.5$. The riprap in test F13P2 failed due to an instability in the dam toe and not direct riprap failure. The placed ripraps in the model tests were denser packed than in the field as reflected by the low P_c values. The packing factor for dumped riprap are in the same range for model and field tests. Dumped riprap failure commenced as soon as overtopping of the riprap structure was achieved leading to sliding on the underlying filter layer. Erosion of the first stone coincided with riprap failure as indicated by $q_s\,q_c^{-1} = 1$. Placed ripraps in general endured the erosion of single stones and hence $q_s\,q_c^{-1} < 1$, except for P03 which was loaded in a single loading step and F13P2 which failed due to toe instability. In some tests, it was not possible to observe removal of the first stone, marked with n/a in Table 1. The maximum displacements Δx_{max} were smallest for test P05–P07 in which the riprap did not fail. The displacements accumulated at the top of the chute forming a gap. Consequently, the riprap stones at the gap gradually lost their interlocking and were prone to erosion. Placed riprap failure initiated with progressive erosion at the top of the chute and rapidly propagated down

Fig. 3. Field site for the large-scale riprap tests. The 3 m high and 12 m wide test dams were situated in the outlet channel (in the center) of the large dam Svartevatn (in the background on the right). (Photo of test F13P1: NTNU)

Table 1. Summary of results from model and field tests. The letter "P" in the test name implies placed riprap and "D" dumped riprap. The field tests are additionally named with "F13" for the series in 2013 and "F15" for 2015. Median stone diameter d_{50}, packing factor P_c, critical discharge at riprap failure q_c, ratio between the discharge at erosion of the first stone q_s and q_c, critical stone-related Froude number $F_{s,c}$ and maximum displacements in flow direction at the marked stone in the top of the chute Δx_{max}.

Test	d_{50} (m)	P_c (-)	q_c (m²s⁻¹)	$q_s q_c^{-1}$ (-)	$F_{s,c}$(-)	Δx_{max} (m)
P01	0.057	0.56	0.24	0.42	5.6	0.110
P02	0.057	0.55	0.36	0.28	8.4	0.106
P03	0.057	0.52	0.25	1.00	5.9	0.066
P04	0.057	0.53	0.40	0.50	9.4	0.108
P05[a]	0.057	0.48	>0.49	n/a	>11.5	0.012
P06[a]	0.057	0.50	>0.49	<0.73	>11.5	0.013
P07[a]	0.057	0.56	>0.49	n/a	>11.5	0.023
P08	0.057	0.55	0.24	0.81	5.6	0.038
D01	0.057	1.05	0.04	1.00	0.9	n/a
D02[a]	0.057	0.83	0.05	1.00	1.2	n/a
F13P1	0.54	n/a	6.5	n/a	5.2	n/a
F13P2	0.54	n/a	2.0	1.00	1.6	n/a
F15P1	0.37	0.75	6.1	0.74	8.7	n/a
F15P2	0.37	0.64	7.5–8.0	0.84–0.79	10.6–11.3	n/a
F15D1	0.37	0.84	0.4–0.8	1.00	0.6–1.2	n/a

[a]Chute length shorter than 1.8 m

the chute. This observation of displacements as failure origin of placed riprap on steep slopes is described in detail in Hiller et al. (2018).

Results from Hiller et al. (2018) describe 1D failure mechanism in placed ripraps on steep slopes exposed to overtopping scenarios. Further, all past experimental research on placed ripraps had focused on analyzing the 1D failure mechanism in placed ripraps exposed to overtopping. A 2D description of the same was unavailable in international literature. Hence, the study of Ravindra et al. (2020) was aimed at past findings describing unidimensional failure mechanisms in placed ripraps to 2D by providing qualitative and quantitative descriptions of 2D displacements of placed riprap stones under overtopping scenarios. Experimental data sets accumulated by Hiller et al. (2018) through physical modelling investigation conducted on model ripraps with toe support constructed with angular stones on a steep slope of $S = 0.67$ were further analysed along with additional experimental data. Figure 4 illustrates 2D displacements analysis results from a single test, test P02 presented in Table 1.

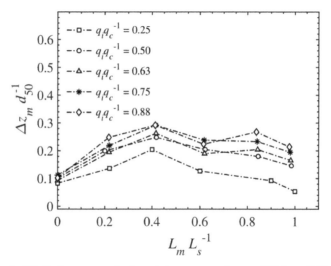

Fig. 4. 2D displacements of selected riprap stones for test P02 from Table 1.

The plot depicts development of 2D stone displacements as a function of applied overtopping discharge magnitudes ($q_i\, q_c^{-1}$). The horizontal axis of the plot ($L_m\, L_s^{-1}$) represents the distance to the respective selected riprap stones from the riprap stone placed adjacent to the fixed toe structure along the x-axis (L_m) normalized over the total riprap length (L_s). The vertical axis of the plot represents the progressive stone displacements along the z-axis normalized over the median riprap stone diameter ($\Delta z_m\, d_{50}^{-1}$). Reference is made to Ravindra et al. (2020) for detailed description of the analysis.

From Fig. 4, it can be observed that the selected riprap stones underwent progressive displacements in the x-z plane as a function of incremental overtopping. The 2D stone displacements were found to be dependent on both the overtopping flow magnitude and distance of the respective stones from the fixed toe structure. The 2D displacements arising due to flow attack lead to buckling of the riprap structure which in turn leads to ultimate riprap collapse. Similar displacement patterns were found for all the tests analysed. The observed buckling mechanism was concluded as a resultant of formation of a bearing structure as a consequence of the interlocking forces setup within placed ripraps. Total riprap failure in the conducted tests was found to be initiated at the upstream section of the riprap when the maximum displacement of riprap stones along the x-axis exceeded the size of the longest axes of the riprap stones, a-axis, at the upstream end of the chute ($\Delta x_{max} \approx a \approx 1.6\ d_{50}$) (Hiller et al. 2018). Cumulative analysis of the 2D stone displacement patterns further revealed that ultimate placed riprap failures were in general found to be achieved when the stone displacements along the z-axis reached magnitudes of the order 0.3–0.35 times the median stone diameter of riprap stones ($0.3\ d_{50} \leq \Delta z \leq 0.35\ d_{50}$). The deformation behaviour resembled buckling of a slender-long column pinned at one end and free at the other. Furthermore, a non-dimensional equation describing the 2D displacements of placed riprap stones was developed employing Euler's Buckling theory and was further calibrated employing experimental data sets.

4 Discussion

Dumped and placed riprap differ in packing density as well as stability and failure mechanism. Dumped riprap failed in the tests conducted as part of this study by sliding of the protection layer. Erosion of single stones from dumped riprap could not be observed. The steep slope of $S = 0.67$ is close to the angle of repose of the material and this was probably the reason as to why dumped riprap failure was characterized by sliding mechanism as opposed to erosion of single stones.

The results about the stability in terms of the critical stone-related Froude number $F_{s,c}$ for placed ripraps scatter more than for dumped riprap. As far as placed ripraps are concerned, erosion of individual riprap stones is possible without resulting in ultimate failure of the entire structure. Arching was observed in the remaining riprap around the erosion hole of single stones, and the interlocking forces between the riprap stones compensated for the loss. Hence, it suggests itself that the closer the stones are packed (low P_c value) the more stable the placed riprap is. This point is also supported by accumulating displacements in flow direction due to small rearrangements of the stones. Consequently, a gap developed in the transition between the horizontal crest and the inclined chute, and the stones adjacent to the gap lost their interlocking and finally were eroded. The critical size of the gap correlated with the longest axes of the riprap stones (Hiller et al. 2018).

The buckling failure mechanism in the placed ripraps described by Ravindra et al. (2020) further extends past findings to 2D. The results add valuable knowledge to the theoretical and simplified considerations of the failure scenario "disruption of the protection layer" by Siebel (2007). The observed buckling process in placed ripraps exposed to overtopping is a direct consequence of the interlocking forces setup between individual riprap stones leading to the formation of a unified structure behaving in a manner predictable by well-established structural theorems. This also offers an explanation for the difference in failure mechanisms of dumped and placed ripraps. Since placed ripraps exposed to overtopping loads form a unified structure as a consequence of the generated interlocking effect, detachment of a single loosely placed stone from the structure does not necessarily entail loss of structural integrity. This is because, the configuration of the neighboring stones can still offer a considerable degree of resistance against the destabilizing force. However, since these interlocking forces are negligible when considering dumped ripraps, dislodgement of individual stone elements lead to exposure of the underlying filter layer in turn resulting in erosion or sliding of the riprap as a whole.

The above described processes with displacements and buckling develop gradually and are hence time-dependent. The loading of accidental overtopping is limited in time as it is related to a certain flood hydrograph and/or in combination with extraordinary events such as a clogged gates or spillway. Considering probabilities for such scenarios and assessing risk by combining probability for overtopping and consequences would be recommended.

Siebel (2007), Dornack (2001), Sommer (1997), Larsen et al. (1986) investigated the 1D behaviour of placed ripraps and stated that the interlocking of riprap stones allows for the transfer of longitudinal forces within the placed ripraps. They concluded that these forces, when large enough, can either cause progressive erosion or rupture of the riprap layer. This paper describes the rupture as a structural collapse. Thus, the failure of placed ripraps exposed to overtopping can be a consequence of either sliding or structural collapse. In case of a constrained toe structure as employed in the experimental model lab tests considered part of this study, the riprap structure is likely to fail as a consequence of structural collapse or buckling as a toe support is provided to avoid sliding of the riprap structure. However, in case of an unrestrained toe, the riprap section could undergo sliding along the steep slope as a result of limited frictional resistance offered at the foundation. Hence, the current investigation results suggest riprap toe support as a key factor influencing overall stability and the underlying failure mechanism.

Further, placed ripraps in the large-scale field tests were not provided with toe support in the same way. The riprap in test F13P2 failed due to toe instability and hence, identifies that the support conditions at the boundaries of placed ripraps are crucial. For the tests conducted in 2015, larger stones were placed at the dam toe and along the abutments to prevent riprap failure along the boundaries and to focus on the riprap itself. The observations indicate that placed riprap is more prone to irregularities and that the stability depends on good foundation conditions than for dumped riprap. Furthermore, detailed survey of existing state of toe support conditions for placed ripraps constructed on several Norwegian rockfill dams conducted by Ravindra et al. (2019) revealed that well-defined toe support measures stabilizing riprap toes are currently not implemented at any of the surveyed rockfill dams. Since toe support condition is considered as a quintessential parameter governing placed riprap stability, further experimental research is required in order to better understand failure mechanism of placed ripraps with realistic toe support conditions. This forms the basis for an ongoing experimental study at NTNU, Trondheim.

5 Conclusion

The results of the presented study on model and field tests of dumped and placed ripraps add valuable data for riprap stability on steep slopes. Toe support, the steep slope and placing the riprap stones in an interlocking pattern lead to a structural behavior in placed riprap. I.e., stability of placed ripraps is dependent more on the interaction between the individual stones leading to a bearing structure. Stability of dumped ripraps on the contrary is governed by erodibility of individual stones. This conclusion is backed by the study results describing 1D and 2D deformations in placed ripraps exposed to overtopping flows. The observed displacements resulted in a gap at the top of the inclined chute as well as in buckling in the middle section of the chute. The displacements in x- and z-direction were hence combined to result in a 2D deformation of the placed riprap structure. Furthermore, the observed deformations were found to be analogous to empirical predictions from the Euler's Buckling theory. The displacements developed gradually, and the process depends hence on time and the applied hydraulic loading.

Acknowledgements. The authors thank their advisors Prof. Leif Lia, Ass. Prof. Fjola G. Sigtryggsdottir and Prof. Jochen Aberle at the Department of Civil and Environmental Engineering at NTNU for their support. The financial support of the collaborators within the project "Placed riprap on rockfill dams" (*PlaF*) coordinated by Energy Norway and the Research Council of Norway [project no. 235730] as well as the support of HydroCen, Norway is kindly acknowledged.

References

Dornack, S.: Überströmbare Dämme-Beitrag zur Bemessung von Deckwerken aus Bruchsteinen [Overtopping dams-Design criteria for riprap] (Ph.D. thesis). Technische Universitat Dresden (2001)

Hiller, P.H.: Large-scale overtopping test with placed riprap (2015). https://www.iahrmedialibrary.net/the-library/applied-hydraulics/hydraulic-structures/overtopping-test-ntnu-sira-kvina/1035

Hiller, P.H., Lia, L.: Practical challenges and experience from large-scale overtopping tests with placed riprap. In: Toledo, M.Á., Moran, R., Onate, E. (eds.) Dam Protections against Overtopping and Accidental Leakage, pp. 151–157. CRC Press/Balkema, London (2015)

Hiller, P.H.: Riprap design on the downstream slopes of rockfill dams, (Ph.D. thesis). Norwegian University of Science and Technology NTNU, Trondheim (2017)

Hiller, P.H., Aberle, J., Lia, L.: Displacements as failure origin of placed riprap on steep slopes. J. Hydraul. Res. **56**(2), 141–155 (2018)

Hiller, P.H., Lia, L., Aberle, J.: Field and model tests of riprap on steep slopes exposed to overtopping. J. Appl. Water Eng. Res. **7**(2), 103–117 (2019)

Hiller, P.H.: Kartlegging av plastring på nedstrøms skråning av fyllingsdammer (Report B1-2016-1). NTNU, Trondheim (2016)

International Commission on Large Dams: Dam failures statistical analysis. ICOLD Bulletin 99. Paris: ICOLD (1995)

International Commission on Large Dams: World register on dams, general synthesis (2020). https://www.icold-cigb.org/GB/world_register/general_synthesis.asp. Accessed 5 Feb 2020

Kjærnsli, B., Valstad, T., Høeg, K.: Rockfill Dams: Design and Construction. Norwegian Institute of Technology, Trondheim (1992)

Larsen, P., Bernhart, H.H., Schenk, E., Blinde, A., Brauns, J. Degen, F.P.: Überströmbare Dämme, Hochwasserentlastung über Dammscharten [Overtoppable dams, spillways over dam notches] (Unpublished Report Prepared for Regierungsprasidium Karlsruhe). Universität Karlsruhe (1986)

Linford, A., Saunders, D.: A Hydraulic Investigation of through and overflow rockfill dams (Report No. RR888). British Hydromechanics Research Association (1967)

Ministry of Petroleum and Energ: Forskrift om sikkerhet ved vassdragsanlegg (Damsikkerhets-forskriften) [Dam safety regulation] (2009)

Norwegian Water Resources and Energy Directorate: Database SIV, [Norwegian dam register]. (Database; data extracted 31 October 2016). NVE (2016)

Olivier, H.: Through and overflow rockfill dams-new design techniques. In: Proceedings of the Institution of Civil Engineers, pp. 433–471 (1967)

Orendorff, B., Al-Riffai, M., Nistor, I., Rennie, C.D.: Breach outflow characteristics of non-cohesive embankment dams subject to blast. Can. J. Civ. Eng. **40**(3), 243–253 (2013)

Ravindra, G.H.R., Sigtryggsdottir, F.G., Asbølmo, M.F., Lia, L.: Toe support conditions for placed ripraps on rockfill dams - A field survey. Vann, **3**, 185–199 (2019)

Ravindra, G.H.R., Sigtryggsdottir, F.G., Lia, L.: Buckling analogy for 2D deformation of placed ripraps exposed to overtopping. J. Hydraul. Res. (2020)

Siebel, R.: Experimental investigations on the stability of riprap layers on overtoppable earthdams. Environ. Fluid Mech. **7**(6), 455–467 (2007)

Sommer, P.: Überströmbare Deckwerke [Overtoppable erosion protections] (Unpublished report No. DFG-Forschungsbericht La 529/8-1). Universität Karlsruhe (1997)

Toledo, M.Á., Moran, R., Onate, E.: Dam Protection against Overtopping and Accidental Leakage. CRC Press/Balkema, London (2015)

Towards Physics-Based Large-Deformation Analyses of Earthquake-Induced Dam Failure

Duruo Huang[1], Feng Jin[1(✉)], Gang Wang[2], and Kewei Feng[2]

[1] Department of Hydraulic Engineering, Tsinghua University, Beijing 100084, China
jinfeng@tsinghua.edu.cn
[2] Department of Civil and Environmental Engineering, The Hong Kong University of Science and Technology, Hong Kong SAR, China

Abstract. Earthquake-induced damage to embankment dams have been frequently reported in the past earthquakes. Embankment dam failure triggered by ground shaking is a complicated nonlinear, progressive, large-deformation process. Yet, conventional computational method has significant limitations in modeling the large-deformation failure process due to mesh distortion and numerical issues. In this study, a two-phase hydro-mechanically coupled Material Point Method (MPM) is developed, which provides a new tool to investigate fully nonlinear mechanism of dam failure and post-failure large-deformation behavior under earthquake shaking. The progressive failure process of the embankment dam, including slip triggering and post-failure large-deformation behavior, are studied by the developed MPM. As an example, the case history of Lower San Fernando dam failure during the 1971 M6.6 San Fernando earthquake in California is simulated. Subjected to strong earthquake loading, the dam materials experience significant loss of strength due to liquefaction. The entire failure process of the dam is captured by the model, indicating great promise of the MPM method in dam failure analyses.

Keywords: Earthquake-induced dam failure · Large deformation analysis · Progressive failure

1 Introduction

Earthquake-induced dam failure is one of the most catastrophic effects of earthquakes, as evidenced by many historic events over the past decades. A notable example is the earthquake-induced Lower San Fernando dam failure in California during the 1971 M6.6 San Fernando earthquake [1, 2]. Realistic modelling of earthquake-induced dam failure is crucial for the design of key infrastructure and to protect human lives in seismically active regions. Embankment dam failure triggered by ground shaking is a complicated nonlinear, progressive, large-deformation process. Yet, conventional computational method has significant limitation in modeling the large-deformation failure process due to mesh distortion and numerical issues. It is thus important to develop innovative numerical methods that can model the entire dam failure process, including initiation and large

J.-M. Zhang et al. (Eds.): ICED 2020, SSGG, pp. 117–124, 2020.
https://doi.org/10.1007/978-3-030-46351-9_10

deformation. It could significantly improve our understanding of the complex process and the underlying influential factors.

With the advancement of computational technology, there is an opportunity to develop a physics-based model to study the complex process of earthquake-induced dam failure. In this study, a hydro-mechanically coupled MPM model is developed, which provides a new tool to investigate the fully nonlinear mechanism of dam failure and post-failure large-deformation behavior under earthquake shaking. The progressive failure process of the embankment dam, including slip triggering and post-failure large-deformation behavior, are studied by the developed MPM. As an example, the case history of Lower San Fernando dam failure during the 1971 San Fernando earthquake in California is simulated using the MPM. Subjected to strong earthquake loading, the dam materials experience significant loss of strength due to liquefaction. The entire failure process of the dam is captured by the model, indicating the great promise of MPM method in dam failure analyses.

2 A Hydro-Mechanically Coupled MPM Model

2.1 Governing Equations for Porous Media

Hydrogeological conditions have significant impact on the dynamic dam failure analyses. The degree of saturation and development of matric suction influence the mechanical properties of unsaturated soils by changing the effective stress and shear strength [3]. On the other hand, catastrophic dam failure shall involve seepage forces. Moreover, generation of excessive pore pressure within the soil can induce liquefaction and significantly affect the post-failure behavior [4–8]. Therefore, consideration of the coupled soil-water behavior is essential in dam failure simulation. Although a few studies have modeled soil and water coupling using MPM [9–11], such method has not yet been realized for a dynamic simulation.

In this study, a two-phase hydro-mechanically coupled MPM formulation is developed. Porous media consist of the solid particles and numerous pores, which are filled with water and air. Two sets of material point layers are adopted to represent the solid skeleton and pore water respectively. Note that air is neglected in the current formulation. The motion of each constituent is described by Lagrangian methods. Note that subscripts "s" and "w" denote quantities of solid phase and pore water, respectively. For sign convention, tension in the solid phase and compression in the pore water are assumed to be positive. The momentum balance equation for the solid-fluid mixture is provided in Eq. (1) to describe the motion of solid phase,

$$(1-n)\rho_s a_s + n\rho_w a_w = \nabla \sigma + (1-n)\rho_s b + n\rho_w b \tag{1}$$

where n is porosity, ρ is density, σ is total stress, a_s and a_w are the accelerations of solid grains and pore water respectively, b is the unit body force. For partially saturated porous media, the generalized Darcy's equation is adopted to describe the motion of pore water,

$$\rho_w a_w = \nabla p_w + \rho_w b - \frac{n\rho_w g}{k_{ij}}(v_w - v_s) \tag{2}$$

where k_{ij} is the hydraulic conductivity, g is the gravity scalar, $\boldsymbol{v}_w$ and $\boldsymbol{v}_s$ are the velocities of pore water and solid skeleton respectively. Equations (1) and (2) are used to compute the motion of solid and fluid phases, while the total stress is the summation of effective stress and pore pressure,

$$\boldsymbol{\sigma} = \boldsymbol{\sigma}' + p_w \boldsymbol{I} \tag{3}$$

where the effective stress $\boldsymbol{\sigma}'$ will be used in constitutive modeling of the nonlinear soil.

2.2 Solid-Liquid Coupled MPM Formulation

In the coupled MPM system, the solid skeleton and the pore water are represented by two sets of material points to represent the solid skeleton and pore water, as shown in Fig. 1.

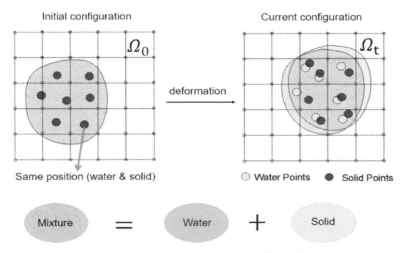

Fig. 1. Configuration of two sets of material points in modelling solid skeleton and pore water

The primary computational procedure for the hydro-mechanical MPM model is summarized as follows. First, project the variables of material points including masses, momentum of the solid/fluid and porosity into the background grid. Second, interpolate the porosity to the fluid points positions, and compute the accelerations of fluid phase and solid phase. Third, update the positions and velocities of material points, and apply the boundary conditions. Fourth, update the positions and velocities of material points and apply the boundary conditions, and subsequently update the pore water pressure of the fluid phase, the soil porosity and soil permeability. Finally, reset the background mesh and a new iteration step starts. Obviously, the numerical procedure takes advantage of both Lagrangian and Eulerian methods to overcome difficulties encountered with a conventional FEM. Pure Lagrangian methods (e.g. FEM) typically result in severe mesh distortion and consequently ill conditioning of the solution scheme, while MPM has a distinct advantage in modeling large-deformation failure mechanics of geomaterial because Lagrangian material points move within a Eulerian background mesh.

3 Case Study of the Lower San Fernando Dam Failure

In this paper, a case study is presented to assess failure of the Lower San Fernando dam due to liquefaction during the 1971 M6.6 San Fernando earthquake. The dam was constructed by hydraulic filling, and the fill soil remained loose after being transported to the dam site through pipeline, making the dam subject to liquefaction. Approximately 80,000 people living in the downstream region were threatened by the very real possibility that the dam would completely fail, inundating the region by a severe flood. The relative density of the hydraulic sand fill was later determined to be between 40 and 70%. During the event, approximately 11 m of freeboard prior to the earthquake were reduced to a 1.5 m. Extensive field investigation demonstrated that liquefaction of the upstream shell near its base was responsible for the slide [1, 2]. The slide mass extended approximately 50 to 60 m into the reservoir.

3.1 The Bounding Surface Hypoplasticity Model and Input Wave

Figure 2(a) shows the dam material zonation used in the dynamic analysis. The dam height is 45 m. Dam material primarily consists of five zones, namely rolled fill, ground shale, clay core, hydraulic fill and rigid plate. To realistically characterize the soil behavior during earthquakes as well as its liquefaction resistance, the bounding surface hypoplasticity sand model is adopted in the numerical analysis [12]. The bounding surface model has been extensively validated in the past to simulate the nonlinear dynamic response of soils, the simultaneous build up and dissipation of pore water pressure during

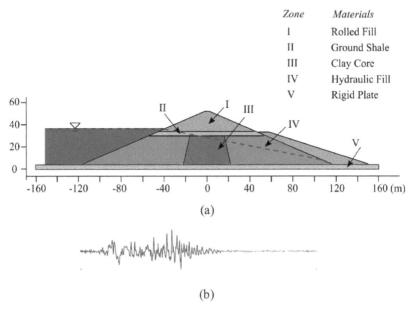

(a)

(b)

Fig. 2. (a) Numerical model of the Lower San Fernando Dam used in MPM analyses, (b) incident ground motion with PGA of 0.6 g

earthquake shaking [13], and the permanent deformation during and after earthquake shaking. Note that one of the fundamental issues in modeling the stress-strain behavior of sand in liquefaction analyses is the correct description of soil dilatancy. Parameters used in the bounding surface model is provided in Table 1 [12].

Note that the epicenter of the San Fernando earthquake was located approximately 10 km northeast of LSFD at a focal depth of approximately 9 km. Based on an analysis recorded ground motions at the LSFD site, the peak ground acceleration in the rock foundation underlying the dam was approximately 0.6 g, while researchers concluded there was no significant amplification between the foundation and the crest of the dam. Therefore, an accelerogram with a PGA of 0.6 g is adopted as the input motion at dam rigid rock base in the analysis of this study. The motion was recorded on the abutment of nearby Pacoima Dam approximately 5 km east of San Fernando, as illustrated in Fig. 2(b).

Table 1. Material parameters in the bounding surface hypoplasticity model

Zone	Description	Void ratio	$\phi(^o)$	k(m/s)	G_0	h_r	k_r	d	R_p/R_f	b
I	Rolled fill	0.56	37	1.8e-7	250	0.15	100	100	1	2
II	Ground shale	0.56	37	1.8e-7	303	0.2	100	100	1	2
III	Clay core	0.6	20	1.8e-8	200	1.2	100	100	1	2
IV	Hydraulic fill	0.56	37	1.8e-7	250	0.15	0.5	2.1	0.75	2

3.2 Analysis Results of the Dam

Figure 3 illustrates evolution of deviator strain of the dam during earthquake shaking, wherein initial failure, its development and the post-failure process can be clearly observed. Figure 3(a) captures a surface shallow slide at onset of the simulation, while the dam maintains stable given sufficient soil shear strength. A deep shear band gradually develops along dam shell due to ground shaking, as can be seen in Fig. 3(b) and (c). A progressive failure mode can be faithfully captured. Shear deformation propagates backward and upward gradually towards surface of the slope, resulting in a catastrophic failure of the embankment dam.

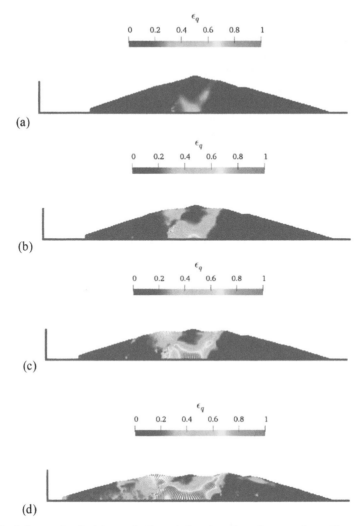

Fig. 3. Evolution of deviator plastic strain in the Lower San Fernando dam from (a) $t = 6$ s, (b) $t = 10$ s, (c) $t = 15$ s, (d) $t = 30$ s

The displacement pattern of the post-earthquake analyses by Seed [2] is shown in Fig. 4(a). For fair comparison, the deformation pattern of dam material zonation and deviatoric plastic strain simulated in this study are presented in Figs. 4(b) and (c). These displacements are in reasonable agreement with field observations over the entire dam zones, and indicate that post-earthquake strength loss due to liquefaction could have been a main factor in the observed large deformations.

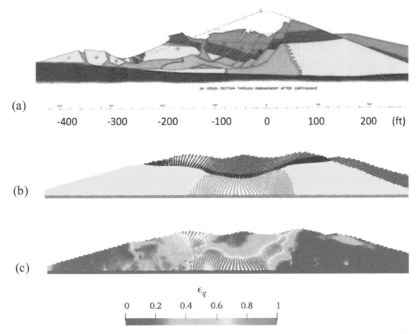

Fig. 4. (a) Comparison of slides in the San Fernando dam between analysis by Seed [2], (b) deformation pattern of dam material zonation, and (c) deviatoric strain simulated in this study

4   Conclusions

Adequately quantifying post-failure large deformation is essential in the design and remediation of earth dams in seismically active regions. As the configuration and soil profile of earth dams are usually complicated, fully coupled analyses are recommended for evaluating large deformation as well as other effects. This paper presents a set of dynamic analyses on the lower San Fernando dams under earthquake shaking using a fully coupled hydro-mechanical MPM model. The model is featured by utilizing a u-U formulation to model the water and soil phases as well as their interaction. Two sets of material points are adopted to represent solid skeleton and pore water, respectively. The generation of excessive pore pressure within the dam material is modelled through the soil-water coupled formulation. A bounding surface hypoplasticity model is also implemented in the MPM to simulate fully nonlinear dynamic behavior of the soil under earthquake loading.

A case study about the Lower San Fernando dam failure during the 1971 San Fernando earthquake in California is conducted using the coupled hydro-mechanical MPM model. Numerical simulations clearly illustrate that a surface failure appears within the saturated zone, and thereafter a deep shear band gradually propagates inside the dam with continued ground shaking. The retrogressive failure mode inside the dam can be faithfully captured in the simulation, which highlights the true advantages of the MPM approach in dealing with large deformation problems. More importantly, more case history studies shall be conducted to sufficiently validate the numerical model in the future, and to provide more insights into this topic.

Acknowledgements. The authors acknowledge support from Joint Research Fund for Overseas Chinese Scholars and Scholars in Hong Kong and Macao (Grant No. 51828902) from National Natural Science Foundation of China, General Research Fund grant No. 16214519 from Hong Kong Research Grants Council, and research fund 2019-KY-02 from State Key Laboratory of Hydroscience and Engineering.

References

1. Seed, H.B., Lee, K.L., Idriss, I.M., Makdisi, F.: Analysis of the slides in the San Fernando dams during the earthquake of Feb. 9, 1971. Earthquake Engineering Research Center Report No. EERC 73-2, University of California, Berkeley (1973)
2. Seed, H.B., Seed, R.B., Harder, L.F., Jong, H.L.: Re-evaluation of the lower San Fernando dam in the earthquake of Feb. 9,1971. Earthquake Engineering Research Center Report No. UCB/EERC-88/04, University of California, Berkeley (1988)
3. Nuth, M., Laloui, L.: Effective stress concept in unsaturated soils: clarification and validation of a unified framework. Int. J. Numer. Anal. Meth. Geomech. **32**(7), 771–801 (2008)
4. Huang, D., Wang, G., Du, C.Y., Jin, F., Feng, K.W., Chen, Z.W.: An integrated SEM-Newmark model for physics-based regional coseismic landslide assessment. Soil Dyn. Earthq. Eng. **132**, 106066 (2020)
5. Wei, J., Wang, G.: Discrete-element method analysis of initial fabric effects on pre-and post-liquefaction behavior of sands. Géotech. Lett. **7**(2), 161–166 (2017)
6. Wei, J., Huang, D., Wang, G.: Microscale descriptors for particle-void distribution and jamming transition in pre- and post-liquefaction of granular soils. J. Eng. Mech. **144**(8), 04018067 (2018)
7. Wei, J., Huang, D., Wang, G.: Fabric evolution of granular soils under multi-directional cyclic loading. Acta Geotech. (2020). https://doi.org/10.1007/s11440-020-00942-8
8. Ye, J., Huang, D., Wang, G.: Nonlinear dynamic simulation of offshore breakwater on sloping liquefied seabed. Bull. Eng. Geol. Env. **75**, 1215–1225 (2016)
9. Bandara, S., Soga, K.: Coupling of soil deformation and pore fluid flow using material point method. Comput. Geotech. **63**(1), 199–214 (2015)
10. Soga, K., Alonso, E., Yerro, A., Kumar, K., Bandara, S.: Trends in large-deformation analysis of landslide mass movements with particular emphasis on the material point method. Géotechnique **66**(3), 248–273 (2016)
11. Yerro, A., Alnoso, E.E., Pinyol, N.M.: The material point method for unsaturated soils. Géotechnique **65**(3), 201–217 (2015)
12. Wang, Z.L., Dafalias, Y.F., Shen, C.K.: Bounding surface hypoplasticity model for sand. J. Eng. Mech. **116**(5), 983–1001 (1990)
13. Wang, G., Xie, Y.: Modified bounding surface hypoplasticity model for sands under cyclic loading. J. Eng. Mech. ASCE **140**(1), 91–101 (2014)

A Cylindrical Erosion Test Apparatus for Erosion Tests on Samples from Yigong Landslide Dam

Lin Wang[1(✉)], Xingbo Zhou[2], and Qiang Zhang[1]

[1] College of Water Resources and Hydropower Engineering, Xi'an University of Technology, Xi'an 710048, China
ruoshuiya@163.com, zhang_qiang333@163.com
[2] China Renewable Energy Engineering Institute, Beijing 100120, China
zhou_xingbo@126.com

Abstract. To quantitatively measure the erodibility of soil including the critical shear stress and the erosion rate, a new measurement system is proposed. The measurement system contains a cylindrical erosion test apparatus (CETA) and a method for calculating shearing stress. The CETA was designed to measure the relationship between the velocity and erosion rate, which was mainly composed of a structure system, a transmission system, an erosion test system, a desilting circulatory system, and a control and data acquisition system. On the basis of the proposed new method, single-size soil samples from the Yigong landslide dam were collected for erosion tests. It is noted that the erosion rate from the apparatus is reliable, which is on the same order as that of the experimental results performed by Briaud. This apparatus can be utilized to analyze the shear stress. The hyperbolic relationship between erosion rate and shear stress is found by proposing the accurate calculation method of shear stress. The hyperbolic model of erosion rate versus shear stress is verified.

Keywords: Erosion · Critical erosion shear stress · Cylindrical erosion test apparatus

1 Introduction

Landslide dam, if out of control, will leading to a great threat to the downstream [1]. Erosion is one of the most important factors for the landslide dam breach, and quantitative analysis for erosion is one of the most critical challenges. Chen et al. [2] takes the Tangjiashan dam as an example, and selects four different erosion models, namely Einstein Brown, Englund Hensen, Du Boys and Meyer Peter Muller, to analyze the breach. It is found that the difference of breach discharge caused by different erosion models can reach 2 or 3 time. Therefore, the analysis method of erosion rate for landslide dam needs to be reevaluated [3, 4].

© Springer Nature Switzerland AG 2020
J.-M. Zhang et al. (Eds.): ICED 2020, SSGG, pp. 125–135, 2020.
https://doi.org/10.1007/978-3-030-46351-9_11

A variety of testing apparatus have been developed as reported by Foster et al. [5], Temple [6], Shaikh et al. [7], Hanson and Simon [8], Wan and Fell [9], Zhu et al. [10], Chang and Zhang [11], and Wu [1]. Wahl [12] finds that the erosion rate of jet apparatus is one magnitude higher than that of HET, while the critical shear stress is two or more orders magnitude lower. Therefore, different erosion models can be obtained by using different measuring apparatuses. At present, there is no erosion calculation method for different soil characteristics. The relationship between erosion rate and shear stress takes on diverse forms, and the analysis results of breach are unstable, sometimes even not convergent. Based on the measured erosion rate of the Tangjiashan dam, Chen et al. [13] proposes a hyperbolic model between erosion rate and shear stress. Chen considers that there is no infinite "strength" when soil materials resist erosion. However, the research on hyperbolic erosion model for landslide dam is still in the ambiguous stage, so it is urgent to carry out in-depth research.

A Cylindrical Erosion Test Apparatus (CETA) is presented for measuring erosion rate by the authors' team. Compared to the above describing apparatuses, the developed CETA has the following advantages. First, this apparatus can provide the highest water velocity to 7 m/s, which can meet the requirements in altered conditions. Second, the system can perform tests on coarse materials (sand, gravel). Third, the apparatus can conduct erosion tests on the soil samples with a maximum particle size of 10 cm.

In this study, a detailed description of the CETA is first introduced. Then, erosion tests are conducted on the soil samples collected from the Yigong landslide dam. The erosion rates with the Yigong samples of different particle sizes are studied and the relationship between erosion rate and shear stress is established.

2 Description for Cylindrical Erosion Test Apparatus

The developed CETA is composed of a struture system, a transmission system, an erosion test system, a desilting circulatory system, and a control and data acquisition system shown in Fig. 1.

The transmission system includes a motor and an impeller in Figs. 2 and 3. In the experiment the motor rotates and drives the impeller, which in turn drives the water in the steel cylinder to wash the sample. The speed regulating motor has a rated voltage of 220 V and a maximum speed of 1400 R/min. The water recycling could be achieved by the desilting circulatory system composed of a desilting pool, a pump and a water-spraying unit. After the soil particles are washed away, they flow into the desilting pool with the water. The water in the desilting pool returns to the pump and the water-spraying unit, while the soil particles remain in the desilting pool. Three observation windows made of plexiglass are specially arranged on the steel cylinder wall, which can be utilized for observing and photographing the testing progress. The size of each observation window is about 40 cm × 28 cm. In addition, considering that the vortex in the cylinder might affect erosion, a cylindrical lamp ring is added at the bottom of the cylinder, which is utilized to provide illumination for the observation of the erosion process during the test. Therefore, the cylinder bottom could be approximated as an annular water flume to reduce the vortex effect and improve observations and photographing performance. The inner diameter of the cylinder is 104 cm, the diameter of the lamp ring around the central axis is 35.4 cm, and the height of the lamp ring is 30 cm.

The principle of the developed CETA is that the motor drives the impeller to rotate, and the impeller rotates to make the water flow. After the water flow starts, it will wash the sample placed at the bottom of the cylinder. Once the sample moves, the speed at this time is the critical velocity. The control and data acquisition system records the water velocity and the process of sample washing. The maximum design speed of the impeller is 11 m/s, and the range of test water depth is 0.4–0.7 m.

Fig. 1. Photo of the CETA

Fig. 2. Motor   **Fig. 3.** Impeller

The apparatus could only acquire the impeller rotation speed. Therefore, the water velocity under a certain impeller rotation speed should be calibrated. Utilizing a pitot tube and the FP111 direct-reading velocity meter, the reel rotating rate has been calibrated to the flow velocity on the soil/water surface (refer to Ma [14]). The calibration curve of the velocity with the impeller rotation speed is:

$$y = 0.679x - 0.0103 \qquad (1)$$

where: y is the central bottom water velocity (m/s); x is the impeller central velocity (m/s).

3   Erosion Testing

Erosion rate is usually defined as the height of soil sample washed away in unit time, expressed in cm/s. In this experiment, it is found that if the soil particles are placed in the soil box of the lifting system, the soil will be washed out in a short time and the remaining soil mass cannot be recorded. The test apparatus cannot lift soil samples like EFA (refer to Fig. 4), so the test in the lifting soil box cannot meet the requirements.

Fig. 4. Erosion function apparatus

Therefore, the new erosion rate measurement method is used to carry out erosion research, and the relationship between erosion rate and shear stress is researched through Yigong soil samples.

3.1   Setup and Testing Procedure

(1) Lay the soil in the cylinder, and keep the mass and area of the sample consistent each time, as shown in Fig. 5. After tiling, fill water quickly to 55 cm of the cylinder. The test shall be conducted immediately after water filling to avoid the separation of particles in the soil. Therefore, it is very important to inject water and carry out the test quickly.

(a)   Lifting cylinder   (b) Tiling sample

Fig. 5. Experiment procedure

(2) Due to the sensitivity of the erosion rate to water, it is extremely important to keep the sample flush with the bottom of the cylinder.

(3) The speed of the instrument is modified to control the starting velocity, and the erosion time is recorded. The data to be recorded in this stage are instrument speed, erosion time and weight of soil sample.

(4) Calculate the remaining dry weight of soil sample. If there is too much residual soil sample after the test, part of the soil can be put into three sample boxes. Then, sample boxes shall be dried. The average moisture content shall be calculated. The dry weight of the remaining soil can be calculated by the average moisture content. The data to be recorded are sample box weight, wet weight before drying and dry weight after drying. The data to be measured are moisture content of the sample box, average moisture content, residual dry weight, and dry density and area.

(5) The erosion height can be obtained by erosion dry weight divided by density and area. The erosion rate can be obtained by the height change divided by time duration.

In the previous erosion test, it is difficult to determine the erosion starting time. How to define the initial stage of erosion test has a great impact on the results of erosion rate. This method does not need to determine the starting time of soil particles, but only needs to record the total time of erosion and the quantity of the soil being scoured. This new method solves the problem of determining the erosion starting time in the previous erosion test. The apparatus can measure the erosion rate of large-size soil at present, and the maximum particle size can reach 4 cm.

3.2 Typical Results

As shown in Fig. 6, two soil materials from Yigong have $D_{50} = 8$ mm and $D_{50} = 10$ mm, respectively. The natural density values of the samples in the test are 1.845 g/cm^3 and 1.5852 g/cm^3, respectively, and the water depth is 55 cm. The erosion rate is defined as the height of the soil sample washed away in unit time, and the soil mass is controlled at 4 kg and 3.72 kg. The area of the sample is controlled as 1075 cm^2 and 1085 cm^2, and the results are shown in Fig. 7.

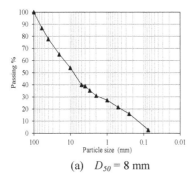

(a) $D_{50} = 8$ mm

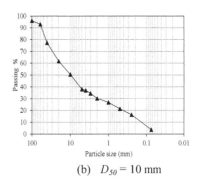

(b) $D_{50} = 10$ mm

Fig. 6. Gain size distributions of two Yigong landslide dam materials

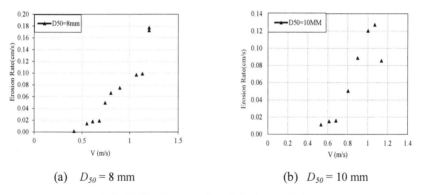

(a) $D_{50} = 8$ mm   (b) $D_{50} = 10$ mm

Fig. 7. Erosion rate of particle size gradation

Erosion Rate

Briaud et al. [15] uses EFA to research the erosion rate of clean coarse sand. D_{50} of this soil sample is 3.375 mm, and the maximum erosion rate is 12000 mm/h = 1200 cm/3600 s = 0.33 cm/s. As shown in Fig. 8, the relationship between the critical velocity and erosion rate is hyperbolic from Briaud. The test results are shown in Fig. 7. The relationship between velocity and erosion rate can be obtained after regression:

$D_{50} = 8$ mm:

$$y = 0.21095 - \frac{0.10569}{v} \tag{2}$$

$D_{50} = 10$ mm:

$$y = 0.21567 - \frac{0.11944}{v} \tag{3}$$

where: v is the velocity (m/s) and y is the erosion rate (cm/s).

It can be seen from Eqs. (2) and (3) that the relationship between critical velocity and erosion rate is hyperbolic. The maximum value of erosion rate in the tests is 0.17 cm/s. The magnitude of test results is consistent with that of clean coarse sand in Fig. 8.

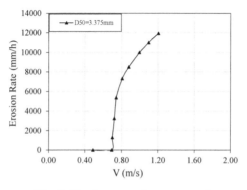

Fig. 8. Erosion curve for coarse sand

Briaud et al. [15] consider the same average velocity in EFA is 1 m/s, and the erosion rate of sand is about 1000 times that of clay, indicating that the erosion rate of different soils is different. The experimental results performed by Briaud show that the numerical magnitude of the erosion rate of clean sand and gravel is 10^4 mm/h, which is consistent with the test results in this paper.

In conclusion, the test results show that the relation between critical velocity and erosion rate can be fitted with a hyperbolic curve. It is believed that the CETA is reliable, and the reliability of the erosion rate test results is also verified, which can be utilized to analyze the shear stress.

Shear Stress

Crowley et al. [16] believe that the most effective alternative method is to use the Moody Chart to estimate the shear stress when it is impossible to measure the shear stress directly.

The calculation formula of shear stress is as follows:

$$\tau = \frac{1}{8} f \rho_w V^2 \tag{4}$$

where τ is shear stress (N/m^2), ρ_w is the density of water (1000 kg/m^3), V is the average speed, f is friction coefficient. According to three parameters from the Moody Chart (Fig. 9): f, Re, $\frac{\varepsilon}{D}$ (relative roughness, which is the average height of the rough part protruding/pipe diameter) can give the value of f.

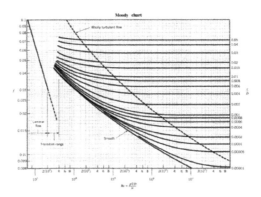

Fig. 9. Moody Chart

The Reynolds number Re is calculated as

$$Re = VD/\upsilon \tag{5}$$

where: D is the hydraulic diameter, υ is the kinematic viscosity of water (10^{-6} m^2/s in 25 °C), the hydraulic diameter D is calculated as 4 times of the hydraulic radius R, which R is defined by the flow area divided by the wetted perimeter.

This formula is expressed as:

$$D = 2ab/(a+b) \tag{6}$$

The hydraulic diameter is:

$$D = \frac{2ab}{(a+b)} = \frac{2329}{214} = 0.1088 \text{ m} \tag{7}$$

When $Re > 100000$, find the corresponding friction coefficient f directly in the Moody Chart.

When $Re < 100000$:

$$f = \frac{0.316}{Re^{1/4}} \tag{8}$$

This equation is the Blasius equation. Crim [17] recommended the use of this equation to calculate the friction coefficient f.

According to the EFA calculation formula, the shear stress is obtained by Eqs. (4), (5), (6), (7) and (8). The relationship between the critical velocity and the shear stress is found through the regression equation. The relationship between the critical velocity and the shear stress is shown in Tables 1 and 2.

Table 1. Relationship between velocity and critical shear stress in $D_{50} = 8$ mm

Velocity (m/s)	Erosion rate (cm/s)	Shear stress (N/m²)
0.553	0.014	0.773
0.682	0.019	1.116
0.805	0.066	1.489
0.417	0.001	0.472
0.743	0.050	1.297
0.614	0.018	0.929
0.900	0.075	1.810
1.069	0.097	3.859
1.198	0.173	4.846
1.198	0.177	4.846
1.130	0.099	4.313

Table 2. Relationship between velocity and critical shear stress in $D_{50} = 10$ mm

Velocity (m/s)	Erosion rate (cm/s)	Shear stress (N/m²)
1.137	0.086	4.688
1.069	0.127	4.145
1.001	0.120	3.635
0.900	0.089	1.810

(*continued*)

Table 2. (*continued*)

Velocity (m/s)	Erosion rate (cm/s)	Shear stress (N/m^2)
0.682	0.016	1.116
0.614	0.015	0.929
0.805	0.050	1.489
0.533	0.011	0.724

The Model of Erosion Rate Versus Shear Stress

Chen et al. [13] proposed a hyperbolic model for evaluating the erosion rate related to shear stress. Einstein and Krone [18], Partheniades [19], Van Prooijen and Winterwerp [20] all consider that the erosion rate can be expressed by the function related to shear stress. Slagle [21] also considers that the method similar to EFA-SRICOS [15] or the Miller-Sheppard model [22] can be used to simulate the relationship between erosion rate and shear stress with the empirical formula. Therefore, the empirical formula can be utilized to fit the relationship.

With regression analysis, the relationship between erosion rate and shear stress is:
$D_{50} = 8$ mm:

$$y = 0.13571 - \frac{0.08491}{\tau} \tag{9}$$

$D_{50} = 10$ mm:

$$y = 0.13167 - \frac{0.10153}{\tau} \tag{10}$$

where: τ is shear stress (N/m^2), and y is the erosion rate (cm/s).

It is confirmed that the relationship between the erosion rate and the shear stress is hyperbolic in Fig. 10. When the shear stress is low, there is a strong hyperbolic function relationship between the erosion rate and the shear stress.

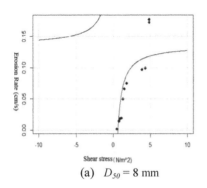 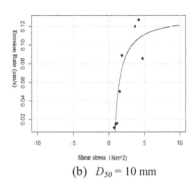

(a) $D_{50} = 8$ mm   (b) $D_{50} = 10$ mm

Fig. 10. Relationship between critical shear stress and erosion rate

4   Conclusions

This paper reports the effort in developing a new method of soil erosion measurement. The soil erosion can be characterized by the erosion rate and shear stress induced by the water flow, and a soil erosion function between the erosion rate and the shear stress. This paper introduces a cylindrical erosion test apparatus (CETA), which is a measure of the relationship between erosion rate and water velocity. It is found that the present test results agree well with experimental results performed by Briaud. The apparatus can be utilized to analyze the shear stress. Through establishing shear stress calculation model, the relationship between erosion rate and shear stress for Yigong soil samples is hyperbolic. The hyperbolic model of erosion rate versus shear stress is verified, which can be utilized for breach analysis and simulation.

References

1. Wu, W., Altinakar, M.S., Song, C.R.: Earthen embankment breaching. J. Hydraul. Eng. **137**(12), 1549–1564 (2004)
2. Zhou, X.B., Chen, Z.Y., Wang, L.: Comparison of sediment transport model in dam break simulation. J. Basic Sci. Eng. **6**, 1097–1108 (2015). (in Chinese)
3. Morris, M.W. IMPACT Project: Final Technical report – January 2005. IMPACT Project, HR Wallingford (2005). http://www.impact-project.net
4. Coleman, S.E., Andrews, D.P., Webby, M.G.: Overtopping breaching of noncohesive homogeneous embankments. J. Hydraul. Eng. **128**(9), 829–838 (2004)
5. Foster, G.R., Meyer, L.D., Onstad, C.A.: An erosion equation derived from basic erosion principles. Trans. ASAE **20**(4), 678–0682 (1977)
6. Temple, D.M.: Stability of grass lined channels following mowing. Trans. ASAE **28**(3), 750–0754 (1985)
7. Shaikh, A., Ruff, J.F., Abt, S.R.: Erosion rate of compacted Na-montmorillonite soils. J. Geotech. Eng. **114**(3), 296–305 (1988)
8. Hanson, G.J., Simon, A.: Erodibility of cohesive streambeds in the loess area of the midwestern USA. Hydrol. Process. **15**(1), 23–38 (2001)
9. Wan, C.F., Fell, R.: Investigation of rate of erosion of soils in embankment dams. J. Geotech. Geoenviron. **130**(4), 373–380 (2004)
10. Zhu, Y.H., Lu, J.Y., Liao, H.Z., Wang, J.S., Fan, B.L., Yao, S.M.: Research on cohesive sediment erosion by flow: An overview. Sci. China Ser. E **51**(11), 2001–2012 (2008)
11. Chang, D.S., Zhang, L.M.: Simulation of the erosion process of landslide dams due to overtopping considering variations in soil erodibility along depth. Nat. Hazard Earth Syst. **10**(4), 933–946 (2010)
12. Wahl, T.L.: Laboratory investigations of embankment dam erosion and breach processes. Rep. T032700–0207A, CEA Technologies, Inc. (CEATI), Montréal (2007)
13. Chen, Z.Y., Ma, L.Q., Yu, S., Chen, S.J., Zhou, X.B., Sun, P., Li, X.: Back analysis of the draining process of the Tangjiashan barrier lake. J. Hydraul. Eng. **141**(4), 05014011 (2015)
14. Ma, L.Q.: Flood analysis of landslide dam breach. Post-Doctoral Dissertation, China Institute of Water Resources and Hydropower Research, Beijing (2014, in Chinese)
15. Briaud, J.L., Ting, F.C.K., Chen, H.C.: Erosion function apparatus for scour rate predictions. J. Geotech. Geoenviron. **127**(2), 105–113 (2011)
16. Crowley, R.W., Bloomquist, D.B., Shah, F.D.: The sediment erosion rate flume (SERF): A new testing device for measuring soil erosion rate and shear stress. Geotech. Test. J. **35**(4), 649–659 (2012)

17. Crim, S., Jr.: Erosion functions of cohesive soils. M.S. thesis, Draughon Library, Auburn University (2003)
18. Einstein, H.A., Krone, R.B.: Experiments to determine modes of cohesive sediment transport in salt water. J. Geophys. Res-Atmos. **67**(4), 1451–1461 (1962)
19. Partheniades, E.A.: Erosion and Deposition of Cohesive Soils. World J. Biol. Psychiatry, Official J. World Fed. Soc. Biol. Psychiatry **2**(4), 190–192 (1965)
20. Prooijen, B.C.V., Winterwerp, J.C.: A stochastic formulation for erosion of cohesive sediments. J. Geophys. Res-Atmos. **115**(C1), 104–118 (2010)
21. Slagle, P.: Correlations of Erosion Rate-Shear Stress Relationships with Geotechnical Properties of Rock and Cohesive sediments. FDOT Report No. BD-545, RPWO # 3, University of Florida, Gainesville (2006)
22. Sheppard, D.M., Bloomquist, D., Marin, J., Slagle, P.: Water Erosion of Florida Rock Materials. FDOT Report No. BC354 RPWO #12, Florida Department of Transportation, Tallahassee (2005)

Determination of Non-uniform Input Ground Motion for High Concrete Face Rockfill Dams

Yu Yao[1], Rui Wang[2(✉)] (iD), Tianyun Liu[2], and Jian-Min Zhang[2]

[1] China Renewable Energy Engineering Institute, Liupukang North Street 2, Xicheng District, Beijing 100120, China
[2] Tsinghua University, Beijing 100084, China
wangrui_05@tsinghua.edu.cn

Abstract. A method is developed in this study to determine the non-uniform input ground motion based on seismic records for the seismic analysis of high concrete face rockfill dams (CFRDs). The acceleration time histories recorded during the 2011 Tohoku earthquake at the Haga station is used as an example to determine the potential functions of the seismic waves. The recorded surface acceleration time history and calculated non-uniform input are then used to conduct analysis of the seismic response of the Gushui CFRD under both uniform and non-uniform input motions. Under the condition that the acceleration time histories on the surface of the free field are assumed to be the same for non-uniform and uniform seismic input, the seismic response of the CFRD under non-uniform input is found to be in general significantly smaller, while the local dynamic stresses around the edges of the concrete face slab are greater. The analysis results suggest that non-uniformity of the ground motion input has important effects on the seismic response of high CFRDs, and should be considered in the seismic design of CFRDs.

Keywords: Seismic · Potential functions · Non-uniform input · High CFRD · Seismic response

1 Introduction

Current design and evaluation practice for high concrete face rockfill dams (CFRDs) assume a uniform input ground motion at the base (e.g. ICOLD 2001). However, due to site effects and the large size of such structures, the actual base input motion tends to be highly non-uniform. Currently, a number of high CFRDs are being planned and constructed in high seismicity zones in China, such as Gushui (maximum height of 245 m), Cihaxia (maximum height of 253 m), Maji (maximum height of 277.5 m), etc. Therefore, there is an urgent need to understand the influence of such input motion non-linearity on the seismic response of CFRDs.

Various methods have been developed to simulate the seismic response of geotechnical structures under non-uniform input motions, including (1) different kinds of local artificial boundaries, for instance, the viscous boundary (Lysmer and Kulemeyer 1969), the visco-elastic boundary (Deeks and Randolph 1994), and the transmitting boundary (Liao et al. 1984), which can be applied in the finite element method (FEM) or the

© Springer Nature Switzerland AG 2020
J.-M. Zhang et al. (Eds.): ICED 2020, SSGG, pp. 136–145, 2020.
https://doi.org/10.1007/978-3-030-46351-9_12

finite difference method (FDM) analysis; and (2) methods to solve the global wave field equations, for instance, the boundary element method (BEM) (Hall 1994), the scaled boundary finite element method (SBFEM) (Song and Wolf 1997), the wave function expansion method (WFEM) (Sanchez-Sesma et al. 1985), the Green function method (GFM) (Wong 1982), and the wave function combination method (WFCM) (Yao et al. 2016, 2019). The common fundamental basic requirement for these methods is the determination of the incident waves. Spatial variability of the input motion is often considered based on fitting existing earthquake records to various functions (e.g. Dibaj and Penzien 1969; Shen and Xu 1983; Tian 2003). Researchers have further developed seismic wave propagation models from the source to the ground surface to determine the non-uniform input motion (e.g. Zerva et al. 1986; Beck 1978). However, the wave scattering at the ground surface are not appropriately considered by such approaches. To solve this problem, a new method is proposed in this study to determine the potential functions of the seismic waves in the near field based on actual seismic records. Using the proposed method, the acceleration time histories at two different depths recorded during the 2011 Tohoku earthquake at the Haga station is used to determine the potential functions of the seismic waves as an example. The response of the Gushui CFRD on Lancang River is then analyzed under both uniform and non-uniform input, using the Haga station input motion. The characteristics of the response of high CFRDs subjected to non-uniform input is investigated by the comparison between the non-uniform and uniform input motion cases.

2 Method to Determine Incident Waves

To determine the incident waves, we assume that all of the incident waves are propagated from the same earthquake focus, and consist of P waves, S waves and Rayleigh waves.

First, based on the refraction law, the angle of refraction is dependent on the velocity of the waves. Therefore, we assume that the incident angles are determined by five parameters, which are the horizontal incidence angle (the angle between the projection of the incidence wave on the horizontal plane and the East direction) of P waves θ_P^h, the vertical incidence angle (the angle between the incidence wave and the upward direction) of P waves θ_P^v, the horizontal incidence angle of S waves θ_S^h, the vertical incidence angle of S waves θ_S^v, and the horizontal incidence angle of Rayleigh waves θ_R^h. The maximum range of the horizontal incidence angle is from 0° to 360°, and the maximum range of the vertical azimuth angle is from 0° to 90°.

With these incidence angles assumed, we need to determine the potential functions of P waves, SV waves, SH waves, and Rayleigh waves in the frequency domain, denoted as φ_P, φ_{SV}, φ_{SH} and φ_R. The acceleration of any recording location in the frequency domain can be expressed as:

$$-\omega^2 \cdot \begin{bmatrix} U_{Pxi} & U_{SVxi} & U_{SHxi} & U_{Rxi} \\ U_{Pyi} & U_{SVyi} & U_{SHyi} & U_{Ryi} \\ U_{Pzi} & U_{SVzi} & U_{SHzi} & U_{Rzi} \end{bmatrix} \begin{pmatrix} \varphi_P \\ \varphi_{SV} \\ \varphi_{SH} \\ \varphi_R \end{pmatrix} = \begin{pmatrix} \hat{a}_{xi} \\ \hat{a}_{yi} \\ \hat{a}_{zi} \end{pmatrix} \tag{1}$$

where ω is the circular frequency; U stands for the displacement generated by the unit potential function in frequency domain corresponding to ω, and the subscripts stand for the type of the wave, the direction, and the index of the recording location, respectively; $\hat{a}$ stands for the recorded acceleration transformed to frequency domain corresponding to ω. If there are two recording locations or more, Eq. (1) will be an overdetermined equation, and an optimal least square solution can be obtained.

The relative error of the least square solution can be assessed as:

$$ERR_r = \frac{\left\| A - \tilde{A} \right\|_2}{\left\| \tilde{A} \right\|_2} \tag{2}$$

where A is a matrix which consists of the acceleration time history for three directions calculated by the least square solution, i.e.

$$A = \begin{bmatrix} a_{xi} \ a_{yi} \ a_{zi} \end{bmatrix} \tag{3}$$

$\tilde{A}$ is a matrix which consists of the recorded acceleration time history for three directions, i.e.

$$\tilde{A} = \begin{bmatrix} \tilde{a}_{xi} \ \tilde{a}_{yi} \ \tilde{a}_{zi} \end{bmatrix} \tag{4}$$

When calculating the least square solution, the horizontal incidence angle and the vertical incidence angle should be assessed within their respective ranges. The combination of incident angles corresponding to the smallest ERR_r is taken as the actual incident angles, and the corresponding least square solution is taken as the coefficients of potential functions. Following this same procedure as the incidence angle, other unknown parameters, such as the Young's modulus of the base rock, can also be determined.

3 Example for Incident Wave Determination

The acceleration time histories at two different locations recorded during the 2011 Tohoku earthquake at the Haga station (National Research Institute for Earth Science and Disaster Prevention website 2015) are used to determine the incident wave using the proposed method as an example. The records include the acceleration time histories at the surface and 112 m depth underground (Fig. 1).

The site at Haga station is mostly flat. Thus, we can regard the site as a half space. For simplicity, the half space is assumed to be isotropic elastic, with a Poisson's ratio of 0.22 and a density of 2700 kg/m³. The Poisson's ratio remains constant, thus the refractive index is the same for P waves and S waves, suggesting that the horizontal incidence angles are the same for different kind of incident waves and the vertical incidence angles are the same for different kind of incident body waves (P and S waves). Unknown parameters include the horizontal incidence angle θ_1, the vertical incidence angle θ_2, and the Young's modulus of the site E.

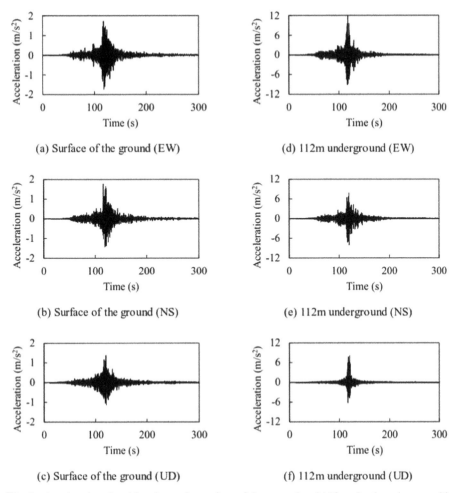

Fig. 1. Acceleration time histories at the surface of the ground and 112 m depth underground in the east-west (EW), north-south (NS), and vertical (UD) directions

The earthquake focus is located at 142.860° east longitude, 38.103° north latitude, and the Haga station is located at 140.075° east longitude, 36.548° north latitude. The angle between the direction from the epicenter to the station and east is 215°. The relative error ERR_r obtained using the proposed method for various θ_1 around 215°, and various θ_2 and E are shown in Fig. 2. It can be seen that when θ_1 equals to 155°, θ_2 equals to 40° and E equals to 6 GPa the relative error ERR_r reaches its minimum value of 0.293. The corresponding least square solution can be taken as the coefficients of potential functions, thus the incident waves are determined.

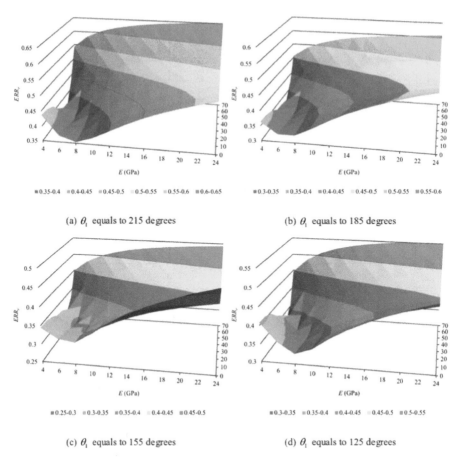

(a) θ_1 equals to 215 degrees

(b) θ_1 equals to 185 degrees

(c) θ_1 equals to 155 degrees

(d) θ_1 equals to 125 degrees

Fig. 2. The relative error ERR_r for different values of θ_1, θ_2 and E at Haga station

4   Dynamic Analysis of Gushui CFRD

The Gushui CFRD is located on Lancang River. The maximum height, length and width of the dam is 245 m, 710.6 m and 396 m, respectively. The direction of the dam axis is 23°48' from south to east. The upstream water depth is 228 m. The peak ground acceleration is 2.81 m/s².

Figure 3 shows the finite element mesh for the dam, which consists of 10729 nodes and 10318 hexahedron elements. The concrete face slab is modeled as linear elastic, with a Young's modulus of 30 GPa, a Poisson's ratio of 0.167, and a density of 2400 kg/m³. An equivalent visco-elastic model (Shen and Xu 1996), which has accumulated much application experience in the seismic design and analysis of rockfill dams in China over the past several decades, is adopted in this study as the constitutive model of the rockfill material in the dynamic analysis. The parameters used for the rockfill material of the CFRD are listed in Table 1.

Both the non-uniform and the uniform seismic input are used for the dynamic analysis. For non-uniform input, the acceleration records of the Haga station from the 110 s

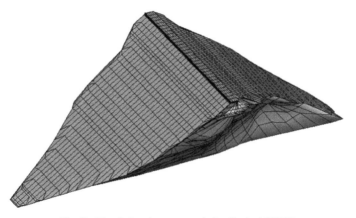

Fig. 3. The finite element mesh for Gushui CFRD

Table 1. Dynamic calculation parameters

Material	k_2	λ_{max}	μ_d	k_1	n	c_1	c_2	c_3	c_4	c_5
Cushion zone	3223	0.16	0.2	40	0.455	0.0145	0.97	0	0.0491	0.51
Transition zone	3828	0.24	0.2	42	0.345	0.0145	0.97	0	0.0491	0.51
Main rockfill zone	2660	0.21	0.2	39	0.444	0.0084	0.81	0	0.1082	0.62
Secondary rockfill zone	2000	0.24	0.2	39	0.47	0.0084	1.62	0	0.1082	0.62
Drainage zone	3223	0.16	0.2	40	0.455	0.0145	0.97	0	0.0491	0.51

to the 130 s scaled to the PGA at the Gushui site are used. For uniform input, the surface ground motion at Haga station is scaled to the PGA at the Gushui site. This way, the acceleration time histories on the surface of the free field are the same under non-uniform and uniform input.

Figures 4(a) and (b) show the contours of the peak acceleration in the direction along the river on the middle section of the dam for non-uniform input and uniform input, respectively. It can be observed that the peak acceleration under non-uniform ground motion input is overall smaller. Figures 5(a) and (b) show the contours of the peak dynamic compressive stress perpendicular to the dam slope within the concrete face slab under uniform and non-uniform input, respectively. In similar fashion, Figs. 6(a) and (b) show the contours of the corresponding peak dynamic tensile stress. Under non-uniform input, both the compressive and the tensile stress perpendicular to the dam slope within the concrete face slab are in general smaller, which is to be expected considering the difference in the acceleration of the CFRD in Fig. 4. However, Figs. 5(c) and 6(c) show the difference of the peak dynamic stress of the face slab between the non-uniform input case and the uniform input case, it can be seen that around the edges of the concrete face slab, the stress of the concrete face slab under non-uniform input is greater. This increase of local stress under non-uniform input could be caused by the phase and amplitude difference between the input motions at different positions. The difference in stress due to non-uniform input motion should be considered in design as a potential threat to the safety of the seepage control system, as such locations are weak links of the seepage control system where failures are especially detrimental and difficult to repair.

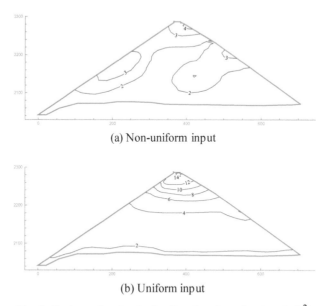

(a) Non-uniform input

(b) Uniform input

Fig. 4. Peak acceleration in the direction along the river (m/s^2)

Figures 7(a) and (b) show the residual settlement contours on the cross section in the middle of the dam for non-uniform input and uniform input, respectively. Similar to the acceleration results, the residual settlement under non-uniform input is smaller than that under the corresponding uniform input.

Table 2 compares the maximum peak acceleration along the river, peak dynamic compressive and tensile stress of the concrete face, and post-earthquake compressive and tensile stress of the concrete face under uniform and non-uniform input. The results show around 10% to 60% relative difference. These results indicate that current design and analysis approaches of using uniform input could result in seismic responses to deviate significantly from reality. Under the conditions of study in this paper, the dynamic response and residual deformation induced by uniform input is in general larger than those induced by non-uniform input, suggesting current design and analysis methods to be conservative. However, the stress results shown in Fig. 5(c) and Fig. 6(c) also suggest that the seismic response obtained under uniform input is not globally conservative, where the dynamic stress around the edges of the concrete face slab could be larger under non-uniform input compared with that under uniform input. The greater dynamic stress at the edges could provide explanations for some of the damage observed by Zhang et al. near the abutment of the Zipingpu CFRD after the Wenchuan earthquake (Zhang et al. 2015).

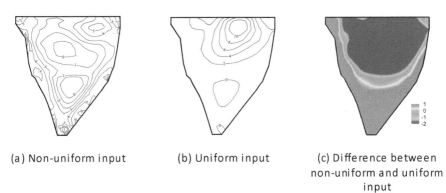

(a) Non-uniform input   (b) Uniform input   (c) Difference between
                                                                      non-uniform and uniform
                                                                      input

Fig. 5. Peak dynamic compressive stress in the concrete slab perpendicular to the dam slope (MPa)

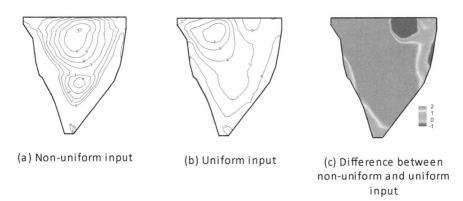

(a) Non-uniform input   (b) Uniform input   (c) Difference between
                                                                      non-uniform and uniform
                                                                      input

Fig. 6. Peak dynamic tensile stress in the concrete slab perpendicular to the dam slope (MPa)

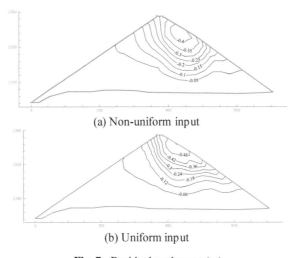

(a) Non-uniform input

(b) Uniform input

Fig. 7. Residual settlement (m)

Table 2. The peak response of the dam under uniform input and non-uniform input

Items	Uniform input	Non-uniform input	Percentage of the difference
a_y(m/s^2)	14.12	5.93	58%
s(m)	0.51	0.45	12%
σ_{dxt}(MPa)	−8.94	−7.59	15%
σ_{dxp}(MPa)	10.57	9.41	11%
σ_{dst}(MPa)	−10.06	−7.06	30%
σ_{dsp}(MPa)	12.29	7.04	43%
σ_{pxt}(MPa)	−2.29	−1.42	38%
σ_{pxp}(MPa)	15.30	10.98	28%
σ_{pst}(MPa)	−3.66	−2.40	34%
σ_{psp}(MPa)	11.70	8.27	29%

Note: a_y - Acceleration along the river
s - Residual settlement
σ_{dxt} - Dynamic tensile stress of the face along the axis of the dam
σ_{dxp} - Dynamic pressure stress of the face along the axis of the dam
σ_{dst} - Dynamic tensile stress of the face along the slope of the face
σ_{dsp} - Dynamic pressure stress of the face along the slope of the face
σ_{pxt} - Post-earthquake tensile stress of the face along the axis of the dam
σ_{pxp} - Post-earthquake pressure stress of the face along the axis of the dam
σ_{pst} - Post-earthquake tensile stress of the face along the slope of the face
σ_{psp} - Post-earthquake pressure stress of the face along the slope of the face.

5 Conclusions

A new method is developed in this study to determine the potential functions of the seismic waves at a specific site based on seismic records, which can consider the scattering of the ground surface for better description of spatial variability. This method is adopted to determine the incident waves at the Haga station for the 2011 Tohoku earthquake, and the resulting non-uniform ground motion is used for the dynamic analysis of Gushui CFRD.

The dynamic analysis for Gushui CFRD shows that when the acceleration at the surface of the free field for dynamic simulations with uniform and non-uniform input are kept consistent, the response of CFRDs under non-uniform input is in general significantly smaller, suggesting that the current design and analysis methods tend to be conservative. However, the dynamic stress around the edges of the concrete face slab could be larger under non-uniform input compared with that under uniform input, and pose non-negligible threat to the seepage control system.

The example analyzed in this study is used to showcase the input motion determination method and the influence of non-uniform input on the seismic response of CFRDs.

The recorded ground motions at the specific site of the CFRD should be used, but unfortunately such an ideal site was not available. It should also be noted that the non-uniform input motion calculation method is developed under the assumption that the bedrock is a homogeneous isotropic linear elastic medium, and the modulus is much larger than that of the rockfill dam.

Acknowledgements. The authors would like to thank the Tsinghua University Initiative Scientific Research Program (2019Z08QCX01) and National Natural Science Foundation of China (No. 51678346 and No. 51708332) for funding this work.

References

ICOLD: Design features of dams to effectively resist seismic ground motion. Committee on Seismic Aspects of Dam Design, Bulletin 120, ICOLD, Paris (2001)

Lysmer, J., Kulemeyer, R.L.: Finite dynamic model for infinite media. J. Eng. Mech. **95**, 759–877 (1969)

Deeks, A.J., Randolph, M.F.: Axisymmetric time-domain transmitting boundaries. J. Eng. Mech. **120**(1), 25–42 (1994)

Liao, Z.P., Wong, H.L., Yang, B.P., et al.: A transmitting boundary for transient wave analysis. Sci. Sinica **27**(10), 1063–1076 (1984)

Hall, W.S.: Boundary Element Method. Springer, Dordrecht (1994)

Song, C., Wolf, J.P.: The scaled boundary finite-element method—alias consistent infinitesimal finite-element cell method—for elastodynamics. Comput. Methods Appl. Mech. Eng. **147**(3–4), 329–355 (1997)

Sanchez-Sesma, F.J., Miguel, A.B., Ismael, H.: Surface motion of topographical irregularities for incident P, SV, and Rayleigh waves. Bull. Seismol. Soc. Am. **75**(1), 263–269 (1985)

Wong, H.L.: Effect of Surface Topography on the Diffraction of P, SV and Rayleigh waves. Bull. Seismol. Soc. Am. **72**(4), 1167–1183 (1982)

Yao, Y., Liu, T., Zhang, J.: A new series solution method for two-dimensional elastic scattering by a canyon in half-space. Soil Dyn. Earthq. Eng. **89**, 128–135 (2016)

Dibaj, M., Penzien, J.: Response of earth dams to traveling seismic waves. J. Soil Mech. Found. Div. **95**(2), 541–560 (1969)

Shen, Z., Xu, Z.: Seismic response analysis of geotechnical structures considering the traveling wave. J. Hydraul. Eng. **11**, 37–43 (1983)

Tian, J.: Earth dam's response to multi-point input seismic incitation and relative researching method, Ph.D. thesis, College of Water Resources and Hydropower Engineering, Hohai University, Nanjing (2003)

Zerva, A., Ang, A.H.S., Wen, Y.K.: Development of differential response spectra for lifeline seismic analysis. Probab. Eng. Mech. **1**(4), 208–218 (1986)

Beck, J.L.: Determining models of structures from earthquake records. EERL Report No. 78-01, California Institute of Technology, Pasadena, California (1978)

National Research Institute for Earth Science and Disaster Prevention website. http://www.kyoshin.bosai.go.jp/kyoshin/quake/index_en.html. Accessed 1 Jan 2015

Shen, Z., Xu, G.: Deformation behavior of rock materials under cyclic loading. J. Nanjing Hydraul. Res. Inst. **2**, 143–150 (1996)

Zhang, J.M., Yang, Z.Y., Gao, X.Z., Zhang, J.H.: Geotechnical aspects and seismic damage of the 156-m-high Zipingpu concrete-faced rockfill dam following the Ms 8.0 Wenchuan earthquake. Soil Dyn. Earthq. Eng. **76**, 145–156 (2015)

Yao, Y., Wang, R., Liu, T.Y., Zhang, J.M.: Seismic response of high concrete face rockfill dams subjected to non-uniform input motion. Acta Geotech. **14**(1), 83–100 (2019)

Case Histories of Failure of Embankment Dams and Landslide Dams

Effects of Dam Failure Mechanisms on Downstream Flood Propagation

Francesco Federico$^{(\boxtimes)}$ and Chiara Cesali

University of Rome Tor Vergata, Rome, Italy
fdrfnc@gmail.com, cesali@ing.uniroma2.it

Abstract. The careful mapping of areas that might potentially be flooded following a dam failure is necessary to understand, reduce and manage the relevant risks. These ones, in turn, strongly depend upon both the dam failure mechanisms and the characteristics of the areas potentially subjected to inundation (i.e. topography, presence of infrastructures, …). Specifically, referring to the embankment dam on Sciaguana River (Sicily, Italy), 2D unsteady flow simulations of the possible dam break due to *(a)* overtopping or *(b)* piping failure have been carried out by running the HEC RAS code. The effects of an assigned breach formation and progression mechanisms on flood propagation have been parametrically analysed. Inundation mapping results obtained through the proposed dam failure model have been compared with the potentially flooding areas predicted by the local government Authorities. Some breach prediction methods available in technical literature have been furthermore applied; the corresponding results have been discussed and compared with the numerical (HEC RAS) simulations. The proposed analyses may contribute to a better and more reliable delimitation of the flood hazard areas as well as the definition of relevant mitigation countermeasures.

Keywords: Dam break · Flood mapping · HEC RAS

1 Introduction

The failure of a dam triggers a flood wave that advances in the area behind the dam or through the valley below the dam. The effects of a dam break wave may be disastrous and may cause numerous fatalities/calamities as well as great financial losses exceeding many times the price of the hydraulic structure itself. It is thus necessary to take into account that no dam design can ensure the "absolute" protection of potentially endangered areas [1].

The main causes of failure of a dam are [2] (Fig. 1):

overtopping due to insufficient spillway capacity, improper manipulation, a landslide in the reservoir [3], an earthquake or when the design discharge is exceeded (48% of accidents in the world);

internal erosion phenomena (e.g. backward piping erosion) through the dam body and/or its foundation soils (46% of accidents in the world).

These last phenomena are practically unavoidable in earthen structures due to heterogeneity of the grain size of quarried materials, inappropriate compaction, discontinuities

© Springer Nature Switzerland AG 2020
J.-M. Zhang et al. (Eds.): ICED 2020, SSGG, pp. 149–162, 2020.
https://doi.org/10.1007/978-3-030-46351-9_13

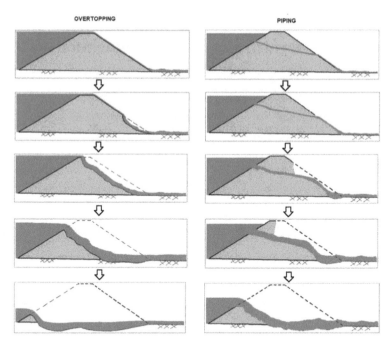

Fig. 1. Progression of dam breach due to overtopping and piping erosion

of displacements, dynamic or cyclic effects. To this purpose, reliable monitoring systems (e.g. distributed thermal, fiber optics, sensors [4]) are under development and application.

However, due to the unpredictability of these phenomena as well as the recent accidents (e.g. *Dicle Dam failure*, Turkey, 2018; *Brumadinho Dam failure*, Brasil, 2019), advanced models must be elaborated in order to predict the effects of a dam break (i.e. inundation of the valley below the dam) as well as the corresponding risks by taking into account the chain of critical events that contribute to the occurrence of an ultimate limit state of the dam [5].

In the paper, some breach prediction (empirical and analytical) methods available in literature are recalled and described. These models are applied to a documented case, also numerically analyzed (through HEC RAS code) by imposing different failure mechanisms (i.e. overtopping and piping erosion).

The comparison between the analytical and numerical results finally allows to evaluate their limits, reliability and application fields.

2   Breach Prediction Models

Breach prediction models available in literature can be distinguished in [6]:

- empirical (or analytical) methods
- semi-physically based models
- physically based models

The following description of these breach prediction methods (and the application of some of them) is intended to highlight their main features and to possibly guide users towards the most appropriate breach model, on the basis of phenomena to be simulated and of application of obtained results.

2.1 Empirical Methods

Empirical methods are based on historic dam failure data, collected together and statistically analyzed by using regression formulas.

These methods provide parametric equations describing dam breaching parameters as a function of simple reservoir properties (e.g. breach width, depth, side slope angle, formation time and peak out flow).

The main advantage of these approaches is their quick applicability. The main input parameters typically include: volume of water above final breach bottom (V_w); total volume of reservoir (V_r); height of water above final breach bottom (h_w); height of dam (h_d); height of breach (h_b). In the paper, the following (Tables 1, 2 and 3) empirical (or analytical) methods to estimate the average breach width (B), the time to failure (t_f), the peak outflow (Q_p) and the outflow hydrograph ($Q(t)$) are considered and afterwards applied [7–12].

Table 1. Empirical methods to evaluate the breach width (B)

Reference	Equation
U.S. Bureau of Reclamation [7, 8]	$B = 3 \cdot h_w$ (1)
Froehlich [9]	$B = 0.23 \cdot k_0 \cdot V_w^{1/3}$ (2)[*]

[*] $k_0 = 1.5$ and 1.0 for overtopping and piping failure, respectively.

Table 2. Empirical methods to evaluate the time to failure (tf)

Reference	Equation
U.S. Bureau of Reclamation [7, 8]	$t_f = 0.011 \cdot B$ (3)
MacDonald and Langridge-Monopolis [10]	$t_f = 0.0179 \cdot V_{er}^{0.364}$ (4)[*]
Froehlich [9]	$t_f = 60\sqrt{V_w / \left(g h_b^2 \right)}$ (5)[**]

[*] V_{er} is the eroded volume evaluable as follows: $V_{er} = 0.0261 \cdot (V_w h_w)^{0.769}$ for earthfill dams, and $V_{er} = 0.00348 \cdot (V_w h_w)^{0.852}$ for rockfill dams; [**] g is the gravity acceleration.

Referring to the outflow hydrograph ($Q(t)$), water flowing over an embankment dam can be determined by applying a broad-crested weir flow equation (Ponce and Tsivoglou 1981) [13]. The flow could be considered as uniform flow along the downstream shell

Table 3. Empirical methods to evaluate the peak flow (Q_p).

Reference	Equation
U.S. Bureau of Reclamation [7, 8]	$Q_p = 19.1 \cdot (h_w)^{1.85}$ (6)
MacDonald and Langridge-Monopolis [10]	$Q_p = 1.154 \cdot (V_w h_w)^{0.412}$ (7)
Froehlich [9]	$Q_p = 0.0175 \cdot k_1 \cdot k_H \sqrt{\dfrac{g V_w h_w h_b^2}{W}}$ (8)[*]
Singh and Snorrason [11]	$Q_p = 13.4 \cdot h_d^{1.89}$ (9)
Xu and Zhang [12]	$\dfrac{Q_p}{\sqrt{g V_w^{5/3}}} = 0.133 \cdot \left(\dfrac{V_w^{\frac{1}{3}}}{h_w}\right)^{-1.276} e^{C_4}$ (10)[**]

[*] $k_1 = 1.85$ and 1.0 for overtopping and piping failure, respectively; $k_H = 1$ and $(h_b/6.1)^{1/8}$ for $h_b < 6.1$ m and $h_b \geq 6.1$ m, respectively; g = gravity acceleration; W = average embankment width. [**] $C_4 = b_4 + b_5$; $b_4 = -0.788$ and -1.232 for overtopping and piping failure, respectively; $b_5 = -0.089, -0.498$ and -1.232 for high, medium and low erodibility of dam materials, respectively; g = gravity acceleration.

with a large channel but a small slope. The breach discharge can be thus calculated as (e.g., Singh and Scarlatos 1988) [14]:

$$Q(t) = C \cdot [B_b(t) + (H(t) - Z(t)) \tan(\alpha)](H(t) - Z(t))^{\frac{3}{2}} \qquad (11)$$

$B_b(t)$ = breach bottom width; $Z(t)$ = elevation of the breach bottom; α = angle of the side slope; $H(t)$ = elevation of the reservoir water surface; C = breach weir coefficient (generally equal to 1.7). The reservoir water level can be obtained by applying the mass balance equations:

$$A\frac{dH(t)}{dt} = Q_{in} - Q_{out} \qquad (12)$$

A being the lake surface area; Q_{in}, the flow rates into the reservoir (evaluable according to hydrological analyses); $Q_{out} = Q_b + Q_s$, the flow rates out of the reservoir (Q_b = breach discharge; Q_s = seepage discharge through the dam; this last is generally set to null value because it is neglegible with respect to Q_b or its measures are not available).

2.2 Semi-physically Based Models

With respect to empirical methods, the semi-physically models add elements of a physical process to a dam breach simulation, by minimising the computational requirements Morris [15].

An outflow hydrograph will be thus generated according to the input breach geometry parameters. The breach process is essentially defined and the semi-physically based models simply calculate the flow that would occur through such as the breach, simulating the flood propagation along the valley below the dam. Among these, the HEC RAS code, developed by the Hydrologic Engineering Centre's River Analysis System, (USACE, United States Army Corps of Engineers), is a one-dimensional and bi-dimensional flow routing model which allows to simulate a breach process by means two options for the failure mode: *piping* and *overtopping*.

The corresponding breach hydrographs can be evaluated according to the following methods: *User Entered Data* (this option is the simplest and allows the user to enter the pre-determined breach geometry parameters. These will have been determined using one of the empirical methods previously described or something similar. The user may also define the relationship between time and the breach progression); *Simplified Physical*: this option is more complex, but takes account of material properties somewhat. Users enter the maximum bounding breach width and height and do not define a breach formation time. Thus, relationships between the velocity of water and the down cutting and widening rates must be defined/assigned [6].

2.3 Physically Based Models

Physically based models combine key hydraulic, structural and geotechnical properties to analytically or numerically predict the breaching process for an embankment dam. Many of these models have been developed in the last few decades [6]. Among these, EMBREA developed at HR Wallingford [16]; AREBA developed under the FRMRC program [17]; DL Breach proposed by Wu [18]; WinDAM C developed by the Agricultural Research Service; Macchione Breach developed at ETH Zurich (Swiss Federal Institute of Technology); BASEMENT (Basic Simulation Environment) elaborated at ETH Zurich. The main overall advantage of these methods is the increased accuracy; many factors and parameters allow to closely predict the behavior and characteristics of an embankment dam breach, including erosion, sediment transport and slope stability.

Existing technical papers can provide information about the relative performance of these physically based breach models in comparison to each other. However, the comparisons and results are often based on the use of older (more easily available) breach models or the inappropriate application of more recent models. It is therefore important to consider this when interpreting any results and ideally to look for independent validation of breach model performance if possible [6].

3   A Case Study: Sciaguana Dam Break Simulation

The Sciaguana dam is located in the Licari - Di Marco district in the municipal territories of Agira and Regalbuto (Enna - Sicily, Italy).

The embankment of the dam, built between 1984 and 1992 along the homonymous river, tributary of the Dittaino River (in turn tributary of the Simeto River), is composed by loose zoned-type materials, with the upstream side of alluvial gravel-sandy and limestone quarry materials and the central core consisting of silt-sandy alluvial materials. The downstream shell is formed by tout-venant materials with a gritty-gravelly grain size. Transition filters are interposed between the core and the flanks, while three draining mats are interposed in the valley side. The reservoir is used for irrigation purposes (Fig. 2).

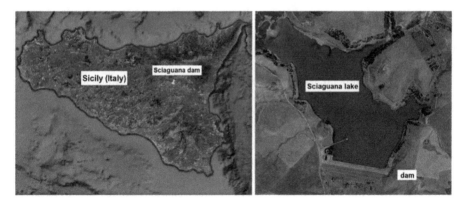

Fig. 2. Sciaguana dam (Sicily, Italy)

The geometrical characteristics of the dam are summarized in the following Table 4.

Table 4. Sciaguana dam: geometrical characteristics

Parameter	Value
Watershed basin (km^2)	64.89
Crown elevation (m asl)	266.00
Maximum reservoir water level (m asl)	260.57
Maximum operation water level (m asl)	257.10
Minimum reservoir water level (m asl)	241.75
Dam height (m)	34.57
Crown length (m)	540
Crown width (m)	30
Reservoir water volume (m^3)	$1.57 \bullet 10^7$
Operation water volume (m^3)	$9.9 \bullet 10^6$
Flood lamination reservoir volume (m^3)	$3.8 \bullet 10^6$

The flow rate from the spillway, under the hypothesis of the maximum stored water volume, is 960.0 m^3/s; from the bottom outlets is 99.0 m^3/s.

No dam break event has ever occurred. However, according to the local territorial plans, a possible failure (whose mechanism is not specified) of the Sciaguana dam would cause the inundation of the areas shown in Fig. 3. Due to the lack of detailed information about the available local studies and the corresponding results, several numerical simulations (through the HEC RAS code) of possible dam breaches, also modelling the consequent flood propagation along the Sciaguana River, have been thus carried out. Additionally, the above described empirical methods have been also applied.

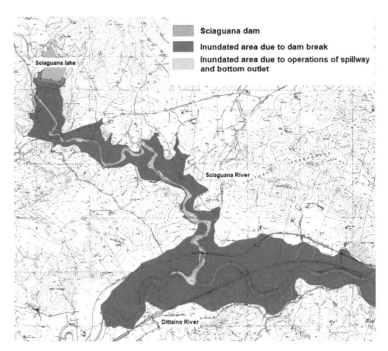

Fig. 3. Sciaguana dam: inundated areas due to dam break, according to local territorial plans

3.1 Numerical Analyses by Using HEC RAS Code

A (pure) 2D unsteady flow model of a possible Sciaguana dam break due to *(a)* overtopping or *(b)* piping failure have been implemented through the HEC RAS code.

DTM (Digital Terrain Model) with spatial resolution 2 × 2 m has been used as topographic input data. The reservoir was modeled through the "*storage area*" element (available in HEC RAS code), to which the "*elevation vs volume*" values deduced from information reported in Table 1, were assigned.

The domain of 2D calculation ("*2D Flow Area*") allows to simulate the potential flooded areas along the Sciaguana River valley. "*Storage area*" (i.e. resevoir) and "*2D Flow Area*" (i.e. Sciaguana River valley) are connected through the "*SA/2D Area Connection*" element, representing the embankment dam (Fig. 4).

Through this element it is possible to define the "*dam break*" properties; as previously specified, the "*dam break*" was simulated through the "overtopping" and "piping erosion" options (according to *User Entered Data* tool).

According to "overtopping" failure mode, the water surface overtops the entire dam and erodes its way back through the embankment or when flow going over the spillway causes erosion that also works its way back through the embankment. The breach formation and progression occur according to a proportional horizontal and vertical increments, precautionally (as default in HEC RAS code, Fig. 5).

The "piping erosion" failure mode allows to model the dam break due to seepage through the embankment, which causes erosion, which in turn causes more flow through the dam, favouring more erosion. A piping failure grows slowly at first but tends to

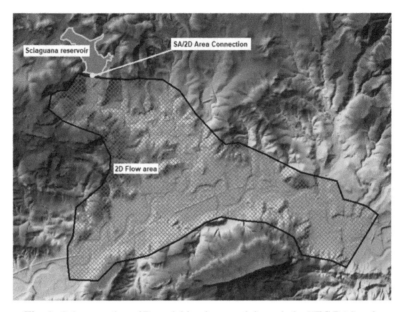

Fig. 4. Sciaguana dam: 2D model implemented through the HEC RAS code

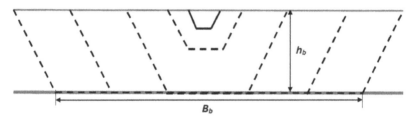

Fig. 5. Overtopping erosion characterized by equal horizontal and vertical progressive increments of the breach (adapted by [19])

pick up speed as the transversal area of the opening (or channel/pipe) begins to enlarge. The definition of two parameters is required to run the code according to this option: "piping coefficient" and "initial piping elevation". The orifice (piping) coefficient is used to calculate the flow through the breach opening during the piping erosion progression; once the embankment above the opening (or pipe) sloughs, and the water is open to the atmosphere, a weir equation is used to compute the breach flow. The following scenarios have been simulated.

(1) *Complete dam failure due to overtopping*, input breach parameters in HEC RAS: center station = 320 m; final bottom width = 430 m; final bottom elevation = 241.75 m asl; left side slope = 3; right side slope = 2.5; trigger failure at WS = 260.57 m asl; breach formation = 1 h (estimated through Eq. (4)); breach progression = linear;

(2) *Complete dam failure due to overtopping*, input breach parameters in HEC RAS: center station = 320 m; final bottom width = 430 m; final bottom elevation =

241.75 m asl; left side slope = 3; right side slope = 2.5; trigger failure at WS = 260.57 m asl; breach formation = 5 h (estimated through Eq. (3)); breach progression = linear;

(3) *Complete dam failure due to overtopping*, input breach parameters in HEC RAS: center station = 320 m; final bottom width = 430 m; final bottom elevation = 241.75 m asl; left side slope = 3; right side slope = 2.5; trigger failure at WS = 260.57 m asl; breach formation = 1 h (estimated through Eq. (4)); breach progression = non linear;

(4) *Partial dam failure due to piping erosion*, input breach parameters in HEC RAS: center station = 320 m; final bottom width = 40 m (average breach width = 60 m, derived from Eqs. (1), (2)); final bottom elevation = 241.75 m asl; left side slope = 1; right side slope = 1; trigger failure at WS = 260.57 m asl; breach formation = 0.75 h (average value between those ones estimated through Eqs. (3), (5)); breach progression = sinusoidal (typically assigned in HEC RAS code to simulate a dam breach due to piping); piping coefficient = 0.5; initial piping elevation = 250.0 m asl.

Referring to the breach progression (according to equal horizontal and vertical progressive increments), the following failure evolution curves have been applied (Fig. 6). Definitively, the main input data of the simulated scenarios can be summarized as follows (Table 5).

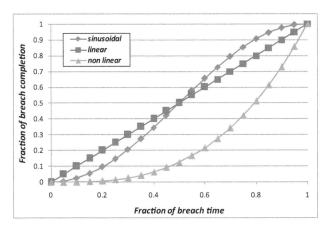

Fig. 6. Breach progression: assigned evolution failure curves

Referring to the initial and boundary conditions ("*unsteady flow data*" tool), the following assumptions are considered:

maximum reservoir water level, as initial condition, applied to the "*storage area*";
normal depth applied to the "*2D flow area*" mesh boundary;
no flow rate along the Dittaino River.

Manning coefficient equal to 0.06 s/m$^{1/3}$ has been assigned to the *2D flow area* mesh, according to local territorial plans studies by Sicily Region government authorities.

Table 5. Sciaguana dam break simulation: considered scenarios

Scenario	Description	Breach formation time	Breach progression
1	Complete dam failure due to overtopping	1 h	Linear
2	Complete dam failure due to overtopping	5 h	Linear
3	Complete dam failure due to overtopping	1 h	Non linear
4	Partial dam failure due to piping erosion	0.75 h	Sinusoidal

The obtained results in terms of (1) inundation areas along the Sciaguana river valley (with the inundation areas derived from the local Authorities as red lines), (2) water level at a specific transversal section along Sciaguana river (at the distance of 1000 m downstream the dam), (3) peak outflow and (4) the corresponding outflow hydrograph, referred to the different simulated scenarios, are shown in the following figures. Sensible differences between the various simulated scenarios and the local territorial planning can be observed. With respect to the inundation maps from the local territorial planning, these differences are mainly due to the topographic data (i.e. DTM) used in the computations.

Referring to the different considered scenarios, wider (and more similar to those ones derived from local Authorities) inundation areas along the Sciaguana river valley are obtained by simulating the break dam due to overtopping (Fig. 7). The differences between the two considered breach mechanisms (i.e. overtopping and piping erosion) are considerable also in terms of water level at a transversal river section (Fig. 8).

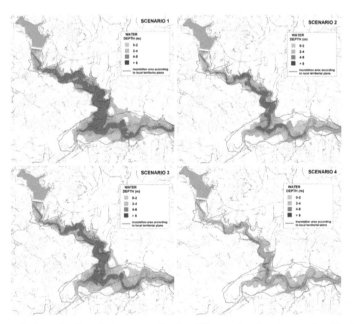

Fig. 7. Numerical results: inundation areas for the different simulated scenarios (the inundation areas derived from local government Authorities are shown as red lines)

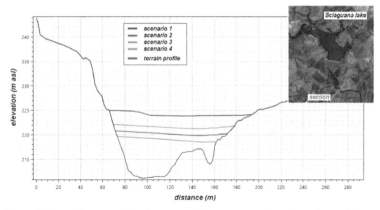

Fig. 8. Numerical results: water level at a specific transversal section along Sciaguana river (at the distance of 1000 m downstream the dam), for the different simulated scenarios

It is worth observing that the inundation maps from local territorial planning can be interpreted and reproduced only by assuming a break dam due to overtopping.

The maximum peak outflow (9562.2 m³/s) is obtained in the scenario 1 (complete dam break due to overtopping, duration of formation = 1 h; breach progression = linear); the minimum peak outflow (3909.4 m³/s) is obtained in the scenario 4 (partial dam break due to piping erosion, duration of formation = 0.75 h; breach progression = sinusoidal) (Fig. 9). Referring to scenario 4, the reduction of the reservoir water level is slower than the other scenarios (Fig. 10).

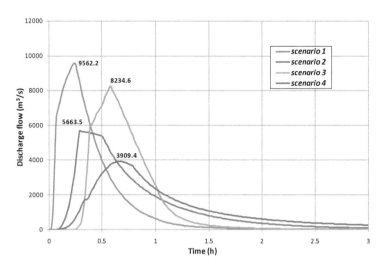

Fig. 9. Numerical results: outflow from the breach dam vs time

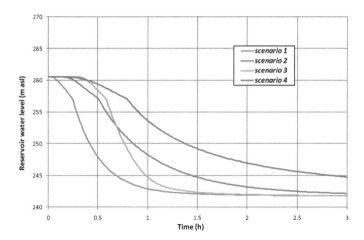

Fig. 10. Numerical results: reservoir water level vs time

3.2 Application of Empirical Breach Prediction Methods

As previously specified, the empirical formulas, Eqs. (1)–(5), have been applied to estimate the average breach width and the time to failure, assumed in the numerical simulations through HEC RAS code.

Therefore, the Eqs. (6)–(11) are applied here to the case study of the Sciaguana break dam and compared with the above numerical results (HEC RAS code) in terms of peak outflow (Table 6) and outflow hydrograph (Fig. 11).

Table 6. Sciaguana dam break simulation: numerical results vs empirical methods

Scenario	Q_p (mc/s)					
	HEC RAS code	Equation (6)	Equation (7)	Equation (8)	Equation (9)	Equation (10)
1	9562.2	4355.9	3565.0	2072.89	10845.4	6326.3
2	5663.5					
3	8234.6					
4	3909.4					

It is firstly worth observing that the Eqs. (6)–(10) to estimate the peak outflow don't take into account the time to failure and the type of breach mechanism (e.g. overtopping, piping erosion).

Equations (6) and (7) provide results similar to those ones numerically obtained in the scenario 4; Eq. (9) similar to those ones numerically obtained in the scenario 1; Eq. (10) similar to those ones numerically obtained in the scenario 2. Equation (8)

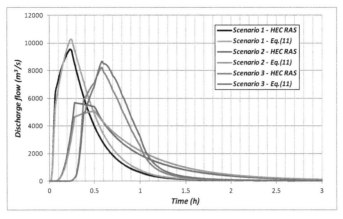

Fig. 11. Outflow hydrographs, for the different simulated scenarios: numerical results (HEC RAS) vs Eq. (11)

provides smaller values of peak outflow than the remaining formulas. Thus, Eqs. (6)–(8) seem to be applicable to the analysis of breach dam due to piping erosion; while Eqs. (9), (10) seem to be applicable to the analysis of breach dam due to overtopping. Equation (11) can be only applied to the breach dam processes due to overtopping. Thus, the results computed by Eq. (11) have been compared with the numerically simulated scenarios 1, 2, 3. Specifically, $B_b(t)$, $H(t)$ and $Z(t)$ (figuring in Eq. (11)) have been derived from the HEC RAS code, for each considered scenarios ($\tan(\alpha) = 1$ and $C = 1.44$ are assigned in Eq. (11)).

It is observed the correspondence, in terms of both form and peak of the outflow hydrograph, between the obtained numerically (HEC RAS) and analytically (through Eq. (11)) results (Fig. 11).

4 Concluding Remarks

The paper provides a synthetic overview of available methods in technical literature to model an embankment dam breach.

Advantages and disadvantages of some breach prediction methods have been described. Specifically, empirical models are simple, easy to apply to determine the potential size of a breach and its peak outflow. These last can be estimated from simple reservoir and dam geometry data, which is often easier to obtain than the more complex soil data required for physically based models. However, the necessity of mapping areas that might potentially be flooded following a dam failure calls for the application of numerical (hydraulic, semi-physical based) models allowing to map the resulting flood waves. To this purpose, HEC RAS code is applied.

Referring to the case study of the Sciaguana break dam, the HEC RAS code has been implemented to analyze the effects associated with the changes of the dam break parameters (e.g. time to failure, mechanisms of breach, breach progression,….) on the inundation areas. The obtained results have been also compared with the potentially flooded areas derived from the studies carried out by the local government Authorities. The analysis through some empirical models of the examined case study has finally allowed to define their limits of application and to evaluate their reliability.

References

1. Jandora, J., Riha, J.: The failure of embankment dams due to overtopping. Translation & review c František Aujesky & Roger Turlan (2008)
2. ICOLD. Internal erosion of existing dams, levees and dikes, and their foundations. Bulletin n.164, Vol. 1: "Internal erosion processes and engineering assessment" (2013)
3. Tessema, N.N., Sigtryggsdottir, F.G., Lia, L., Jabir, A.K.: Case study of dam overtopping from waves generated by landslides impinging perpendicular to a reservoir's longitudinal axis. J. Mar. Sci. Eng. **7**, 221 (2019). https://doi.org/10.3390/jmse7070221
4. Cesali, C., Federico, V.: Coupled thermal and piezometric heads monitoring to detect permeability defects within embankment dams and levees. In: Third International DAM WORLD Conference 2018, Foz do Iguassu, Brazil, 17–21 September 2018 (2018)
5. Federico, F., Musso, A.: Progetto allo stato limite di contatti e transizioni nelle dighe di terra. (in italian) Tech. J. "L'Ingegnere" 1–4, 49–56 (1990)
6. West, M., Morris, M., Hassan, M.: A guide to breach prediction. Editor: Craig Goff, HR Wallingford (2018)
7. U.S. Bureau of Reclamation. Guidelines for defining inundated areas downstream from Bureau of Reclamation dams. Reclamation Planning Instruction No. 82-11, U.S. Department of the Interior, Bureau of Reclamation, Denver (1982)
8. U.S. Bureau of Reclamation. Downstream hazard classification guidelines. ACER Tech. Memorandum No. 11, U.S. Department of the Interior, Bureau of Reclamation, Denver (1988)
9. Froehlich, D.C.: Empirical model of embankment dam breaching. In: International Conference on Fluvial Hydraulics (River Flow) (2016)
10. MacDonald, T.C., Langridge-Monopolis, J.: Breaching characteristics of dam failures. J. Hydraul. Eng. **110**(5), 567–586 (1984)
11. Singh, K.P., Snorrason, A.: Sensitivity of outflow peaks and flood stages to the selection of dam breach parameters and simulation models. J. Hydrol. **68**, 295–310 (1984)
12. Xu, Y., Zhang, L.M.: Breaching parameters for earth and rock fill dams. J. Geotech. Geoenviron. Eng. **135**(12), 1957–1970 (2009)
13. Ponce, V.M., Tsivoglou, A.J.: Modeling gradual dam breaches. J. Hydr. Eng. Div. ASCE **107**(HY7), 829–838 (1981)
14. Singh, V.P., Scarlatos, P.D.: Analysis of gradual earth-dam failure. J. Hydraul. Eng.-ASCE **114**(1), 21–42 (1988)
15. Morris, M.W.: Modelling breach initiation and growth. FLOODsite Report T06-08-02 (2009)
16. Mohamed, M.A.A.: Embankment breach formation and modelling methods. Ph.D. thesis, The Open University, England (2002)
17. Van Damme, M., Morris, M.W., Hassan, M.A.M.: A new approach to rapid assessment of breach driven embankment failures. FRMRC Research Report, WP4.4 (2012)
18. Wu, W.: Simplified physically based model of earthen embankment breaching. J. Hydraul. Eng. **139**(8), 837–851 (2013)
19. HEC-RAS Hydraulic Reference Manual

Stability Assessment of the Embankment Dam Systems Under Different Seismic Loads in Southern California and Evaluation of the Current Design Criteria

Mehrad Kamalzare[✉] and Hector Marquez

Civil Engineering Department, California State Polytechnic University, Pomona, CA, USA
{mkamalzare,hmarquez}@cpp.edu

Abstract. The integrity of system of embankment dams and levees is a crucial component in ensuring the safety of protected communities in any country. The failure of such systems due to natural or man-made hazards can have monumental repercussions, sometimes with dramatic and unanticipated consequences on human life, property and the economy of the states and the country. For highly seismic areas such as Southern California, it is critical to investigate and study the seismic response of embankment dams and levees for the afore mentioned reasons. While experimental studies of embankment dams under seismic loads is expensive, very time consuming, and limited, numerical studies usually suffer from lack of legitimate real data for verification of the developed models. However, organizations such as the California Strong Motion Instrumentation Program (CSMIP) instrument lifeline structures such as earth dams and levees with accelerometers and actively collect strong-motion data. The data obtained from CSMIP accelerometers is then processed by the Center for Engineering Strong Motion Data (CESMD) and made public for earthquake engineering applications. In this study, numerical models of existing earth embankment dams verified with site specific CESMD data are created in order to analyze their stability for a future earthquake, for post-earthquake response purposes. The seismic fragility of the modelled dams was assessed, providing insight for decision makers regarding priority areas important for matters such as maintenance, dam retrofit, or first-aid response locations for a hypothetical major earthquake. Society can be better prepared for a potential catastrophic seismic event.

Keywords: Embankment design · Slope stability · Seismic monitoring

1 Introduction

The integrity of the state and national system of embankment dams and levees is a crucial component in ensuring the safety of protected communities in any country. Levees are constructed along water courses to provide protection against floods while dams are constructed to form reservoirs to store water for urban, industrial or agricultural consumptions. The failure of such systems due to natural or man-made hazards can have

© Springer Nature Switzerland AG 2020
J.-M. Zhang et al. (Eds.): ICED 2020, SSGG, pp. 163–174, 2020.
https://doi.org/10.1007/978-3-030-46351-9_14

monumental repercussions, sometimes with dramatic and unanticipated consequences on human life, property and the economy of the states and the country. The failure of dams and levees during Hurricane Katrina in 2005, which led to the catastrophic flooding of the city of New Orleans, USA, is a highly illustrative example. About 2,000 people lost their lives due to the failure of the levees that were protecting the city, and the property damage was estimated at $81 billion (2005 USD) [1]. There are several other examples that reveal the critical role of embankment dams and levees, and the impacts of their failure on people's lives and properties. There are nearly 14,000 miles of levees under U.S. Army Corps of Engineers (USACE) jurisdiction in the US; but it does not include what is believed to be more than 100,000 additional miles of levees not covered by the USACE safety program. Some are little more than mounds of earth piled up more than a century ago to protect farm fields. Others extend for miles and are made of concrete and steel, with sophisticated pump and drainage systems. They shield homes, businesses and infrastructures such as highways and power plants [2].

Figure 1a shows that 881 counties with a total population of 160 million in the United States are protected by these dams and levees. Figure 1b presents a closer look at the levees (black lines) and the areas protected by them in Southern California, specifically, the Los Angeles metropolitan area. As it is illustrated in Fig. 1b, there are large areas of Orange County between the Los Angeles River and Santa Ana River, which are heavily populated and are being protected by levees. Although Southern California has a relatively lower risk of experiencing hurricanes or typhoons compared to cities such as New Orleans, Louisiana or Houston, Texas, the existence of a large number of active faults, and the high likelihood of earthquakes, makes the assurance of a healthy and reliable dam and levee system a very important matter to the State of California. In the case of an earthquake, the induced seismic forces, failure of the slopes, and the ground rupture would be the main failure mechanisms. In the case of a hurricane or flood that happens relatively quickly, seepage and overtopping would be the most dominant and most probable failure mechanisms. While other failure mechanisms require more time to significantly damage a dam or levee, seismic loads would apply large deformation to the dams or levees in a relatively short time, and eventually lead to dam failure.

The overall stability of levees and embankment dams is a very complicated matter and depends on several multidisciplinary factors such as stability of slopes (Geotechnical Engineering), characteristics and impacts of flooding events (Water Resources Engineering), and erosion properties of the surface and covers (Construction Engineering), among others. Therefore, it is critical to investigate and study the behavior of the system of levees and embankment dams in Southern California using a multidisciplinary research team. This can help to more realistically identify the locations with most critical problems in the levee system and accordingly reevaluate the current existing seismic design criteria in regards to the embankment dam systems. Precisely modeling the structure of dams under seismic loads would help engineers to be able to predict the most probable failure sections, and take the appropriate actions to minimize the risk of failure.

One method to model the seismic response of earth dams is through shaking table tests of scaled models. One of many examples is the work performed by Yuan et al. [3]. The large size of the dams and levees would generally create a great limitation on the experimental and laboratory studies of these structures. Accurate construction of

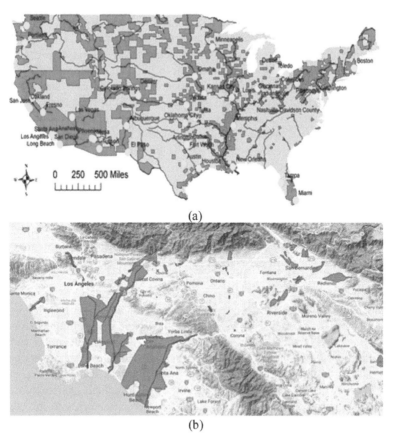

Fig. 1. Areas protected by levees, (a) United States counties protected by levees (shown in red color) and major cities (shown in yellow color), (b) Los Angeles greater metropolitan area with the endanger flood zones shown in purple color.

the laboratory models, lack of precise control on the boundary conditions, difficulties of performing tests with various seismic loads, and large number of required stress and strain sensors, among others are some of the main challenges of experimental investigations of embankment dams and levees [4, 5].

Numerical models, on the other hand, can overcome almost all of the mentioned limitations of the experimental studies, although a thorough verification of the results is an essential part of any numerical study [6, 7]. Alberti et al. [8] analyzed the seismic performance of the San Pietro dam in Southern Italy using a numerical method. The dam was modeled and analyzed through dynamic 2D finite difference analyses using the computer code FLAC 2D. Crosshole tests were performed on various portions of the dam to obtain small strain shear modulus (*Go*) values to model the dam. Prior to the seismic analyses, a static analysis was performed to simulate the dam construction and reproduce the total and effective state of stress at the end of the dam construction. The input motions were obtained from several accelerograms from a worldwide database including a record from the 1994 Northridge earthquake in California. Permanent deformations smaller than 50 cm (20 in) were calculated, based on the input parameters.

Rampello et al. [9], performed a set of finite element analyses to evaluate the behavior of the Marana Capacciotti earth dam in Southern Italy, under seismic load. A constitutive model capable to reproduce soil non-linearity, and calibrated against laboratory measurements of the stiffness of small strains, was used for their investigations. The models were developed in Plaxis software and both artificial and real accelerograms were considered for seismic input values. With respect to the real accelerogram data, the finite element analysis only considered data from a single accelerogram from the 1976 Friuli earthquake in northeast Italy. Prior to seismic analyses, a static model of the construction of the dam was also simulated to produce initial state of stress conditions. The static model was checked with observed settlements during and after construction from extensometers installed on the dam. Material property inputs were obtained from results of recent in situ investigation. Ultimately, the seismic analyses returned acceptable results specifically due to computed settlements at the crest being considerably smaller than the service freeboard. Other researchers have performed seismic analyses of earth dams using numerical methods in recent years to investigate various aspects of embankment design [10–14].

While the studies mentioned above provide important information regarding seismic numerical analysis procedures for earth dams, they were mainly limited to historical seismic input values that have occurred at other sites. However, due to the complexity of the interactions of various sections of embankment dams, it would be very beneficial to analyze a dam using seismic parameters previously experienced by the specific dam in order to verify that the model responds similarly to the actual occurrence.

Zeghal and Abdel-Ghaffar [15], performed numerical analyses to investigate the behavior of the Long Valley earth dam in California, using data from 22 accelerographs instrumented on the dam, primarily to address existing methods of seismic modeling of earth dams. The authors noted that it was a complex task to choose a model for a real structure, especially under seismic conditions. Using the accelerograph data, the dam was determined to behave nonlinearly and having seismic wave propagation at its boundaries. The study also found that constitutive hysteretic models are insufficient to account for dam dissipation mechanisms. The study highlighted the benefit of having strong-motion data to produce information not available by other means.

More recently, Castelli et al. [16] modeled the Lentini earth dam in southeast Sicily, Italy with strong-motion data from a nearby accelerometer recorded during the 1990 Santa Lucia earthquake, which had caused notable damage to the dam. Using Plaxis, a 1D analysis was preformed resulting in the maximum horizontal acceleration versus depth.

Strong-motion earthquake data is constantly being collected for various structures in the State of California, USA. The primary reason for collecting strong-motion earthquake data is that society could greatly benefit from an increased understanding of how certain structures would respond to specific strong-motion values or seismic events [17]. This is especially true for lifeline structures in Southern California. Figure 2 shows the locations of the dams in southern California that are currently being monitored by the Center for Engineering Strong Motion Data (CESMD) and its partners, and the behavior of these dams have been recorded during the past earthquakes in the region. In lack of accurate laboratory work, the available data can be a great source to verify and validate developing numerical models.

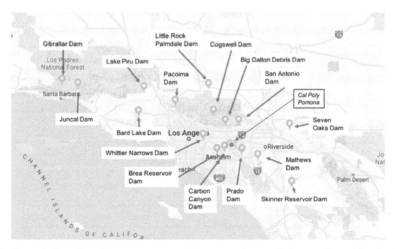

Fig. 2. Dams in Los Angeles greater metropolitan area that are being monitored by the Center for Engineering Strong Motion Data (CESMD) and its partners.

The objective of this investigation is to develop sets of numerical models that simulate different failure mechanisms of these dams under seismic loads. The results can reveal the areas of the dams and levees with higher risks in respect to overall stability, which would eventually lead to the measurement of potential impacts on properties and lives in affected areas. These could lead to the development of action plans for remediation of the system of the dams and reduce the risk of failure in the case of an earthquake or other natural and man-made catastrophes [18].

2 Problem Definition

One of the main goals of this investigation is to revisit and improve the seismic design criteria of embankment dams and levees. This goal can be achieved by developing precise models of embankments, using site specific soil characteristics, and considering the overall behavior of dams under previous seismic loads.

2.1 Input Motions

Site specific accelerometer data is used for this study. There are currently a few organizations such as the California Geologic Survey (CGS), Department of Conservation that use a large number of instruments to continuously record the responses of select structures since 1972. More than 125 structures instrumented with accelerometers by the California Strong Motion Instrumentation Program (CSMIP) are lifeline structures [19]. The data obtained from CSMIP accelerometers is then made public by the Center for Engineering Strong Motion Data (CESMD). The CESMD is a cooperative center established by the United States Geological Survey (USGS) and the CGS to integrate earthquake strong-motion data from the CSMIP. The CESMD provides raw and processed strong-motion data for earthquake engineering applications [20]. In order to analyse an existing structure for post-earthquake response, significant earthquakes that have occurred in Southern California are applied to the model, such as the 1994 Northridge earthquake (6.7 Mw). The CESMD provides records from numerous accelerometers with data from this earthquake and many others. Figure 3 displays a typical output of processed accelerometer data displaying the acceleration, velocity, and displacement during the 1994 Northridge earthquake from an accelerometer located on the Pacoima Dam in California.

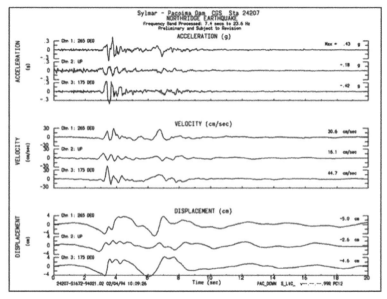

Fig. 3. Sample CESMD processed accelerometer data output (Center for Engineering Strong Motion Data: https://strongmotioncenter.org).

2.2 Site and Dam Characteristics

To investigate the behavior of embankment dams under seismic loads, a few existing dams in Southern California were analyzed. For the purposes of this study, the selected dams were limited to homogeneous earth embankments, currently well instrumented with CSMIP accelerometers, and with existing subsurface investigation for model parameters. Figure 4 displays the Puddingstone dam cross section as an example obtained from the CESMD. The dam is located in the city of San Dimas, California.

The dam presented in Fig. 4 was constructed with locally available weathered shale bedrock. The embankment fill is composed of compacted sandy silty clay (CH-ML)

	Dam No. 1	
Crest Elevation		983.5 ft
Freeboard		7.1 ft
Height Above Streambed		148.0 ft
Crest Length		1,085.0 ft
Crest Width		25.0 ft
Upstream Slope	(Elev. 982 to 912)	2.5 H:1V
	(Elev. 912 to 887)	3 H:1V
	(Elev. 887 & below)	3.5 H:1V
Downstream Slope	(Elev. 982 to 885)	2.5 H:1V
	(Elev. 885 & below)	3 H:1V
Material		CH-MH

(a)

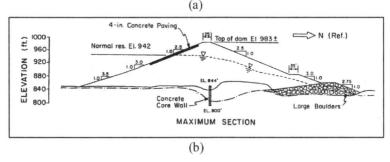

(b)

Fig. 4. Puddingstone dam cross section (a) Properties of the dam; (b) Dam's geometry (not drawn in scale) [20].

with weathered shale fragments. The dam also has a toe drain on the downstream (north) portion made up of large boulders. The dam was completed in 1928 and is now operated by the Los Angeles County Flood Control District.

3 Finite Element Modelling

The numerical models are developed using the finite element modeling software RS2 by Rocscience [21]. The selected dams were modeled geometrically based on the actual conditions of the respective dams. Material properties were then introduced based on the performed subsurface investigations, and obtained values from field and laboratory experiments. The dams were modeled using an appropriate constitutive model. A strain-hardening model with non-linear stiffness was found to be the most appropriate constitutive model for these investigations [9, 11–13, 16], and was used in this study. Subsequently, the appropriate boundary conditions and a uniform mesh were applied to the models. Figure 5 illustrates the cross section of a typical modeled dam with mesh and boundary conditions. Prior to running any analysis, the strong-motion data obtained from the CESMD must be properly applied. In RS2, similar to other finite element models, the seismic motions are applied at the base of the model. However, the strong-motion values obtained from the CESMD are the recorded motions experienced at the location of the accelerometer. To properly evaluate the validity of the models, the strong-motion data must be deconvoluted such that the motions inputs at the base of the embankment dam, results in the motion recorded by the accelerometer at the actual location of the accelerometer. Afterwards, the strong-motion data was carefully modified so that the high frequency components, which do not provide significance would be eliminated from the analyses without impacting the overall properties of the dynamic loading. This is done to reduce computing time. Finally, appropriate Rayleigh damping coefficients, α_M and β_K, were computed for the model.

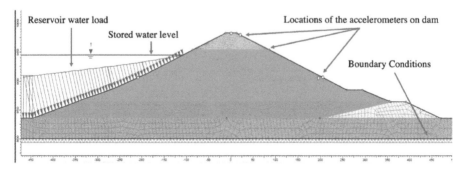

Fig. 5. Typical RS2 model with mesh and boundary conditions

The analysis generally consisted of two major stages. First, the model was analyzed under static conditions (gravity) to achieve an existing (after construction) state of stress. The next stage was the dynamic stage where displacements were calculated based on the input parameters.

3.1 Model Verification

Prior to running an analysis on any probable, significant, earthquake, the developed model must be verified to have confidence in the results. The benefit of picking a dam well instrumented with accelerometers is the possibility of using the recorded data from smaller earthquakes in the past. The developed models in this research were validated with CESMD data recorded by select accelerometers on the dam. The actual displacements of the dam at those selected locations were known from the processed data. The motions input for verification were the recent, previously occurred earthquakes of less magnitude, and the majority of the recorded displacement values were less than 1 cm (0.4 in). The water level of the dams on the days the input motion was experienced was also obtained and modelled accordingly in an effort to create the most representative models possible.

4 Results and Discussions

After obtaining a verified model, the idea is that the model is valid for any possible motion applied to it. To assess the areas of the dam that are at higher risk regarding stability, arbitrarily high strong-motion data obtained from CESMD accelerometers located in Southern California was obtained and modeled for Puddingstone Dam. Conclusions could then be determined based on the obtained results.

Two different embankment dams with various soil properties were investigated in this paper using a developed finite element model. The numerical models were verified using site specific recorded seismic responses. The behaviors of these dams were then studied under predicted future seismic events. Figure 6 shows a typical deformation response of one of the modeled dams. The weaker sections of the dams were identified and checked against the current seismic design criteria. The developed model can also be used to analyze the responses of other earth embankment dams with similar soil properties in the case of a major earthquake event, and assess the possibility of future failure in existing conditions. Displacements throughout the dam are measured and assessed. The crests of the dams are specifically an area of interest in assessing the stability of dams as large deflections of the crest would greatly affect the service freeboard.

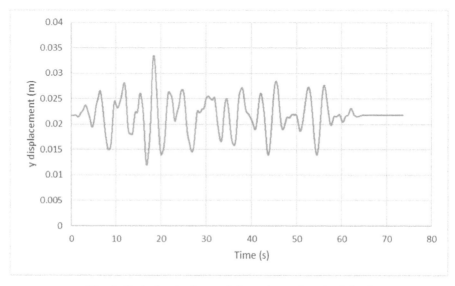

Fig. 6. Typical embankment deformation under seismic load

Seismic fragility analysis, with fragility curves as the outcome, is an efficient approach for seismic risk analysis of engineering structures. However, as far as embankment dams in seismically active regions are concerned, there still lacks a well-established method. As a result of this study, fragility curves for the subject dams were also generated. Figure 7 presents a typical fragility curve result. The seismic vulnerability for the dams under the proposed study were determined by combining the probabilities of various levels of the seismic hazard at the dam location, with the damage probabilities to the dam corresponding to the seismic hazard levels at the site.

Seismic hazard is represented by the peak ground acceleration (PGA) using a number of near field earthquake records in the CSMIP database. The records include different PGA levels and frequency contents that represents a creditable earthquake in active seismic sites. The analyses were carried out using the calibrated models described in the section above. The fragility curves will assist a decision maker for section priorities for dam retrofit and maintenance, and it will be used as a tool for first aid respondents

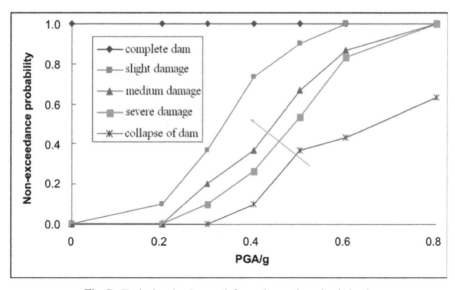

Fig. 7. Typical embankment deformation under seismic load.

to mobilize their resources more effectively at sites susceptible to the risk of higher damage. The ultimate goal is to expand this study in future research for a series of earth dams and levees in a region such as the Los Angeles Basin to identify the key structures that are in emanate risk.

5 Conclusions

In this paper, a set of numerical models were developed that simulate different failure mechanisms of embankment dams and levees under seismic loads in Southern California, USA. The results revealed the areas of the dams and levees with higher risks in respect to overall stability, which would eventually lead to the measurement of potential impacts on properties and lives in affected areas. These could lead to the development of action plans for remediation of the system of the dams and reduce the risk of failure in the case of an earthquake or other natural and man-made catastrophes. The following specific conclusions can be drawn from the study:

1. The finite element modeling method was performed to precisely simulate the existing, at risk dams in Los Angeles metropolitan area, using the available recorded data of the response of the same dams during previously occurred seismic activities. The results showed a good consistency with the previous earthquakes.
2. The data produced over the course of this research validated the behavior predicted by the numerical models. Although the deformations and stresses were recorded during previous earthquakes in the region, the developed model in this research can be used to predict the dam's responses in the case of future earthquakes.
3. The analyses showed that the maximum settlement (vertical deformation) would happen under the crest of the embankment. This can be due to the fact the crest is the

highest elevation on the dam, and the underlying soil would carry the largest stresses in the structure.

4. Maximum lateral (horizontal) deformation would occur about mid-height of the embankment on the slope face and away from the center on either side. As the embankment settles, it extends outward laterally.

5. This study serves as an example for other earth embankment dams to be numerically modelled using strong-motion data to assess the need for post-earthquake response.

Acknowledgments. The authors would like to thank the Geotechnical Engineering Laboratory staff of California State Polytechnic University, Pomona, for their help during this project. The authors would also like to thank the California State for providing partial funding for this research.

References

1. Townsend, F.F.: The federal response to Hurricane Katrina: lessons learned. The White House, Washington, D.C., Tech. Rep. PREX 1.2:K 15 (Feb 2006)
2. Knabb, R.D., Rhome, J.R., Brown, D.P.: Tropical cyclone report: Hurricane Katrina. Nat. Hurricane Center, Miami, FL, Tech. Rep. TCR-AL122005_Katrina (Aug 2005)
3. Yuan, L., Liu, X., Wang, X., Yang, Y., Yang, Z.: Seismic performance of earth-core and concrete-faced rock-fill dams by large-scale shaking table tests. Soil Dyn. Earthq. Eng. **56**, 1–12 (2014)
4. Seed, R.: Letter from Ray Seed, Professor of Civil and Environmental Eng. to the president, American Society of Civil Engineers (October 2007). http://www.lasce.org/documents/RaySeedsLetter.pdf
5. The Full Wiki.: New Orleans: map (May 2009). http://maps.thefullwiki.org/New_Orleans
6. Boyd, E.: Large map of USA counties with levees (October 2009). http://levees.org/large-map-of-u-s-counties-with-levees/
7. Natale, P.J: Report card for America's infrastructure (February 2009). http://www.asce.org/pplcontent.aspx?id=2147484137
8. Aliberti, D., Cascone, E., Biondi, G.: Seismic performance of the San Pietro dam. Procedia Eng. **158**, 362–367 (2016)
9. Rampello, S., Cascone, E., Grosso, N.: Evaluation of the seismic response of a homogeneous earth dam. Soil Dyn. Earthq. Eng. **29**, 782–798 (2009)
10. Andrianopoulos, K.I., Papadimitriou, A.G., Bouckovalas, G.D., Karamitros, D.K.: Insight into the seismic response of earth dams with an emphasis on seismic coefficient estimation. Comput. Geotech. **55**, 195–210 (2014)
11. Wu, C., Ni, C., Ko, H.: Seismic response of an earth dam: finite element coupling analysis and validation from centrifuge tests. J. Rock Mech. Geotech. Eng. **1**, 56–70 (2009)
12. Elia, G., Rouainia, M.: Seismic performance of earth embankment using simple and advanced numerical approaches. J. Geotech. Geoenviron. Eng. **139**, 1115–1129 (2013)
13. Puentes, J., Rodríguez, L., Rodríguez, E.: Numerical models for seismic response of "El Buey" dam. In: GeoCongress (2006)
14. Akhtarpour, A., Naeini, M.: Numerical analysis of seismic stability of a high centerline tailings dam. Soil Dyn. Earthq. Eng. **107**, 179–194 (2018)
15. Zeghal, M., Abdel-Ghaffar, A.M.: Analysis of behavior of earth dam using strong-motion earthquake records. J. Geotech. Eng. **118**, 266–277 (1992)
16. Castelli, F., Lentini, V., Trifarò, C.A.: 1D seismic analysis of earth dams: the example of the Lentini site. Procedia Eng. **158**, 356–361 (2016)

17. Leon, R.T.: Improving the seismic performance of existing buildings and other structures. In: Second ATC SEI Conference (2015)
18. Kamalzare, M., Zimmie, T.F., Cutler, B., Franklin, W.R.: New visualization method to evaluate erosion quantity and pattern. Geotech. Test. J. ASTM **39**, 431–446 (2016)
19. California Strong Motion Instrumentation Program. http://www.conservation.ca.gov/cgs/smip
20. Center for Engineering Strong Motion Data, Center for Engineering Strong Motion Data. http://www.strongmotioncenter.org/
21. RS2 9.0 [Computer software]. Toronto, ON, Rocscience Inc.

Piping Risk Reduction Measures for Mengkuang Dam Upgrading Project

C. H. Khor[✉] and C. K. Toh

Angkasa Consulting Services Sdn. Bhd., 47620 Subang Jaya, Selangor, Malaysia
kch@acssb.com.my

Abstract. Construction of the Mengkuang dam upgrading project was exposed to potential piping risk at several key areas. The variability in site geology, foundation conditions and construction materials added to the difficulty in predicting piping risk of the project. In some cases the risk factors were only uncovered during construction stage after the dam foundation was exposed. Changes to the original design, construction procedures and contract specifications had to be made in order to mitigate the risks. The techniques applied and measures implemented for risk reduction varied in each case in order to suit the actual site conditions and nature of risk encountered. This paper discussed the techniques developed and the issues related to the piping risk reduction measures implemented for the dam upgrading project. The techniques developed for alleviating piping risk was effective, judging from behavior of the dam and dam instrumentation records that were within expected ranges.

Keywords: Earthfill dam · Piping risk · Plastic concrete · Cutoff wall

1 Introduction

Other than overtopping, internal erosion and piping are the main causes of failures and incidents in embankment dams [1]. Despite the fact that piping phenomenon has been recognized for many years, incidents of piping in dams continue to occur [2, 3]. The variability in site geology, foundation conditions and construction materials added to the difficulty in predicting piping risk of the project. In many cases potential piping risks were not identified at the investigation and design stage. The risk factors were only uncovered during construction stage after the dam foundation was exposed. Risk management technique was developed for the construction of the Mengkuang Dam upgrading project [4, 5].

Construction of the Mengkuang dam upgrading project was exposed to potential piping risk at several key areas. Changes to the original design had to be made in order to mitigate the risks. The strategy adopted for the design of foundation cutoff works involved a combination of cut-off trench, plastic concrete cut-off wall, jet grout wall and cement grout curtain in order to suit the varied foundations conditions and presence of erodible materials at the dam foundation and abutment. This paper discussed the techniques developed and the issues related to the piping risk reduction measures implemented for the dam upgrading project.

© Springer Nature Switzerland AG 2020
J.-M. Zhang et al. (Eds.): ICED 2020, SSGG, pp. 175–181, 2020.
https://doi.org/10.1007/978-3-030-46351-9_15

2   Importance of the Dam Upgrading Project

The Mengkuang Dam upgrading construction contract was awarded by the Government of Malaysia to China International Water & Electric Corp. (M) Sdn. Bhd. in 2012. This major water source augmentation project is vital for Dua River Water Treatment Plant which provides over 80% of the potable water supply of Penang State.

The original Mengkuang Dam was an earthfill dam 31 m high and about 1 km long completed in 1985. The project involved raising the existing dam by 11 m and construction of a new dam 45 m high and about 2 km long. The upgraded dam was completed in July 2015.

3   Site Conditions

The dam site is part of the Kulim Granite Formation which is a medium to coarse grained, porphyritic granite and granodiorite. The granite is typically siliceous, containing over 50% quartz, and microcline as its dominant feldspar, and minor mafic minerals that are mainly biotite. The expanded Mengkuang Dam has a crest length of about 3 km long and a dam foundation about 300 m wide. The dam foundation is widely covered with sandy clayey silt of alluvium or slope wash of the Quaternary Period. Residual soil layer varies widely from 2 m to 40 m deep. Highly to moderately weathered granite varying from zero to 8 m in thickness generally overlay the slightly weathered to fresh granitic bedrock. The residual soil and weathered granite have permeability coefficients in the range of 10^{-4} cm/s.

4   Internal Erosion Control Measures

The extended dam is a zoned earthfill dam. In order to protect the clay core against internal erosion and piping risk, a comprehensive filter and internal drainage systems and seepage cut-off measures were provided as shown in Fig. 1. Crushed rocks filter was processed from the granite boulders using quarry facility available at site. A laboratory test proposed was set up for testing the acceptability of the crushed rocks filters for clay core constructed of tropical residual soils derived from granite (see Fig. 2) [6, 7].

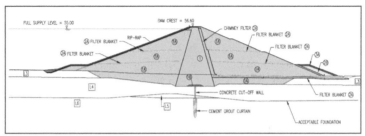

Notes: L3 – Alluvium, L4 – Completely Weathered Rock to Residual Soil, L5 – Moderately to Highly Weathered Rock, L6 – Fresh to Slightly Weathered Rock

Fig. 1. Typical design section of extended dam

Fig. 2. No erosion filter test for Mengkuang Dam

5   Dam Foundation Seepage Cutoff Works

Seepage cutoff work through the dam foundations was provided using a combination of cut-off trench, plastic concrete cut-off wall, jet grouted wall, and consolidation and curtain grouting.

The extended dam on the adjacent valley passed through the pervious to semi pervious weathered granite layer with thickness varying from 2 m to 40 m. The strategy adopted for the design of foundation cutoff works involved a combination of cut-off trench, plastic concrete cut-off wall, jet grout wall and cement grout curtain in order to suit the varied foundations conditions over the 3 km long dam embankment.

Value engineering conducted on options of seepage barrier installation showed that cutoff trench with depth exceeding 10 m was more expensive compared to plastic concrete cutoff wall (see Fig. 3). Other factors such as cutoff trench stability and dewatering issue to enable dry working area during construction period added to the disadvantage of deep cutoff trench option.

Fig. 3. Trenching for plastic concrete wall using clamshell bucket

Jet grout cutoff wall was constructed at the upstream toe of the raised dam and extended up the abutments and the spillway to form a continuous seepage barrier for the raised dam. This option was adopted for the coarser soil foundation at the hilly ground. The jet grouting machine is a lighter machine suitable for ease of maneuvers at sloping ground.

At the base of cutoff trench a 500 mm thick concrete slab was constructed which act as grout cap for consolidation grouting to seal the weathered and fractured granite as shown in Fig. 4.

Fig. 4. Consolidation grouting on moderately weathered rock foundation

Grouting was required to complete the seepage cut-off in the bedrocks which permeability varied between 105.93 Lu and 0.66 Lu. One row of grout curtain was provided for the foundation of both the heightened and the extended embankment, with total length of 3,150 m. The adopted grout curtain seepage criteria was below 3 Lu.

Slope correction excavation or dental concrete were used at areas where the rock profile was uneven (>150 mm) in order to ensure positive contact of the core zone against the rock.

6 Piping Risk Reduction Measures

Due to the presence of erodible materials at the dam foundation and abutment piping risk reduction measures were applied at five key areas of the dam. Some risks impact the safety of existing dam at the construction stage, while others impact on the future performance of the dam.

6.1 Risk of Piping at Left Abutment of Extended Dam

The left abutment of the extended dam contained several bands of hydro-thermally altered granite which is highly erodible materials as shown in Fig. 5. Potential failure mode due to erosion of the weak materials when subject to high pressure seepage flows from the reservoir was deemed unacceptable, after a detailed geological and geotechnical investigation was carried out. The piping risk reduction measure adopted involved construction of three rows of cement grout curtain connected to the grout curtain of the extended dam, radial grout curtain of the outlet tunnel and extended to the left abutment to form a seepage barrier through the fractured zone at the left abutment. The grouting work constituted a variation to the contract with a cost of RM 6.4 million.

Fig. 5. Hydro-thermally altered granite at left abutment and foundation of extended dam

6.2   Risk of Piping at Extended Dam Foundation

At the extended dam foundation, whitish completely weathered granite (which was hydro-thermally altered) was encountered at chainage 400 m to 900 m and intermittently until chainage 1400. This chalk-like materials was highly erodible. Removal of this material was not practical and the cost to be incurred would be prohibitive. Plastic concrete cutoff wall was provided to create a seepage barrier through the erodible foundation [8]. The plastic concrete cutoff wall was designed to achieve its ability to deform without cracking and to resist erosion if it did crack when it is subjected to the embankment loading above.

The hydro-thermally altered granite becomes soft when exposed to rain and tends to form cracks induced by shrinkage when exposed to sunlight. Placing of the clay core was made as soon as the dam foundation preparation was completed in order to control the formation of cracks. The clay core material used for contact with the foundation consisted of plastic clay with plasticity index over 15 and moisture content wetter than optimum up to 4%. Likewise placing of horizontal filter blanket at the downstream shoulder of dam was made soon after foundation surface was ready. A rock-toe with filter was provided at the downstream toe of dam to act as counterweight against uplift pressure of seepage exiting at the dam toe.

6.3   Risk of Hydraulic Fracture of Dam Embankment at River Channel

Construction of the 2 km long dam embankment had proceeded concurrently on both sides of the river channel of Mengkuang River during construction of the diversion tunnel. The gap at the dam embankment in the river channel was completed much later. Differential settlement and crack due to variation in height and time of completion of the embankment was a potential risk. The piping risk reduction measures applied included controlling the steepness of embankment slope at both sides of river channel to not more than 30° in order to avoid soil arching effect. Abrupt variation in ground surface at the edge of river banks was corrected to avoid steps in the profile. This measure was applied to avoid development of low stress zones in which hydraulic fracture can occur as the dam is filled [3].

6.4   Risk of Internal Erosion of Dam Core Adjacent Draw-off Culvert of Existing Dam

Raising of the existing dam involved extending the horse shoe shape draw-off culvert which had experienced high seepage along the wall of the culvert. Potential failure mode of the raised embankment due to internal erosion and piping was a major concern. The conduit facilitates the initiation of internal erosion by causing soil arching adjacent the wall of culvert which lead to low stress zone and hydraulic fracture [3, 9]. The piping reduction measures applied include placing plastic clay with high moisture content (up to 4% above optimum moisture content) to facilitate effective contact between soil and culvert interfaced area. Filter collars together with horizontal filter blanket were placed at the extended culvert of the downstream shoulder of the raised dam.

6.5   Risk of Piping Failure of Existing Dam Due to Dam Raising Works

During construction of dam upgrading works, concentrated leak was observed at about 100 m away from the reservoir. The leaks were discharging from a porous layer of highly weathered granite as shown in Fig. 6. The probability of piping failure of existing dam was assessed using event tree technique to represent the postulated failure mode to assist in decision making on the mitigation measure required [4, 5].

Fig. 6. Piping risk at left abutment of existing dam

The risk reduction option by lowering existing reservoir water level by 10 m was the least cost option. However this option would impact the security of water supply system. An alternative water source from Beris Dam of Kedah state was identified as a backup source. This issue was finally resolved after a long process of negotiation between state government of Penang, Kedah and Federal Ministry of Land Water and Natural Resources (KATS).

7   Conclusions

The measures applied for piping risk reduction had taken into account the actual site conditions and nature of the risk encountered. The techniques developed for alleviating piping risk was effective judging from the behavior of the dam and dam instrumentation records. Monitoring data showed the total seepage recorded in 2019 was less than 9 L/s. Other dam instrumentation readings, e.g. piezometers, inclinometers, extensometers, geodetic surveys, etc. were within the expected ranges.

Acknowledgments. The authors acknowledge the project team members of Water Supply Division (BBA) of Ministry of Land, Water and Natural Resources (KATS), Angkasa Consulting Services Sdn. Bhd. and China International Water & Electric Corp. (M) Sdn. Bhd. who contributed to the development and execution of the piping risk reduction measures for the project.

References

1. Foster, M., Fell, R., Spannagle, M.: The statistics of embankment dam failures and incidents. Can. Geotech. J. **37**(5), 1000–1024 (2000). ISSN 0008-3674
2. Sherard, J.L.: Embankment dam cracking. In: Hirschfeld, R.C., Paulos, S.J. (eds.) Embankment Dam Engineering, pp. 308–312. Wiley, New York (1973)
3. Sherard, J.L.: Hydraulic fracturing in embankment dams. ASCE J. Geotech. Eng. **112**(GT10), 905–927 (1986)
4. Khor, C.H., Shamsuddin, A.H.: Risk management for the Mengkuang dam upgrading and enlargement project. Ingenieur **73**, 52–59 (2018)
5. Khor, C.H., et al.: Risk management for a major water source expansion project. In: Lariyah, M.S., et al. (eds.) 1st International Conference on Dam Safety Management and Engineering 2019, Penang, WRDM, ICDSME 2019, pp. 479–487. MYCOLD, Malaysia (2019)
6. Khor, C.H., Woo, H.K.: An investigation of crushed rock as filters for a dam embankment. ASCE J. Geotech. Eng. **115**(3), 399–412 (1989)
7. Khor, C.H.: Crushed rock filters and leakage control in embankment dams. J. Inst. Eng. Malaysia (IEM) Bil **50**, 5–14 (1992)
8. Electric Power Industry Standard of the People's Republic of China: Specification of Concrete Cut-off Wall Used for Hydropower and Water Conservancy Project. DL/T 5199 – 2004. China Electric Power Press, China (2004)
9. Charles, J.A.: General report, special problems associated with earthfill dams. In: Nineteenth International Congress on Large Dams, Florence, GRQ73, vol. ii, pp. 1083–1198. ICOLD, Paris (1997)

The Outburst Flood of the 2000 Yigong Landslide Dam Based on Limited Information

Danyi Shen[1,2], Zhenming Shi[1,2], and Ming Peng[1,2(✉)]

[1] Key Laboratory of Geotechnical and Underground Engineering of Ministry of Education,
Tongji University, Shanghai 200092, China
pengming@tongji.edu.cn

[2] Department of Geotechnical Engineering, Tongji University, Shanghai 200092, China

Abstract. A super large rock avalanche occurred along Zhamu Creek, southeast Tibet on April 9, 2000, and triggered the Yigong landslide dam. The dam with a height of 60 m, a length of 2500 m, a width of 2500 m and a lake volume of 2.015 Gm^3, threatened life and property both downstream and upstream. This paper reports a rapid assessment method for dam breaching with limited geometric and hydrological information of landslide dams. The method can be divided into three steps: first, Google map was used to obtain regional DEM data; second, Global Mapper was used to analyze the regional DEM data and obtain the river channel profiles; third, the HEC-RAS 4.1 was used to simulate the outburst discharge and flood evolution process. The results show that the excavation of spillway can decrease the dam breach peak discharge and control the life and property risk. The present method can be used for breach flood assessment of sudden unexpected landslide dams and providing a scientific basis for decision making in landslide dams risk management.

Keywords: Yigong landslide dam · Dam breaching · Peak discharge · Flood routing · Excavate spillway

1 Introduction

A landslide dam is a type of natural dam formed by the lateral obstruction of a river with landslide, avalanche, debris flow and so on (Zhang et al. 2016). The longevity of most landslide dams is short, about 52% lasted less than one week (Peng and Zhang 2012). The breach of landslide dam has brought widespread hazard to life and property in history not only because of their magnitude but also because they usually occur suddenly and unpredictably (Walder and O'Connor 1997). For instance, in 1786, the burst of landslide dam in Luding-Kangding, killed more than 10 thousand people (Fan et al. 2012). Due to the sudden formation of landslide dams, it is usually difficult to obtain the detailed geometric parameters and downstream terrain data quickly. Therefore, prediction of potential peak discharge and resulting hydrograph based on limited data can help to decide mitigation measures and reduce life and property losses.

© Springer Nature Switzerland AG 2020
J.-M. Zhang et al. (Eds.): ICED 2020, SSGG, pp. 182–188, 2020.
https://doi.org/10.1007/978-3-030-46351-9_16

Previously, there are two methods have been used to predicted probable peak discharge from potential dam failure (Walder and O'Connor 1997). One method is to establish the regression equations based on the dam parameters and impounded water volume: dam height, dam volume, lake depth, lake volume, or some combination thereof (Costa 1985; Costa and Schuster 1988; Peng and Zhang 2012). The other method is to establish physical model and calculate the peak discharge and breach size of dams by computer (Fread 1991; Mizuyama 2006; Chang and Zhang 2010; Li 2017). Although, the outburst flood of landslide dams is frequently studied, the first model can get approximate peak discharge, but it is difficult to make a detailed analysis for the dam. The second model needs detailed hydraulics and geological parameters, but the information usually poorly known when a landslide dam suddenly formed.

In this study, an rapid assessment method based on regional DEM data of landslide dams and river channels profiles is proposed to predict the discharge and water level of the 2000 Yigong landslide dam. Additionally, the discharge and flooded area of downstream are compared whether the spillway excavated or not.

2 Methodology of Simulating Landslide Dam Breach Flood

The one-dimensional river hydraulic analysis program HEC-RAS 4.1 has been used to establish the river channel model and simulate the flood evolution. The HEC-RAS model solves the Saint Venant equations formulated for natural channels (U.S. Corps of Engineers 2002)

$$\frac{\partial A}{\partial t} + \frac{\partial Q}{\partial x} - q_l = 0 \tag{1}$$

$$\frac{\partial Q}{\partial t} + \frac{\partial (Q^2/A)}{\partial x} + gA\left(\frac{\partial H}{\partial x} + S_f\right) = 0 \tag{2}$$

where A is cross-sectional area perpendicular to the flow; Q is discharge; ql is lateral inflow due to tributary; g is acceleration due to gravity; H is elevation of the water surface above a specified datum; Sf is longitudinal boundary friction slope; t is temporal coordinate; and x is longitudinal coordinate.

The equations are solved by using the well-known four-point implicit box finite difference scheme.

The establishment of the HEC-RAS model includes three steps: first, using Google Map to obtain the regional DEM data; second, using the Global Mapper to analyze the regional DEM data, and establish the river channel profiles; third, importing the river channel profiles to the HEC-RAS to establish the river channel model. Besides, in order to simulate the flood evolution, the channel geometry, boundary conditions and channel resistance are required. During the calculation, a similar breach formation concept is used. The failure begins at the top of the dam, growing in depth, bottom width, and side slope angle simultaneously (Gary 2003).

3   Flood Simulation of the 2000 Yigong Landslide Dam

3.1   Introduction of the 2000 Yigong Landslide Dam

The Yigong landslide dam was caused by rock avalanche (which was mixed with ice blocks and water) in the mountains of Tibet at 30°10′39″N, 94°56′25″E on 9 April 2000 (Delaney and Evans 2015), Fig. 1 shows the location of the Yigong landslide dam area, the distance to Yigong Town is approximately 26 km, and the location is 50 km from Lulang Town and 95 km from Bomi City (Zhou et al. 2014).

The landslide debris completely blocked the Yigong Zangpo River, with a dam volume of 3×10^8 m^3, a dam height of 60 m, a dam width of 2500 m, a dam length of 2500 m, a lake volume of 2.015 Gm^3. Figure 2 shows the water level-lake capacity curve, the inflow rate was about 518 m^3/s.

By 28 April 2000, the Chinese Army Corps completed the excavation of the spillway, with a length of 1000 m, a top width of 150 m, a bottom width of 30 m, and a depth of 30 m. On 10 June 2000, the dam breached and generated a major outburst flood downstream in the Yarlung Zangpo (Tibet). The peak discharge at the dam site was almost about 124,000 m^3/s. The maximum discharge at Tongmai was about 120,000 m^3/s, 17 km downstream from the dam (Shang et al. 2003). Besides, at a distance of 462 km from the dam site (Fig. 2), the water level began to rise at 12:00 pm on 11 June, and the peak discharge was about 44,200 m^3/s (Tewari 2004).

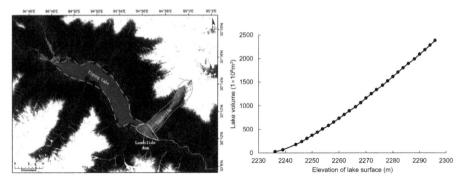

Fig. 1. The 2000 Yigong landslide dam   **Fig. 2.** Water level-lake capacity curve

3.2   Establishment of the Model

The flood routing process of the 2000 Yigong landslide dam can be calculated by HEC-RAS 4.1. Firstly, the Google Earth was used to search the Yigong landslide dam site and river direction to obtain the regional maps based on 30 m DEM data, including longitude and latitude information. Then, the Global mapper was used to obtain the 3D terrain data and river section profiles. According to the topographic characteristic, a total of 37 typical river section are extracted. Finally, the regional DEM data, river section profiles and spillway parameters are input to the HEC-RAS 4.1. The Manning coefficients on the channel and bank of the Yigong landslide dam are 0.035 and 0.045, respectively. Figure 4 shows the section of the dam site (Fig. 3).

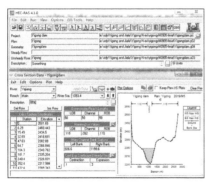

Fig. 3. Interface of HEC-RAS 4.1

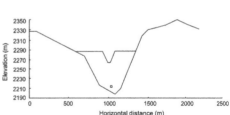

Fig. 4. Section of the dam site

3.3 Flood Evolution Simulation

The lower reaches of the Yigong landslide dam are mostly located in high mountains and canyons. According to Google Earth map, it was found that there are three main areas that may be impacted by the dam breach flood, namely Tongmai bridge, Motuo county and Beibeng village. Among them, the Tongmai bridge was about 17 km away from the dam site and its elevation was about 2,033 m, the Motuo village was about 165 km away from the dam site and its elevation was 1,076 m, and the Beibeng country was about 189 km away from the dam site and its elevation was 859 m.

Figure 5 shows the flood process and the upstream and downstream water level before and after the excavation of spillway at the dam site. It can be found that after the excavation of the spillway, the maximum water level at the dam site decrease obviously, and the peak discharge is decreased from 251,507 m^3/s to 118,284 m^3/s.

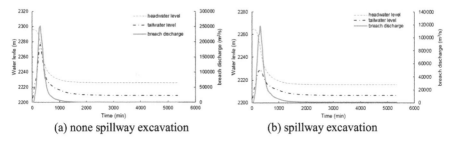

(a) none spillway excavation   (b) spillway excavation

Fig. 5. Water level and breach discharge at the dam site

Figures 6 and 7 show the discharge and flooded area of the Tongmai bridge. The peak discharge at the Tongmai bridge of excavated and non-excavated spillway are 241,911 m^3/s and 116,929 m^3/s, respectively. The water level and breach discharge decreased significantly after the spillway excavated compared to none spillway excavated. Besides, the Tongmai bridge will be flooded after the dam breach whether the spillway excavated or not. This is also consistent with the actual record. Shang et al. (2003) also reported a maximum instantaneous discharge of about 120,000 m^3/s at the Tongmai bridge, and flood the Tongmai bridge.

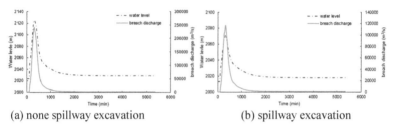

(a) none spillway excavation   (b) spillway excavation

Fig. 6. Water level and breach discharge at the Tongmai bridge

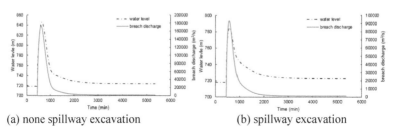

(a) none spillway excavation   (b) spillway excavation

Fig. 7. Flooded area at the Tongmai bridge

Figures 8, 9, 10 and 11 show the discharge and flooded area of the Motuo county and Beibeng village, both the discharge and water level are decreased obviously after the spillway excavated. The Motuo county and Beibeng village are no longer flooded after the excavation of spillway.

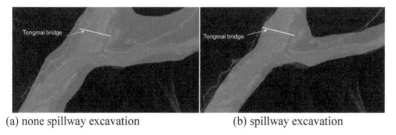

(a) none spillway excavation   (b) spillway excavation

Fig. 8. Water level and breach discharge at the Motuo county

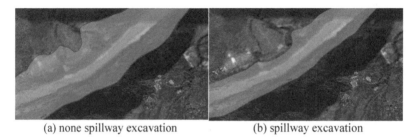

(a) none spillway excavation   (b) spillway excavation

Fig. 9. Flooded area at the Motuo county

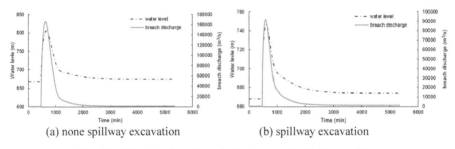

(a) none spillway excavation   (b) spillway excavation

Fig. 10. Water level and breach discharge at the Beibeng village

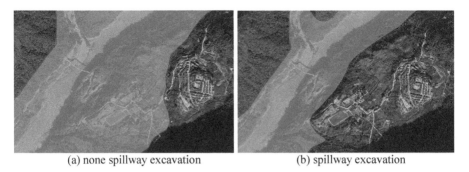

(a) none spillway excavation   (b) spillway excavation

Fig. 11. Flooded area at the Beibeng village

The result of flood evolution simulation shows that the peak discharge and peak water level are greatly reduced with the spillway excavated compared to none spillway excavated. Besides, the excavation of spillway greatly reduce the hazard of the downstream. This is because the excavation of the spillway reduced the maximum reservoir water level of the landslide dam by almost 30 m, and the corresponding maximum reservoir capacity by 1 billion m^3, which was conducive to reducing the water energy and dam breach rate of the landslide dam. Therefore, the excavation of spillway plays an important role in disaster reduction.

4 Results

This paper analyzes the outburst flood evolution of the Yigong landslide dam based on limited information, the main conclusions are as follow:

The three-dimensional geographic information combined with the flood evolution software HEC-RAS 4.1 can realize the rapid assessment of the outburst flood when a landslide dam formed and provide a basis for the emergency management and decision-making of the landslide dams.

After the Yigong landslide dam formed, the peak discharge of dam breach can be as high as 251,507 m^3/s without taking engineering measures (spillway), which will flooded the Tongmai bridge, Motuo village and Beibeng country in the downstream. After the excavation of spillway, the Motuo village and Beibeng country are not flooded, which indicates that the excavation of spillway can reduce the risk of downstream when the dam breach, but give early warning and evacuate the downstream population is still needed.

References

Brunner, G.W.: Dam and levee breaching with HEC-RAS. In: World Water & Environmental Resources Congress 2003, pp. 1–9 (2003)

Costa, J.E.: Floods from dam failures. US Geological Survey (1985)

Costa, J.E., Schuster, R.L.: The formation and failure of natural dams. Geol. Soc. Am. Bull. **100**(7), 1054–1068 (1988)

Chang, D.S., Zhang, L.M.: Simulation of the erosion process of landslide dams due to overtopping considering variations in soil erodibility along depth. Nat. Hazards Earth Syst. Sci. **10**(4), 933–946 (2010)

Delaney, K.B., Evans, S.G.: The 2000 Yigong landslide (Tibetan Plateau), rockslide-dammed lake and outburst flood: review, remote sensing analysis, and process modelling. Geomorphology **246**, 377–393 (2015)

Fan, X., Westen, C.J.V., Xu, Q., Gorum, T., Dai, F.: Analysis of landslide dams induced by the 2008 Wenchuan earthquake. J. Asian Earth Sci. **57**(5), 25–37 (2012)

Fread, D.L.: BREACH, an erosion model for earthen dam failures. Hydrologic Research Laboratory, National Weather Service, NOAA (1988)

Li, X.N.: Study on dam break erosion and flood routing of earth dam. China Institute of Water Resources and Hydropower Research, Beijing (2017)

Mizuyama, T., Satofuka, Y., Ogawa, K., Mori, T.: Estimating the outflow discharge rate from landslide dam outbursts. In: Proceedings of the Interpraevent International Symposium on Disaster Mitigation of Debris Flows, Slope Failures and Landslides, vol. 1, no. 2, pp. 365–377 (2006)

Peng, M., Zhang, L.M.: Breaching parameters of landslide dams. Landslides **9**(1), 13–31 (2012)

Shang, Y., Yang, Z., Li, L., Liu, D., Liao, Q., Wang, Y.: A super-large landslide in Tibet in 2000: background, occurrence, disaster, and origin. Geomorphology **54**, 225–243 (2003)

Tewari, P.: A study of soil erosion in Pasighat town (Arunachal Pradesh) India. Nat. Hazards **32**, 257–275 (2004)

U.S. Corps of Engineers: HEC-RAS River Analysis System, Hydraulic Reference Manual. Hydraulic Engineering Center Report CPD-69, Davis, CA (2002)

Walder, J.S., O'Connor, J.E.: Methods for predicting peak discharge of floods caused by failure of natural and constructed earthen dams. Water Resour. Res. **33**(10), 2337–2348 (1997)

Zhang, L.M., Peng, M., Chang, D.S., Xu, Y.: Dam Failure Mechanisms and Risk Assessment. Wiley, Hoboken (2016)

Zhou, J., Cui, P., Hao, M.: Comprehensive analyses of the initiation and entrainment processes of the 2000 Yigong catastrophic landslide in Tibet, China. Landslides **13**(1), 39–54 (2016)

Estimation and Prediction Method of Water Level, Flow and Bursting Time of Baige Barrier Lake in Jinsha River

Yu Yao[✉] and Fuqiang Wang

China Renewable Energy Engineering Institute, Liupukang North Street 2, Xicheng District, Beijing 100120, China
yaoyutsinghua@126.com

Abstract. The author participated in the emergency treatment of the barrier lake formed by the second landslide of Baige, Jinsha River, and put forward the estimation and prediction methods of the water level, storage capacity and discharged flow process of the barrier lake and the prediction methods of the time when the barrier lake enters the stage of burst. These methods have been applied in the emergency treatment, providing reference information for the emergency response.

Keywords: Baige barrier lake · Emergency response · Burst · Estimation and prediction · Development process

1 Introduction

In the wee hours of October 11, 2018 and on November 3, 2018, two landslides occurred in Baige village, Boluo Town, Jiangda County, Tibet Autonomous Region, blocking the Jinsha River and forming two dangerous cases of barrier lake.

Since Baige landslide formed a barrier to block the river, China Renewable Energy Engineering Institute has set up a technical team for emergency response to the dangerous situation of the barrier lake on the Jinsha River, organized professional forces to predict the possible risks and influence scope, and provided technical support for the national energy administration to formulate emergency plan. In the second emergency response process, China Renewable Energy Engineering Institute carried out basic efforts such as the measurement of the reservoir capacity of the barrier plug and risk judgment, analysis of bursting flood of barrier lake and evolution, and risk dynamic assessment. The Institute rechecked and forewarned the time of burst, updated and forewarned the water conditions along the river in real time after burst. The countermeasures, such as artificial intervention excavation of the barrier, cofferdam breaking of Suwalong Hydropower Station and emergency vacating of Liyuan Hydropower Station, are put forward, which provide technical support for risk prediction and scientific decision-making of emergency response of the barrier lake. As a member of the technical team for emergency response of Jinsha River barrier lake, I participated in the emergency response work of the two barrier lakes. I estimated and predicted the water level and flow information in the development

© Springer Nature Switzerland AG 2020
J.-M. Zhang et al. (Eds.): ICED 2020, SSGG, pp. 189–197, 2020.
https://doi.org/10.1007/978-3-030-46351-9_17

process of the second barrier lake, and successfully predicted the time when the barrier lake entered the stage of burst. In the second part hereof, the dangerous situation and emergency response process [1] of the second Baige barrier lake are briefly introduced; in the third part, the estimation and prediction method of the development process of the barrier lake water level, water storage and discharge is given; in the fourth part, the prediction method of the time when the barrier lake enters the stage of flow burst is given, and the accuracy of the prediction method is analyzed; in the fifth part, this paper is summarized.

2 Overview of the Second Danger Case of Baige Barrier Lake and the Emergency Response

At about 17:40 on November 3, 2018, the second landslide occurred at the original landslide point of Baige barrier lake. The landslide is about 1.6 million m^3, and the slope is about 6.6 million m^3, of which the total is 8.2 million m^3. The last natural discharge channel is blocked. The total volume of the barrier is about 30.2 million m^3, and the top elevation is about 2,966 m, which will form a barrier lake with a storage capacity of 775 million m^3. See Fig. 1 for the barrier conditions after the second landslide.

After the dangerous situation of the barrier lake was researched and determined, it is decided to carry out manual intervention excavation for the barrier. At 14:42 on November 8, the first equipment arrived at the weir top and excavation began; at about 17:00 on November 11, the construction of the barrier drain channel was completed. At 5:00 on the November 12, the water level of the barrier lake rose to the bottom elevation of the manually excavated discharge chute; at 10:50, the manually excavated discharge chute officially and comprehensively overflowed. At 8:00 on November 13, the flow of the burst increased to about 70 m^3/s and entered the bursting stage; at 18:20, the maximum peak flow of the burst was estimated to be about 33,900 m^3/s. See Fig. 2 for flow burst of barrier.

After the burst of the barrier, the flood peak formed rapidly advanced to the downstream, passing through the stations of Yebatan, Lawa, Batang, Suwalong, Benzilan, Shigu and Liyuan in turn. At 20:00 on the November 13, the flood at Yebatan Hydropower Station reached a peak of 28,300 m^3/s; at 1:00 on November 14, the flood at Batang Hydrological Station reached a peak of 20,900 m^3/s; at 3:50 on November 14, the flood at Suwalong Hydropower Station under construction reached a peak of 19,800 m^3/s. From 3:00 on November 15, the inflow of Liyuan Hydropower Station began to increase significantly, and the maximum inflow reached 7,200 m^3/s at 12:30, then decreased slowly.

After the occurrence of the second landslide, through the analysis and evaluation of the impact of the breach, three engineering measures are mainly taken: (1) To reduce the adverse impact of the bursting flood on the power station under construction, the existing power station and the residents along the river, it is needed to manually intervene and excavate the barrier; (2) To reduce the adverse impact of the superposition of the bursting flood and the bursting flood caused by the collapse of the cofferdam of Suwalong Hydropower Station, it is needed to remove the cofferdam of Suwalong Hydropower Station (3). To ensure the safety of concrete faced rock-fill dam of the Liyuan Hydropower Station, it is needed to empty the storage of Liyuan Hydropower Station in emergency.

Fig. 1. Barrier Blocking the River after the Second Baige Landslide

Fig. 2. Burst of Barrier (16:00 on November 13)

3 Estimation and Prediction of Development Process of Water Level, Storage and Discharge Flow of Barrier Lake

The process curve of the water level, storage capacity and discharge of the weir lake can be drawn according to the water level storage capacity curve of Baige barrier lake, the shape, size and elevation information of discharge chute, the released water level information of the barrier lake and the flow information from the upstream of the barrier lake, and the prediction curve can be further drawn based on the existing data.

Before the discharge channel overflows, the measured and predicted curves of the water level and storage capacity of the barrier lake can be drawn on a single map with double coordinates, as shown in Fig. 3. The water level ordinates are arranged according to equal spacing, and the storage capacity (water storage) ordinates are marked after the corresponding storage capacity of each water level is found according to the water level-storage capacity curve. The abscissa is the time (MM-DD).

According to the measured water level time information, the measured curve can be drawn. The predicted curve is drawn through the "add trend line" function of EXCEL.

Note that the current time is t and the corresponding elevation is h. Several (for example, three) times t^i recorded before and corresponding elevation h^i (i = 1, 2, 3) are selected. It is estimated that a prediction curve at the time T corresponding to the elevation H at the top of the barrier is composed of (t^i, h^i), (t, h) and (T, H), and the Curve l of its cubic polynomial function is added. Pay attention to adjust the value of T so that the trend line and the common time segment of the measured line can be reunited and

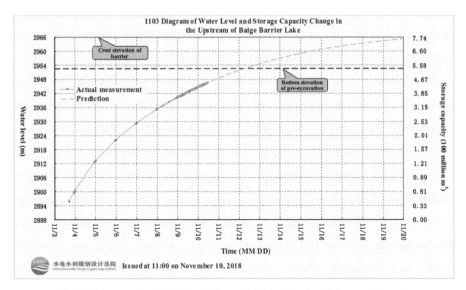

Fig. 3. Measured and Predicted Curve of Water Level and Storage Capacity

smoothly connected as much as possible, and then change the format of Curve l to "no line". In this way, the drawing of the actual measurement and the prediction of water level and storage capacity curve are completed.

After the overflow of discharge chute, the curve of water level, storage capacity and discharge need be drawn. The abscissa of water level storage capacity change curve and discharge change curve is time (MM DD HH). The two curves are arranged up and down to align the abscissa. Before entering the flow breaking stage, the discharge chute size is known, and the discharge flow can be calculated by the formula (2) of broad-crested weir discharge:

$$Q = mB\sqrt{2g}(h - z)^{3/2} \tag{1}$$

In the formula, m is the parameter, and 0.4 is taken; B is the width, and the length of the middle line of the trapezoid section of the discharge chute is taken, changing with the water level; g is the acceleration of gravity, and 9.8 m/s^2 is taken; h is the water level of the upstream of the barrier lake, and z is the elevation of the bottom of the discharge chute. The change curve of water level, storage capacity and discharge flow of the barrier lake before entering the burst stage is shown in Fig. 4.

After entering the burst stage, the discharge flow is calculated by water balance. The average discharge from t_1 to t_2 can be obtained from the measured reservoir capacity $V(t_1)$ and the upstream inflow $q(t_1)$ corresponding to the upstream water level h_1 at t_1, and the measured reservoir capacity $V(t_2)$ and the upstream inflow $q(t_2)$ corresponding to the upstream water level $q(t_2)$ at t_2. The average discharge flow between t_1 and t_2 is:

$$Q = [q(t_1)+q(t_2)]/2 + [V(t_2) - V(t_1)]/(t_2 - t_1) \tag{2}$$

The average flow can be used as the discharge flow at time of $(t1 + t2)/2$. According to a series of data, discharge flow at a series of times can be got, which can be obtained

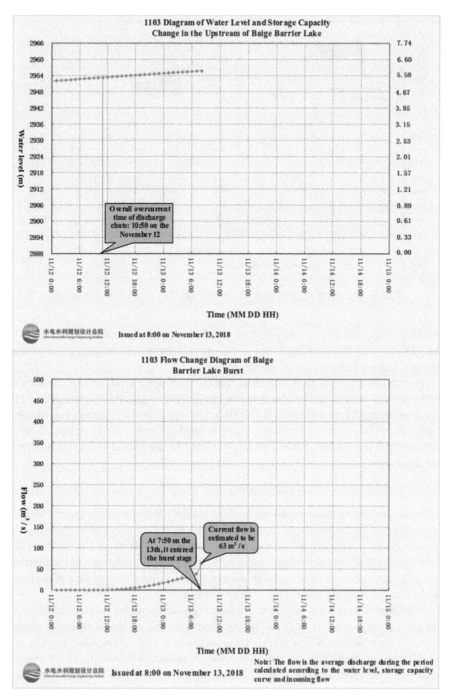

Fig. 4. Water Level Storage Capacity and Discharge Flow Change Curve of the Barrier Lake (Before Entering the Burst Stage)

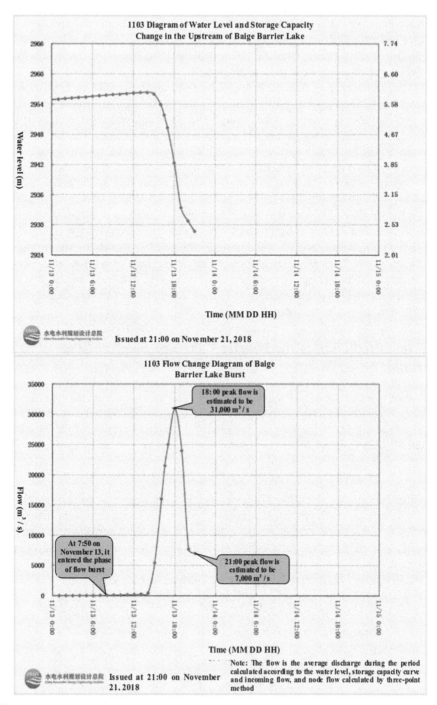

Fig. 5. Water Level- Storage Capacity and Discharge Curve of Barrier Lake (after Entering the Stage of Flow Burst)

by interpolation or extrapolation. The peak discharge needs to be obtained by shortening the time interval as much as possible. The change curve of water level, storage capacity and discharge of the barrier lake after entering the stage of flow burst is shown in Fig. 5.

There are two points need to be explained:

(1) In the water level prediction, the water level information at the time when the estimated water level reaches the weir crest, the current time and the past time was used to fit with the cubic function, and the estimated time was adjusted to make the prediction curve coincide with the actual measurement and trend line in the current time segment as much as possible and connect smoothly. This is to use the change trend from the past to the present to predict the future. The closer the current time is, the more accurate it is. So it is necessary to timely update the predicted curve based on the newly obtained data.

(2) Before entering into the burst stage, the discharge flow was small and the influence on the change of water level was too small. The water balance could not be used to calculate such a small discharge flow, while the shape of the discharge chute was known, and the discharge could be calculated more accurately with the relevant hydraulic formula; after entering the burst stage, the shape of the breach was unknown, and the discharge became large, which was the main factor of the change of water level, so water balance could be used is used to calculate the discharge flow.

4   Time Prediction of the Barrier Lake Entering the Stage of Flow Burst

According to the prediction of water level change and the formula of starting velocity of rock block, the time of Baige barrier lake entering the stage of flow burst can be inferred.

The Formula (3) of the average velocity of the starting section of the rock block caused by local scour is

$$V_c = 0.75\sqrt{(\rho_s/\rho - 1)gd}(D/d)^{1/6} \tag{3}$$

In the Formula, ρ_s, the density of rock block is 2,700 kg/m³; ρ, the density of water is 1000 kg/m³; g is the acceleration of gravity, and 9.8 m/s² is taken; d is the diameter of rock block, and 0.5 m is taken; D is the calculated water depth, the difference between the water level of the barrier lake and the elevation of the bottom of the discharge chute is taken, that is, $D = h - z$, in which z = 2,952 m.

When starting, the section flow meets the Formula that

$$Q = V_c B(h - z) \tag{4}$$

By simultaneous Formula (1), (3) and (4), the following can be got

$$h = z + \left[\frac{0.75}{m}\sqrt{\frac{\rho_s/\rho - 1}{2}}\right]^3 d \tag{5}$$

Substituting in the data, h = 2,954.6 m (starting velocity Vc = 2.85 m/s) can be got. Looking up the latest water level prediction curve, it is can be got that the corresponding time of the water level is about 8:00 am on November 13.

Next, the prediction accuracy of this method was analyzed. The coefficient of rock block diameter d in Formula (5) is calculated to be 5.17, that is, the selection deviation of rock block diameter is 0.1 m, the elevation deviation would be about 0.5 m, and the corresponding time on the prediction curve would be about three hours. It took 21 h for the barrier to pass through the chute completely and enter the flow burst stage. However, the selection of the rock block diameter was very subjective, and the difference of 0.2 m was normal, so the error of the prediction method was about 30%. The initial stage of flow burst developed slowly. There was no obvious time point at the time of entering the flow breaking stage. At 8:00 in the morning, the front personnel came to the scene to observe the flow breaking phenomenon after dawn, so the time of entering the flow burst stage was set as 8:00 in the morning.

5   Conclusion

This paper introduces the second dangerous situation and emergency response process of Baige barrier lake, in which the estimation and prediction method of barrier lake water level, discharge and burst time put forward by the author in the process of emergency response are given. It also analyzes the accuracy of the prediction method of the time that the barrier lake enters burst stage, and confirms the rationality of the successful prediction in the process of emergency response.

References

1. China Renewable Energy Engineering Institute: Summary Report on Emergency Response of Baige Barrier Lake on Jinsha River (2019). (in Chinese)
2. Lili, Z., Feng, W.: Hydraulics. Tsinghua University Press, Beijing (2015). (in Chinese)
3. Ning, M.: On the threshold velocity of slit-sand and gravel-stone. J. Yangtze River Sci. Res. Inst. **28**(1), 6–11 (2011). (in Chinese)

Dam Failure Process Modelling

Modelling of Overtopping and Erosion of Embankment Dams on a Drum Centrifuge

Wenjun Lu[1] and Limin Zhang[1,2(✉)]

[1] Department of Civil and Environmental Engineering, The Hong Kong University of Science and Technology, Hong Kong, China
{wenjunlu,cezhangl}@ust.hk
[2] HKUST Shenzhen Research Institute, Shenzhen, China

Abstract. A drum centrifuge accommodates a long soil model in its drum, hence models distributed geotechnical problems better than a beam centrifuge. A modular package has been designed for the HKUST 870 g-ton drum centrifuge to simulate overtopping and erosion of embankment dams. For the sake of space-saving and convenient operation, 1/6 of the drum model channel is designed as a fluid storage tank. The fluid storage tank is sealed by a top cover, and controlled by a fixed baffle and a movable baffle driven by a three-dimensional centrifuge actuator. After the drum centrifuge spins to the design speed, the movable baffle can be lifted by the actuator to release the fluid into the reservoir of the embankment. The rising rate of the water level can be controlled by adjusting the lifted height of the baffle. The embankment can then be overtopped. High-speed cameras and a PICV system will be employed to monitor the erosion or break process of the embankment. Pore pressure sensors are installed inside the embankment to measure the pore pressure response.

Keywords: Drum centrifuge · Embankment overtopping · Erosion · Flood · Geotechnical hazard

1 Introduction

In the past decades, numerous embankment failures have caused catastrophic loss of life and property in many countries (Liu et al. 2009; Chang and Zhang 2010). Statistics show that overtopping and internal erosion in the earthfill are two most common causes of embankment dam failure (Peng and Zhang 2012, Chanson 2015). However, the understanding of the mechanisms of overtopping failure and internal erosion of the embankment dams is still inadequate due to the lack of an effective research approach. The real-time in situ measurement of the embankment failure is costly and difficult to conduct. The laboratory flume tests are constrained to small scales and confining pressure, so that they cannot properly simulate either the real soil behavior or the flood-soil interaction (Zhu et al. 2006; Balmfroth et al. 2009; Cao et al. 2011; Do et al. 2016). In comparison, a centrifuge model is able to fulfill the identical stress field in a small-scale model, which has been used to study embankment problems (Cheng and Zhang 2011; Li et al. 2018). However, the traditional beam centrifuge modeling has two limitations

© Springer Nature Switzerland AG 2020
J.-M. Zhang et al. (Eds.): ICED 2020, SSGG, pp. 201–205, 2020.
https://doi.org/10.1007/978-3-030-46351-9_18

in simulating the overtopping and internal erosion of embankment dams: (1) limited space of the model container for accommodating gentle-slope embankment dams; (2) boundary reflection of the model container. In order to overcome the limitations of beam centrifuge tests, a drum centrifuge with a much longer model channel has become an alternative and a more suitable experimental approach (Springman et al. 2001; Eichhorn and Haigh 2019).

An 870 g-ton drum centrifuge with a three-dimensional (3D) robot will be installed in the Hong Kong University of Science and Technology (HKUST). A modular package has been designed on this drum centrifuge to simulate overtopping and erosion of embankment dams. This paper aims to introduce the principle and technical specifications of the HKUST drum centrifuge facility, and the model package for embankment dam simulation.

2   Principle of Drum Centrifuge Modeling

The principle of centrifuge modeling (including beam and drum centrifuges) is to use a spinning centrifuge machine to increase the gravitational acceleration of the soil model by the centrifugal force, so that the stress field in the model is identical to that in the prototype (Schofield 1980). The scaling laws for the centrifuge model test are shown in Table 1. Substituting the pore fluid with a high-viscosity fluid resolves the time conflict (Chu and Zhang 2010).

Table 1. Similarity laws for centrifuge model tests.

Centrifugal acceleration (G)	Length	Strain	Stress	Displacement	Time (dynamic)	Time (diffusion)
N	$1/N$	1	1	$1/N$	$1/N$	$1/N^2$

The difference between a beam centrifuge and a drum centrifuge lies in the model container, which is a box on the beam centrifuge, but a circular channel on the drum centrifuge. The much longer model channel enables the drum centrifuge to simulate distributed geotechnical problems, such as long-distance landslides, debris flows, wave-seabed interactions and embankment dam erosion (Fig. 1).

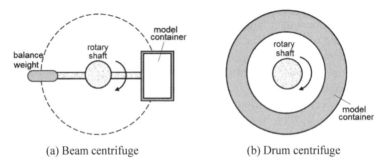

(a) Beam centrifuge   (b) Drum centrifuge

Fig. 1. Top view of beam and drum centrifuges.

3   HKUST Drum Centrifuge

An 870 g-ton drum centrifuge is being developed by Thomas Broadbent and Sons Limited and will be installed in the Hong Kong University of Science and Technology (Fig. 2(a and b)) in April 2020. The drum, with an inner diameter of 2.2 m, is fabricated with carbon manganese steel and a robust protective paint coating. The depth, width and length of the model channel are 0.4 m, 0.7 m and 6.9 m, respectively. The maximum rotation speed is around 450 r/min, which realizes a radial acceleration of 250 g on a test payload of 3480 kg. The HKUST drum centrifuge is the largest of its kind in the world. Besides, it is possible to add the capability of spinning both 30 g-ton beam and 2 g-ton permeameter test environments in the future, which will further augment the modelling power of this facility.

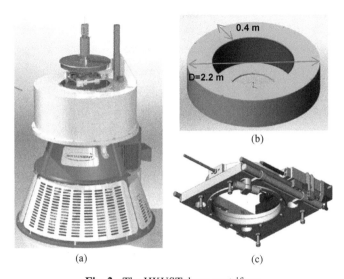

Fig. 2. The HKUST drum centrifuge.

As shown in Fig. 2(c), an actuator that can move in two directions in the plane is attached to the tool table on the HKUST drum centrifuge. The maximum moving velocity and the load capacity are 5 mm/s and ±5 kN respectively. In combination with the tool table rotary actuator integrated into the twin concentric shaft system on the basic centrifuge, an inflight 3D movement of the actuator in the model channel is achieved, which allows fully controllable 3D positioning and load application throughout the whole model in the drum channel. This 3D actuator system can be used as a sand rainer, a soil pourer, a model profiler and a loading device.

A compact high-speed image capture system using a high-speed camera and particle image velocimetry (PIV) technique, and an array of photoconductive sensors and Doppler velocimeters are employed to instrument the drum channel to enable capturing the debris flow, bed erosion and sediment transport processes in the channel. Two independent Ethernet-based 32-channel data acquisition systems, Drum DAS and Tool

Table DAS, are installed at the base platform. The Tool Table DAS utilizes a fiber optic rotary joint for the LAN interface, whereas the Drum DAS utilizes a wireless 150 MN/S interface for the LAS connection due to the presence of the brake and clutch assembly of the Twin Concentric Shaft Drive.

4   Embankment Dam Modular Package

In order to simulate overtopping and erosion of embankment dams, a fluid circulation system has been designed. For the sake of space-saving and convenient operation, 1/6 of the drum model channel is designed as a fluid storage tank. The fluid storage tank is sealed by a top cover, and controlled by a fixed baffle and a movable baffle driven by the 3D centrifuge actuator, as shown in Fig. 3.

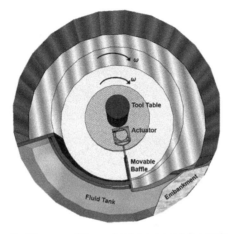

Fig. 3. Diagram of the embankment modular package.

Before the test, a feed pipe together and the 3D actuator system are used for inflight model installation. The drum channel and tool table spin together at a low speed, meanwhile, soil is poured into the channel at a designed location through the feed pipe. Then, the 3D actuator system is utilized for shaping the embankment slopes, levelling the dam rest, and profiling the pre-made breach on the embankment. After the embankment dam model is installed, the drum centrifuge is gradually accelerated to a design rotational speed and maintains this speed for a period of time to facilitate soil consolidation.

During the test, the movable baffle is uplifted by the actuator to let the fluid to flow into the reservoir behind the embankment. The rising rate of the water level can be controlled by adjusting the lifted height of the baffle. The baffle keeps moving upward until the water overtops the embankment. Photoconductive sensors and Doppler velocimetry sensors are used to monitor the flow velocity. High-speed cameras and the PICV system are employed to monitor the erosion or break process of the embankment. Pore pressure sensors inside the embankment are used to measure the change of pore pressure inside the embankment.

5 Summary

This paper provides a brief introduction to the 870 g-ton drum centrifuge in the Hong Kong University of Science and Technology. A modular package to simulate overtopping and erosion of embankment dams is presented in detail.

Acknowledgement. The drum centrifuge facility is supported by the Research Grants Council of the Hong Kong SAR (Grant No. C6021-17EF) and HKUST.

References

Balmfroth, N.J., von Hardenberg, J., Zammett, R.: Dam-breaking seiches. J. Fluid Mech. **628**, 1–21 (2009)

Chang, D.S., Zhang, L.M.: Simulation of the erosion process of landslide dams due to overtopping considering variations in soil erodibility along depth. Nat. Hazards Earth Syst. Sci. **10**(4), 933–946 (2010)

Cao, Z., Yue, Z., Pender, G.: Landslide dam failure and flood hydraulics. Part I: experimental investigation. Nat. Hazards **59**, 1003–1019 (2011)

Cheng, S., Zhang, J.M.: Dynamic centrifuge model test on concrete-faced rockfill dam. J. Earthq. Eng. Eng. Vib. **31**, 98–102 (2011)

Chanson, H.: Embankment overtopping protection systems. Acta Geotechnica**10**, 305–318 (2015)

Do, X.K., Kim, M., Thao Nguyen, H.P., Jung, K.: Analysis of landslide dam failure caused by overtopping. Procedia Eng. **154**, 990–994 (2016)

Eichhorn, G.N., Haigh, S.K.: Imaging of landslide-pipeline interaction in a geotechnical drum centrifuge. In: 2nd International Conference on Natural Hazards & Infrastructure, Chania, Greece, 23–26 June 2019

Li, J., Zhang, J., Xu, J., Wang, F., Wang, B., Li, Q.: Dynamic behavior of polymer antiseepage wall for earth dam by centrifuge test. Int. J. Geomech. **18**(12), 1–17 (2018)

Liu, N., Zhang, J.X., Lin, W., Cheng, W.Y., Chen, Z.Y.: Draining Tangjiashan barrier lake after Wenchuan earthquake and the flood propagation after the dam break. Sci. China Ser. E Technol. Sci. **52**(4), 801–809 (2009)

Peng, M., Zhang, L.M.: Breaching parameters of landslide dams. Landslides **9**(1), 13–31 (2012)

Schofield, A.N.: Cambridge geotechnical centrifuge operation. Geotechnique **20**(3), 227–268 (1980)

Springman, S.M., Laue, J., Boyle, R., White, J., Zweidler, A.: The ETH Zurich geotechnical drum centrifuge. Int. J. Phys. Model. Geotech. **1**, 59–70 (2001)

Zhu, Y.H., Visser, P.J., Vrijling, J.K.: Laboratory observations of embankment breaching. In: Proceeding of the 7th ICHE Congress, Philadelphia, US, September 2006

Displacement Prediction of Concrete Dams: Combining Finite Element Method and Data-Driven Method

Zhenzhu Meng[1], Xueyou Li[1(✉)], Chenfei Shao[2], and Chongshi Gu[2]

[1] School of Civil Engineering, Sun Yat-Sen University, Guangzhou 510275,
People's Republic of China
lixueyou@mail.sysu.edu.cn

[2] State Key Laboratory of Hydrology, Water Resources and Hydraulic Engineering,
Hohai University, Nanjing 210098, China

Abstract. In this study, we gave insights to the displacement prediction of concrete dams, in specific, we integrated a finite element method with a data-driven method named Random coefficient model. Using the finite element method, the coefficients of explanatory variables in the Random coefficient model were of each monitoring point were constrained, in order to model the displacement of these monitoring points simultaneously, and to take the temporal and spatial correlations among monitoring points into account. The proposed model was validated by a case study of the concrete dam at Jinping-I hydropower station. Results indicated that the proposed model outperformed both the classical statistical model and the Random coefficient model without finite element method.

Keywords: Dam displacement · Finite element method · Data-driven approach · Random coefficient model

1 Introduction

Two approaches have been commonly used to predict the displacements of a dam: one is numerical simulation using finite element methods, and the other is statistical model [1]. For statistical models, the modeling methods have been improved from simple linear regression methods to advanced machine learning algorithms. In addition, recent years, many synthetic statistical models which integrate several machine learning methods have been developed [2]. One issue should be noted is that finite element methods can simulate the displacement at an arbitrarily point of the dam, whereas statistical models can merely analyze the displacement data at each monitoring point. Moreover, the data recorded at each monitoring point was often modeled independently without considering the spatial and temporal correlations between each monitoring point. In order to take the structural correlations between each monitoring point into account, in advance of the statistical modelling, recent researches had classified the monitoring points into several groups using clustering methods. For example, Shao et al. [3] clustered the monitoring points based on their spatial characteristics. Hu et al. [4] provided a clustering model which considers the temporal-spatial characteristics. Although the clustering methods

© Springer Nature Switzerland AG 2020
J.-M. Zhang et al. (Eds.): ICED 2020, SSGG, pp. 206–211, 2020.
https://doi.org/10.1007/978-3-030-46351-9_19

can reflect the structural correlations among monitoring points to a certain extent, their constraint ability is still not satisfactory, as they provide constraint for each group rather than for each monitoring point.

In this study, we took the advantage of finite element method that it can simulate the displacement at arbitrary points on the dam, to constrain the spatial and temporal correlations between each monitoring point. Specifically, we investigated the effects of the crucial factors (i.e., water pressure and temperature component) on the dam displacements, by varying them gradually and conducting simulation using the finite element model. As for the statistical model, the random coefficient model was selected, which has a benefit in solving multi-colinearity problem by making the coefficients of each explanatory variable follow one or several Gaussian distributions.

2 Model Development

The displacement of the concrete dam δ consists of three components: water pressure component δ_H, temperature component δ_T and aging component δ_θ:

$$\delta = \delta_H + \delta_\theta + \delta_T \tag{1}$$

The water pressure component δ_H denotes displacement caused by itself δ_{1H}, dam foundation δ_{2H} and rotation of the dam bedrock δ_{3H}. Here, δ_H can be simplified as $\delta_H = \sum_{i=1}^{n} a_i H^i$ ($n = 4$ for arch dams, further information see [4]), where H is the upstream water level, a_i are coefficients related to the dam height h, the downstream slope angle m, the distance from the monitoring point to the dam foundation d. Here, a_i mainly depends on the dam's material properties including elastic modulus of the dam body E_c and elastic modulus of the foundation E_r. Assuming that E_c and E_r have a linear relation: $E_r = \lambda E_c$, δ_H becomes:

$$\delta_H = \frac{1}{E_c} \sum_{i=1}^{n} a_i^0 H^i \tag{2}$$

where a_i^0 is a coefficient determined by h, m, d and λ. With a given designed elastic modulus of the dam body E_c^0 and upstream water level H, we can obtain a numerical solution of water pressure component δ_H^0 using the finite element method. Then, a series of coefficients a_i^1 can be calculated from Eq. (3), by fitting the upstream water level H and numerical solution δ_H^0.

$$\delta_H^0 = \frac{1}{E_c^0} \sum_{i=1}^{n} a_i^1 H^i \tag{3}$$

Substituting a_i^0 by a_i^1 in Eq. (2), δ_H can be expressed as:

$$\delta_H = \frac{E_c^0}{E_c} \frac{1}{E_c^0} \sum_{i=1}^{n} a_i^1 H^i = \frac{E_c^0}{E_c} \delta_H^0 = \xi \delta_H^0 \tag{4}$$

where ξ is the ratio between E_c^0 and E_c.

We assume that the displacement linearly depends on the temperature load $dL = \alpha L dT$, where α is the coefficient of linear expansion, L is the length and T is the temperature. The coefficient of linear expansion α is the only random coefficient in the equation. δ_T can be expressed as:

$$\delta_T = \alpha_c \sum_{i=1}^{n} b_i^0 T^i \tag{5}$$

where b_i^0 represents the coefficients of temperature component, α_c is the actual coefficient of linear expansion of the dam body.

δ_T then can be expressed as:

$$\delta_T = \frac{\alpha_c}{\alpha_c^0} \frac{\alpha_c^0}{1} \sum_{i=1}^{n} b_i^1 T^i = \frac{\alpha_c}{\alpha_c^0} \delta_T^0 = \psi \delta_T^0 \tag{6}$$

where ψ is the ratio from α_c to α_c^0.

δ_θ can be expressed by an empirical formula:

$$\delta_\theta = \sum_{l=0}^{L} \sum_{m=0}^{M} (d_{1lm}\theta + d_{2lm} \ln \theta) x^l z^m \tag{7}$$

where x and z are horizontal and vertical coordinates of the monitoring point, respectively, θ is time, d_{1lm} and d_{2lm} are unknown coefficients.

The radial displacement fitted by the finite element method becomes:

$$\begin{aligned} \delta_i &= \delta_{iH} + \delta_{iT} + \delta_{i\theta} \\ &= \xi_i \delta_H^0 + \psi_i \delta_T^0 + \sum_{l=0}^{L} \sum_{m=0}^{M} (d_{1lm}\theta + d_{2lm} \ln \theta) x_i^l z_i^m + u \end{aligned} \tag{8}$$

where i is the index of monitoring point, u is the random error.

The coefficients of explanatory variables can be solved using a random coefficient model. The monitoring displacement data is a two dimensional panel data contains time series and cross-section panel. Data on one panel represents the displacement data at an indicated time. The regression coefficients of one panel can be expressed by:

$$y_{it} = \sum_{k=1}^{K} \beta_{ki} x_{kit} + u_{it} = \sum_{k=1}^{K} (\beta_k + \gamma_{ki}) x_{kit} + u \tag{9}$$

where y_{it} is the two-dimensional dam displacement data; x_{kit} is the two-dimensional data of explanatory variables; t is the time index; i is the index of section; k is the index of the explanatory variables; the coefficient β_{ki} is independent with time, which can be decomposed into β_k and γ_{ki}; $\beta = (\beta_1, \ldots, \beta_K)'$ is the common mean coefficient vector, $\gamma = (\gamma_{1i}, \ldots, \gamma_{Ki})'$ is the degree of deviation from each individual to the common mean value. The objective is to obtain β_{ki} from the two dimensional data series y_{it} and x_{kit}. Details can be found in [4].

3 Case Study

We selected the radial displacement data of the concrete dam at Jinping- I Hydropower Station as an example. A dataset of 24 monitoring points recorded from June 16, 2013 to August 25, 2015 were selected (see Fig. 1). Each monitoring point has 274 validated data samples.

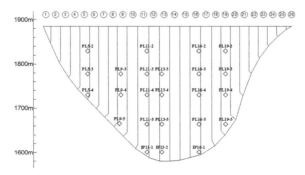

Fig. 1. Distribution of dam sections and the monitoring points from the upstream side view.

We first developed a three-dimensional finite element model based on the actual geological characteristics of the dam including dam body, dam foundation and surrounding mountains. The model of dam body was discretized into 38537 elements and 31941 nodes. The surrounding mountains were set to 2–3 times higher than the dam body. The crest of the dam and the dam body were free to move. The water load was applied to the upstream surface of the dam and the bottom of reservoir. The elastic modulus of dam body and dam foundation were initially set to 30 GPa and 25 GPa, respectively. We first neglected the gravity and temperature load, and varied the water level from 1700 m to 1880 m, and then simulated the associate dam displacement (see Fig. 2).

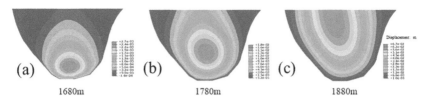

Fig. 2. Radial displacement simulated by finite element method with varying upstream water levels: (a) 1700 m; (b) 1780 m and (c) 1880 m.

Afterwards, we calibrated the coordinates of the monitoring points in real world with the coordinates in the finite element model. Taking the monitoring points at Sect. 5# as an example, Fig. 3(a) and (b) display the displacement versus the upstream water level and temperature, respectively.

The coefficients of the explanatory variables in each component can be fitted with the random coefficient model. The dataset was divided into two groups, 80% of the whole

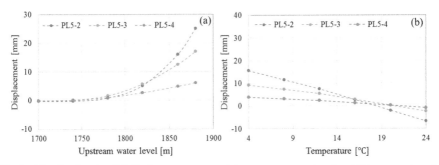

Fig. 3. The simulated radial displacement versus (a) upstream water level and (b) temperature for monitoring points located at vertical line 5#.

data were selected as fitting data, and the other 20% were selected to validate the model. The fitting and predicting results obtained from the proposed synthetic model are shown in Fig. 4. Both the fitting results (left side of the blue line) and validating results (right side of the blue line) fit well with the monitoring data.

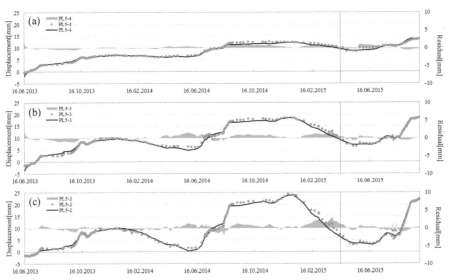

Fig. 4. Modeling results for monitoring points located at vertical line 5#: (a) PL5-4, (b) PL5-3 and (c) PL5-2. The red dots represent the monitoring data and the black line represents the fitting results. The green area denotes the residual.

To evaluate the accuracy of the proposed model, we calculated the coefficient of determination R for data of all the monitoring points, and compared R of the proposed model (FEM-RCM) with the statistical model regressed using ordinary least squares regression (OLS), and with the synthesis model of clustering model and random coefficient model (Clustering-RCM). As shown in Fig. 5, the proposed model overperforms

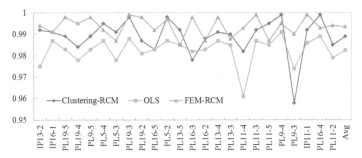

Fig. 5. Comparing R of the FEM-RCM model with Clustering-RCM model and statistical model using OLS regression.

the Clustering-RCM model, and the Clustering-RCM model overperforms the statistical model using OLS regression.

4 Conclusion

In this work, we presented a synthetic model which integrates the finite element method into a statistical model (i.e., random coefficient model), aiming to constrain the structural correlations in statistical model by the temporal and spatial characteristics obtained from finite element model. With this objective in mind, we first used the finite element method to simulate the displacement at each monitoring point, and determined the coefficients of explanatory variables of the water pressure and temperature components of displacement. Then, we implanted the coefficients obtained with the finite element method into the random coefficient model. Finally, we validated the proposed model with a case study. We found that the proposed model overperformed a combined model of random coefficient model and Gaussian mixture clustering model, owing to the benefit that finite element model can provide constraint for each monitoring point, whereas the clustering model merely provides constraint for each class.

References

1. Léger, P., Côté, P., Tinawi, R.: Finite element analysis of concrete swelling due to alkali-aggregate reactions in dams. Comput. Struct. **60**(4), 601–611 (1996)
2. Shao, C., Gu, C., Meng, Z., Hu, Y.: A data-driven approach based on multivariate copulas for quantitative risk assessment of concrete dam. J. Mar. Sci. Eng. **7**(10), 353 (2019)
3. Shao, C., Gu, C., Yang, M., Xu, Y., Su, H.: A novel model of dam displacement based on panel data. Struct. Control Heal. Monit. **25**(1), e2037 (2018)
4. Hu, Y., Shao, C., Gu, C., Meng, Z.: Concrete dam displacement prediction based on an ISODATA-GMM clustering and random coefficient model. Water **11**(4), 714 (2019)

Centrifuge Model Tests on Ecologically Reinforced Soil Slopes Under Vertical Loading

Minhao Xiao, Yiying Zhao, and Ga. Zhang$^{(\boxtimes)}$

Tsinghua University, Beijing, People's Republic of China
zhangga@tsinghua.edu.cn

Abstract. The mechanism of plant roots for slope stabilization is an important issue for the landslide prevention and control. The study of deformation and failure behavior of ecologically reinforced slopes under vertical loading is of great significance to a reasonable slope stability evaluation method. In this paper, centrifugal model tests were conducted to study the deformation and failure behavior of ecologically strengthened soil slopes and pure soil slopes under vertical loading conditions. A coupling effect between deformation localization and local failure was found in the slope failure process. The occurrence and development of deformation localization caused the slope failure. Compared with the pure soil slope, ecological soil slopes could bear higher loads and exhibit smaller deformation. Plant roots could increase stability of the soil slope by reducing the degree of deformation localization in the slope, slowing down the formation process of the slip surface and finally delaying the slope failure.

Keywords: Plant roots · Top loading · Failure characteristics · Reinforcement mechanism

1 Introduction

In the twenty-first century, a large number of water conservancy projects have been built in China, and the protection of reservoir slopes has always been an important issue in the engineering field. In order to reduce landslide disasters, many slope protection projects have been built in China in recent years, and at the same time, local environment has been damaged to some extent. Ecological slope protection uses plant roots as soil anchor to improve the shear resistance of soil, finally achieving the effect of slope consolidation. It also has the function of water conservation, soil erosion reduction and ecological environment improvement, which traditional projects do not have. To maximize the reinforcement effect and minimize the financial outlay of the ecological slope protection, many researchers used the finite element method, physical slope model, field sampling experiment to study the mechanical and hydrological properties of plant roots and slope stability from the aspect of root geometry [1], root architecture [2], root-soil composites [3, 4] and roots with water flow [5]. Due to the complex distribution of plant roots, the centrifuge model tests have an advantage.

© Springer Nature Switzerland AG 2020
J.-M. Zhang et al. (Eds.): ICED 2020, SSGG, pp. 212–220, 2020.
https://doi.org/10.1007/978-3-030-46351-9_20

2 Description of Model Tests

2.1 Devices

The device for the centrifuge model test was the 50 g-t geotechnical centrifuge at Tsinghua University, which could be accelerated to a maximum of 250 g. The slope model container was 500 mm long, 200 mm wide and 350 mm high. The container was made of aluminum alloy with negligible deformation during the test. For clear observation, one side of the metal model container was replaced by a 4-cm-thick organic glass, which was bonded by the silica gel. Compared with centrifugal force, the friction between the slope model and the organic glass was negligible.

A loading device with a maximum load of 10 kN was installed to provide vertical loading through the loading plate (Fig. 1). The loading plate was 200 mm long and 50 mm wide, keeping a distance from the top of the model before tests. The load was uniform in width because of the same wide size of the loading plate and the model container. The loading device was controlled by electric signals. Step load was conducted during the tests to make deformation caused by the load keep stable. The loading plate was rigidly connected to a horizontal device, which provided restrictions on the possible tilting.

2.2 Schemes

Two centrifuge model tests were conducted with the application of vertical load on the top of the slopes. One soil slope was planted with grass as ecological reinforcement, the other was not treated. The physical parameters of the ecologically reinforced soil slope and the pure soil slope were controlled to the same extent, which made plant roots the only variable.

2.3 Model Preparation

Figures 1 and 2 showed the schematic and the photographic views of the model slope. The slope was 300 mm in height with a gradient of 1.5:1. An additional horizontal ground soil layer was set at the bottom to eliminate the influence of the model container, which was 20 mm in depth.

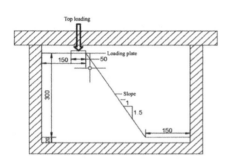

Fig. 1. Schematic view of the model (unit: mm).

Fig. 2. Photographic view of the model.

A type of silty clay with a specific gravity of 2.7 was used in the tests as the model soil. The density and the water content of the soil were controlled around 1.50 g/cm^3 and 18%. Watering was ceased one day before the model test for the ecological soil slope. After the tests, the soil moisture content near the slip surface was 16.9% and 17.1% for the pure soil slope and ecological soil slope, separately. Thus, the water content effect on slope stability could be left out for simplification. The model soil was compacted into the container in layers, each layer 50–60 mm thick. The slope for the test was obtained from the compacted cuboids by oblique cutting and redundant soil removing.

Ryegrass was selected as the experimental plants. Because the root system of ryegrass is well developed and often penetrates deep into the soil. It should be noted that ryegrass grows fast and survives hard conditions, which provides convenience for the experiment.

The seeds of ryegrass were sowed in pits with 5 mm in diameter, which were dug by nails after facture of model slopes. Watering once or twice a day was conducted during the next two weeks. Before the experiment, grass body above slopes was cut off to prevent its dragging effect under centrifugal force. The taproot of the most experiment ryegrass was 1–3 mm long and 0.2–0.3 mm in diameter (Fig. 3).

Fig. 3. Close view of grass roots after the tests.

2.4   Test Process and Measurements

During the test, the model was accelerated gradually to the 50 g-level, 1–2 min keeping for each 5 g to make deformation stable. The loading plate was fixed rigidly on the loading device and did not contact the top of the slope before loading. When the centrifugal acceleration reached the 50 g level, the settlement of the slope top gradually came to stability and then vertical load was steppedly applied. Real-time loading data was monitored by a load transducer to control the loading process, which was installed at the top of the slope.

An image-recording and displacement measurement system was used to record the images of the slope during the tests [6]. A camera was installed on the centrifuge to record pictures and videos of the soil slope through the organic glass. A series of images were obtained by dividing videos. After choosing the measurement points and an image series, an image-correlation analysis algorithm was used to measure the displacement of specific points. The system located measurement points by difference of grey scale, so as to track and obtain the displacement. A number of white granite particles were placed on one side of the slope to create a colorful region with a significant grey scale difference, which could satisfy the measurement requirements and make it easy for observation.

3 Deformation and Failure Behavior of Slopes

3.1 Ultimate Bearing Capacity

After the deformation of soil slopes caused by centrifugal force at 50 g stabilizing, the vertical load was applied on the top of the slope. The load-settlement relationship was shown in Fig. 4. It could be seen that the settlement at the slope top continued to increase with increasing load. The ultimate bearing capacity of the ecological soil slope was around 170 kPa. For the pure soil slope, the ultimate bearing capacity was about 75 kPa. This demonstrated that the plant roots had a significant effect on slope stability.

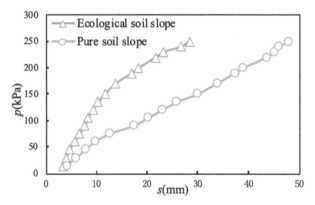

Fig. 4. Load-settlement relationship during the tests.

3.2 Failure Morphology

Through the analysis of a series of images, the slip surface of the slopes under loading conditions could be observed and marked using the black lines in the photographs (Figs. 5 and 6). The slip surface of ecologically reinforced soil slope on the top of slope was deeper than that of the pure soil slope.

3.3 Deformation Behavior

Displacement of some typical points were measured to analyze the deformation and failure characteristics of the soil slope. These points, A, B and C, were in the same location for the reinforced and unreinforced slopes, under the top loading plate, near the slip surface and on the base body, respectively (Figs. 5 and 6).

Fig. 5. Slip surface of the ecologically rein forced soil slope.

Fig. 6. Slip surface of the pure soil slope.

The measurement results showed that the horizontal displacement of the three points exhibited significant difference (Fig. 7). As the load increased, the horizontal displacement of point A and B had a remarkable increase, while the horizontal displacement of point C was relatively small. Further comparison indicated that the displacement of ecologically reinforced slope was significantly smaller than the pure soil slope, demonstrating that plant roots could reduce the deformation of the soil slope.

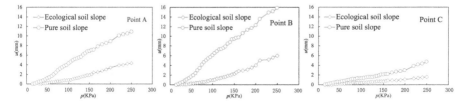

Fig. 7. Horizontal displacement of three typical points.

Three groups of contour points at different elevations were selected to measure the horizontal displacement at several typical moments, and the horizontal distribution curve was obtained as shown in Fig. 8. It could be seen that for the pure soil slope, the displacement of the soil body in the base part was relatively small, and the deformation and failure mainly occurred on the sliding body and the soil body under the load. This showed that the base part of the soil slope could be approximately regarded as a rigid body, which was insignificantly affected by the load.

The same analysis was conducted on the ecologically reinforced soil slope (Fig. 9). It was found that the horizontal displacements of points from the base part and from the sliding part were also quite different, but the horizontal displacement did not increase strictly from the interior to the surface of the slope. It tended to decrease near the surface of the slope. This showed that the degree of sliding deformation was changed by the plant roots. It had been found that the plant roots in the soil not only reinforced soil as soft ribs, but also loosed soil and changed soil water content through transpiration [7]. In the test, the plant roots decreased slope deformation, thus significantly showed its shear resistant ability. But at the same time, it could be found that structure of soil near the slope surface was changed and some complex deformation was produced. Under this experimental parameters, the plant roots turned out the reinforcement effect.

Comparing Figs. 8 and 9, it could be found that the horizontal displacement of the ecologically reinforced soil slope was smaller than that of the pure soil slope at the same location. This indicated that the deformation trend of the ecological soil slope was less intense.

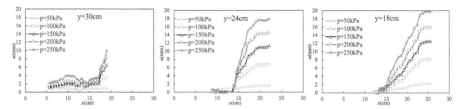

Fig. 8. Horizontal distributions of horizontal displacement for the pure soil slope.

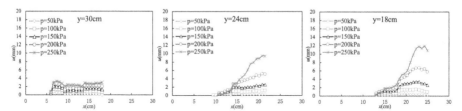

Fig. 9. Horizontal distributions of horizontal displacement for the ecologically reinforced soil slope.

Figure 10 showed the displacement vector diagram of some points on the ecologically reinforced soil slope. Analyze the horizontal distribution of displacement for five rows of typical points on the ecological soil slope. It could be found that the largest displacement appeared at points near but not closest to the slope surface for the upper three rows of points, where the grass was planted, while for the two rows of points below, the largest displacement appeared at points closest to the slope surface. This demonstrated that there was a tendency of adhesion in the soil under the surface of slope. The trend of adhesion was about 4–5 cm and the end of plant roots was generally about 5–6 cm from slope surface. This demonstrated that the plant roots had shear resistant ability and adhesion characteristics in the slope.

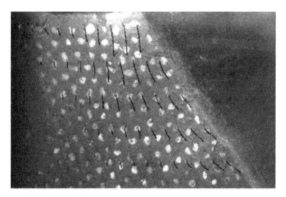

Fig. 10. Displacement vector graph of the ecologically reinforced soil slope.

3.4   Failure and Reinforcement Mechanism

The failure mechanism of soil slopes was discussed based on the measured displacement and integrated analysis of deformation and failure process [8].

The study of failure mechanism showed that the development and localization of deformation was the reason for the occurrence of sliding surface and sliding body [9]. The horizontal distributions of the horizontal strain at the middle height for three different moments were shown in Figs. 11 and 12. For both tests, the horizontal strain distribution was uniform and small at the beginning of the loading. With increasing load, the horizontal strain increased and an evident peak appeared and grew in two certain areas. This demonstrated that the deformation localization occurred during loading and became severe with the rise of load. After the slip surfaces appeared, the deformation localization continued to grow. Thus, it could be concluded that there was a coupling effect between deformation localization and local failure. The development of the deformation localization caused local failure and the local failure promoted new deformation localization. In both tests, the deformation localization caused two slip surfaces. One developed to the slope surface, the other developed to the base part of soil slope (Figs. 5 and 6).

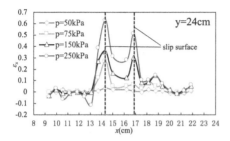

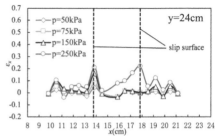

Fig. 11. Horizontal distributions of horizontal strain for the pure soil slope.

Fig. 12. Horizontal distributions of horizontal strain for the ecologically reinforced soil slope.

Analyze the occurrence time of the two slip surfaces by ultimate bearing capacities (the ecological slope: 170 kPa, the pure soil slope 85 kPa). It could be found that for the pure soil slope, the deformation localization developed almost synchronously, both behaved significant change between the load of 50 kPa and 150 kPa. For the ecologically reinforced soil slope, two slip surface appeared under more different load (Fig. 12). The one developing to the base part of the soil slope appeared severe deformation between the load of 50 kPa and 150 kPa, while the one developing to the slope surface appeared severe deformation between the load of 150 kPa and 250 kPa. Thus, it could be concluded that plant roots could reinforce the soil slope by reducing the degree of deformation localization, restraining and delaying formation of the slip surface, and finally delay the slope failure. This was the plant roots reinforcement mechanism.

4   Conclusion

Comparative centrifuge model tests were conducted to study the deformation and failure characteristics of the slope with plant roots. Based on the observations and measurement results, the main conclusions could be drawn as follows:

(1) The plant roots could increase stability level and reduce deformation of the soil slope.
(2) Plant roots could change the original soil structure. Moderate amount of plants could restrain sliding deformation and show shear resistant ability. The slip surface curve for the ecological soil slope was less smooth and behaved a more changeable curvature.
(3) There was a coupling effect between deformation localization and local failure in the slope failure process for the ecological reinforced and unreinforced slopes. The slope failure mechanism could be attributed to the development of deformation localization.
(4) Plant roots could reduce the degree of the deformation localization in the slope, and thus delayed the slope failure.

Acknowledgements. The study is supported by National Key R&D Program of China (2018YFC1508503) and Tsinghua University Initiative Scientific Research Program.

References

1. Ng, C.W.W., Kamchoom, V., Leung, A.K.: Centrifuge modelling of the effects of root geometry on transpiration-induced suction and stability of vegetated slopes. Landslides, 1–14 (2015)
2. Li, Y., Wang, Y., Ma, C., Zhang, H., Wang, Y., Song, S., et al.: Influence of the spatial layout of plant roots on slope stability. Ecol. Eng. **91**, 477–486 (2016)
3. Wang, X., Hong, M.M., Huang, Z., Zhao, Y.F., Ou, Y.S., Jia, H.X., et al.: Biomechanical properties of plant root systems and their ability to stabilize slopes in geohazard-prone regions. Soil Tillage Res. **189**, 148–157 (2019)
4. Zhang, L., Yue, S.Y., Xue, J.: Study and analysis on the protection mechanism of river vegetation slope protection. Henan Water Cons. South-to-North Water Diversion **2017**(1), 84–84 (2017)

5. Liu, F.Y., Liu, L.S., Wang, J.S., Liu, L.: Water flow test and application of different vegetation ecological slope protection in Jingjiang River. Port Waterway Eng. **546**(09), 18–23 + 46 (2018)
6. Zhang, G., Hu, Y., Zhang, J.M.: New image analysis-based displacement-measurement system for geotechnical centrifuge modeling tests. Measurement **42**(1), 87–96 (2009)
7. Wu, H.W.: Atmosphere-plant-soil interactions: theories and mechanisms. Chin. J. Geotech. Eng. **39**(1), 1–47 (2017)
8. Zhang, G., Hu, Y., Wang, L.: Behaviour and mechanism of failure process of soil slopes. Environ. Earth Sci. **73**(4), 1–13 (2015)
9. Zhang, J.H., Zhang, Y., Pu, J.L., Wei, Z.X., Gao, Y.L.: Centrifuge modeling of soil nail reinforcements. Chin. Civil Eng. J. **42**(1), 76–80 (2009)

Numerical Simulation of Landslide Dam Formation and Failure Process

Shengyang Zhou[1] and Limin Zhang[1,2,3]

[1] Department of Civil and Environmental Engineering, The Hong Kong University of Science and Technology, Hong Kong, China
szhouaq@connect.ust.hk, cezhangl@ust.hk
[2] State Key Laboratory of Hydraulics and Mountain River Engineering, College of Water Resource and Hydropower, Sichuan University, Chengdu, China
[3] HKUST Shenzhen Research Institute, Shenzhen, China

Abstract. Landslide dams are typically formed by loose landslide deposits which can be vulnerable to overtopping failure. Subsequent flash flood can pose great threat to downstream area. Hong Kong is subject to subtropical weather and the terrain is hilly. Due to the changing climate, extreme rainfall is becoming more and more frequent which can cause a large number of landslides. Therefore, it is essential to evaluate possible landslide dam formation and failure process, especially on Hong Kong Island which is densely populated. In this paper, a physically-based distributed cell model, EDDA 2.0, is adopted which is capable to predict rainfall-induced landslides, debris flows, erosion and deposition process, and flash floods. Possible landslide dam formation and failure are predicted under an extreme rainfall condition (65% of the 24-h PMP, Probable Maximum Precipitation). Although the magnitude of the landslide and the landslide dam is not very large, the subsequent flash flood can pose much greater threat to people's lives and properties especially in urban area such as Hong Kong Island where the population density is rather high.

Keywords: Landslide dam · Dam break · Flash flood · Rainstorm

1 Introduction

Due to the subtropical weather and the hilly terrain, Landslides and debris flows are common natural hazards in Hong Kong. Hong Kong is experiencing extreme rainfall more and more frequently under the changing climate. If a large-scale landslide happens, loose deposits can be deposited in the channel. With large amount of surface water entering the channel, a landslide dam can be formed. If the dam breaks, downstream area can be seriously flooded, posing great threat to people's lives and properties. This can be especially serious when happening on Hong Kong Island, where the terrain is rather hilly, and the population density is high.

There have been numerous studies on landslide dam break simulation. For example, Valiani et al. simulated Malpasset Dam break with a two-dimensional finite volume method [1]; Fan et al. simulated dam-breach flood of the Tangjiashan landslide dam

© Springer Nature Switzerland AG 2020
J.-M. Zhang et al. (Eds.): ICED 2020, SSGG, pp. 221–225, 2020.
https://doi.org/10.1007/978-3-030-46351-9_21

with two models: the BREACH model and the SOBEK 1-D-2-D model [2]; Chang and Zhang simulated the erosion process of landslide dam due to overtopping with a physically-based breach model [3]. Several physically-based breach model have been developed to simulate dam break, e.g., DAMBRK[4], BREACH [5], BEED [6], SIMBA [7], and etc. However, these models can only simulate breaching process.

This paper aims to simulate the full process of landslide dam formation, dam failure, and subsequent flooding conditions with a physically-based model (EDDA 2.0). Infiltration analysis is conducted first to predict likely landslides. Then landslide movement and deposition is calculated. After that, water accumulation and erosion of landslide dam is simulated. Finally, the erosion process and flood routing analysis is carried out.

2 Methodology

In this study, a physically-based distributed cell model, EDDA 2.0, proposed by Shen et al. (2017) is adopted to predict likely rainfall-induced slope failures, landslide movements, landslide dam formation, dam break and subsequent flash floods over a small catchment located at Central area on Hong Kong Island (see Fig. 1).

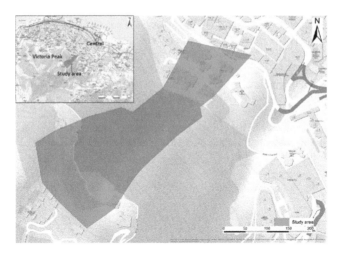

Fig. 1. Study area

2.1 Prediction of Landslide

To predict rainfall-induced landslides, infiltration process has to be analyzed first. This paper adopts the Richards equation to describe the infiltration process:

$$\frac{\partial}{\partial z^*}\left(k\frac{\partial \Psi}{\partial z^*}\right) + \frac{\partial k}{\partial z^*}cos\beta = \frac{\partial \theta}{\partial t} \tag{1}$$

where k is the coefficient of permeability, Ψ is pore-pressure head, z^* is the layer thickness; β is the slope angle; θ is the volumetric water content; and t is time.

In this model, landslide is treated as a flow material, and the dynamics of the flow material can be described by depth-averaged mass conservation equations and momentum conservation equations:

$$\frac{\partial h}{\partial t} + \frac{\partial (hv_x)}{\partial x} + \frac{\partial (hv_y)}{\partial y} = i[C_{v*} + (1 - C_{v*})s_b] + A[C_{vA} + (1 - C_{vA})s_A] \qquad (2)$$

$$\frac{\partial (C_v h)}{\partial t} + \frac{\partial (C_v hv_x)}{\partial x} + \frac{\partial (C_v hv_y)}{\partial y} = iC_{v*} + AC_{vA} \qquad (3)$$

$$\frac{\partial v_x}{\partial t} + v_x \frac{\partial v_x}{\partial x} = g\left[-\text{sgn}(x)S_{fx} - \frac{\partial (z_b + h)}{\partial x}\right] - \frac{v_x\{i[C_{v*} + (1 - C_{v*})s_b] + A[C_{vA} + (1 - C_{vA})s_A]\}}{h} \qquad (4)$$

where h is the flow depth; t is time; v_x and v_y are the average flow velocities in the x and y directions, respectively; i is the rate of erosion or deposition, which can be expressed as $i = -\partial z_b/\partial t$; A is the entrainment rate; C_{v*} is the volume fraction of solids in erodible bed; C_{vA} is the volume fraction of solids in the entrained materials; s_b and s_A are the degrees of saturation in the erodible bed and the entrained materials, respectively; C_v is the volumetric sediment concentration of the mixture; g is the gravitational acceleration; S_{fx} is the flow resistance slopes in the x direction; z_b is the bed elevation; and the sgn function is used to make sure the direction of resistance is opposite to the flow direction.

2.2 Prediction of Landslide Dam Formation and Failure

Erosion process and deposition process are of great importance when simulating landslide dam formation and dam break. When the shear stress is greater than the critical shear stress value and the volumetric sediment concentration is smaller than a critical value, bed erosion occurs. When the flow mixture moves to a flatter area, where the volumetric sediment concentration is larger than the equilibrium value and the flow velocity is smaller than a critical value, deposition occurs. The erosion process can be described by the following equation:

$$i = K_e(\tau - \tau_c) \qquad (5)$$

The deposition process is expressed as:

$$i = \delta_d\left(1 - \frac{V}{pV_e}\right)\frac{C_{V\infty} - C_V}{C_{v*}}V \qquad (6)$$

where i is the erosion/deposition rate, K_e is the coefficient of erodibility; τ is the shear stress, τ_c is the critical erosive shear stress; δ_d is a coefficient of deposition rate; V is the flow velocity; V_e is the critical flow velocity; $Cv\infty$ is the equilibrium volumetric sediment concentration; p is a coefficient of location difference.

Through the above analysis, the landslide dam formation and dam break process can be described.

3   Case Study on Hong Kong Island

Since there has been no landslide dam formation in Hong Kong, the case in this paper is assumed. The study area is relatively small, and the landslide scale is small. Due to the location of the study area, once a landslide dam is formed, it can be very dangerous because the down-stream area is densely populated.

Figure 2 shows the location of the landslide and landslide dam. Firstly, a major landslide event occurs with the heavy rainfall. The landslide volume is as high as $6,000\,\text{m}^3$ (see Fig. 2(a)) After the failure of the slope, the landslide material runs down along the channel and finally deposited somewhere downstream (see Fig. 2(b)). The deposit of landslide blocks the main stream and a landslide dam is formed when the surface water accumulates. When the water depth is larger than the dam height, water runs downstream on top of the dam, and thus, erosion occurs. And then due to overtopping, the landslide dam begins to fail.

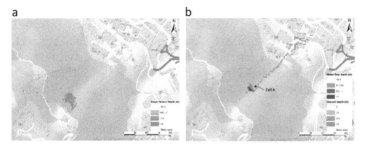

Fig. 2. (a) Slope failure inventory; (b) landslide dam

Figure 3 shows the discharge history of a specific cell A (see Fig. 2(b)). After the landslide dam formation, the discharge becomes 0, which means the water is blocked in the dam. Then when dam failure occurs, the flow discharge becomes larger suddenly and the peak discharge can be as high as $40\,\text{m}^3/\text{s}$. Then after the major failure, the discharge becomes smaller and smaller.

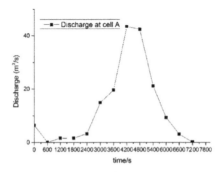

Fig. 3. Flow discharge at cell A

4 Conclusions

This paper presents a physically-based distributed cell model for predicting the full process of landslide occurrence, landslide dam formation, and dam break due to overtopping erosion. This case is an assumed case since there is no actual landslide dam breaking case in Hong Kong. The hazard scale is relatively small, but it shows the capability to simulate the full process of the model, EDDA 2.0. And therefore, this model can be used for simulating landslide dam formation and failure process, and subsequent risk assessment.

Acknowledgement. The authors acknowledge the financial support from the National Key Research and Development Program of the Ministry of Science and Technology of China (Project No. 2018YFC1508600).

References

1. Valiani, A., Caleffi, V., Zanni, A.: Case study: malpasset dam-break simulation using a two-dimensional finite volume method. J. Hydra. Eng. **128**(5), 460–472 (2002)
2. Fan, X., Tang, C.X., van Westen, C.J., Alkema, D.: Simulating dam-breach flood scenarios of the Tangjiashan landslide dam induced by the Wenchuan Earthquake. Nat. Haz. Earth Syst. Sci. **12**(10), 3031 (2012)
3. Chang, D.S., Zhang, L.M.: Simulation of the erosion process of landslide dams due to overtopping considering variations in soil erodibility along depth. Nat. Haz. Earth Syst. Sci. **10**(4), 933–946 (2010)
4. Fread, D.L.: The development and testing of a dam-break flood forecasting model. In: Proceedings of Dam-Break Flood Routing Model Workshop, Bethesda, MD, 164–197 (1977)
5. Fread, D.L.: BREACH: an erosion model for earth dam failures, National Weather Service (NWS) Report, NOAA, Silver Spring, Maryland, USA (1988)
6. Singh, V.P., Quiroga, C.A.: A dam-breach erosion model: I. Formulation Water Res. Manag. **1**(3), 177–197 (1987)
7. Temple, D.M., Hanson, G.J., Neilsen, M.L., Cook, K.R.: Simplified breach analysis model: part I, Background and model components. In: 25th Annual USSD Conference USSD Proceedings: Technologies to enhance dam safety and the environment, Salt Lake City, Utah, pp. 151–161 (2005)

Site Amplification Effects of Canyon on the Post-earthquake Profile of Earth Dam

Zheng Zhou, Fei Zhang$^{(\boxtimes)}$, Ning Zhang, and Yu Zhang

Key Laboratory of Ministry of Education for Geomechanics and Embankment Engineering,
Hohai University, Nanjing, China
feizhang@hhu.edu.cn

Abstract. Most predictions of the seismic displacement of dam slope are often based on Newmark method but neglects the site amplification effects of the canyon. The purpose of this study is to investigate the effects on the dam deformation using Nested Newmark Model (NNM), which can assess the post-earthquake profile of the dam slope. The seismic amplification is analytically obtained from the homogeneous half-space solutions of seismic waves of the scattering induced by a symmetrical V-shaped canyon, and then involved into NNM to calculate the relative displacements from the toe to the crest. An example is presented here to make a comparison of the seismic displacements of the dam with or without the site amplification effects of canyon. The comparison indicates the significant influences of the canyon-shaped site amplification on the dam slope stability. The presented method provides a way to predict the post-earthquake profile of earth dam in a canyon.

Keywords: Earth dams · Earthquake · Permanent displacement · Amplification effects · Canyon

1 Introduction

Previous experiences (e.g., [1, 2]) with earth dams located in a canyon demonstrate a higher incidence of performance problems during earthquakes. The permanent slope displacements after seismic event are very important for the safety of the dam. The classic Newmark method [3] is widely used and improved to determine the seismic displacement. However, the site amplification effect of the canyon has not been involved in Newmark method. Zhang et al. [4] present a rigorous solution of wave functions for the plane SH waves scattering induced by a symmetrical V-shaped canyon and obtain a high amplification of displacements at the wave. The narrow V-shaped canyon could result in seismic amplifications and then could make the permanent displacement much larger. The purpose of this study is to include the derived site amplification effects of canyon into Nested Newmark Model (NNM), which is proposed by Leshchinsky [5] and can enable assessment of a post-earthquake slope profile. Through an example, the site amplification effect is further investigated on the seismic displacement of dam slope.

© Springer Nature Switzerland AG 2020
J.-M. Zhang et al. (Eds.): ICED 2020, SSGG, pp. 226–230, 2020.
https://doi.org/10.1007/978-3-030-46351-9_22

2 Methodology and Formulations

The two-dimensional (2D) canyon model is depicted in Fig. 1. The normalized displacements of the surface of canyon model have been obtained by Zhang et al. [4]. In fact, the normalized displacements represent the transfer factor of the site response at a specific frequency. The transfer functions at a specific location can be obtained by combining the transfer factors at various frequencies. The input Fourier spectrum of the incident seismic acceleration time history can be obtained with the fast Fourier transform technique (FFT), Afterwards, the corresponding Fourier spectrum of the input velocity time history should be multiplied by the transfer functions to obtain the response Fourier spectrum at the targeted locations. Finally, through the inverse fast Fourier transform, the earthquake response in the time domain can be obtained accordingly. In this way, the amplification or de-amplification of the wave at each frequency is taken into account in the site responses.

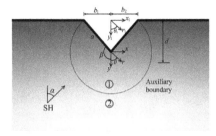

Fig. 1. 2D model of a symmetrical V-shaped canyon redrawn after Zhang et al. [4]

In the framework of the nested Newmark method, Leshchinsky [5] employs the LE approach to determine the post-earthquake profile of the slopes subjected to the translational and rotational failure, respectively. Figure 2 represents a homogeneous soil slope model subjected to the pseudo-static seismic forces on the verge of failure. The inclination of the slope is described by angle β; the height of the slope is defined as H; and the soil is characterized by internal friction angle φ, cohesion c, and unit soil weight γ. First, discretizing the lower slope into n blocks and conducting the pseudo-static approach can obtain the yield acceleration ky_i and the corresponding slip surface. With the acceleration-time history of the input motion obtained by the aforementioned method, the acceleration of the block i could be derived as follow:

$$k_{Hi}(t) = [k(t) - ky_i]g \tag{1}$$

which can subsequently be integrated over time increments, Δt, to attain the relative velocity of a given block:

$$v_i(t) = v_i(t_0 + \Delta t) + \int_{t_0+\Delta t}^{t} k_{Hi}(t)dt \tag{2}$$

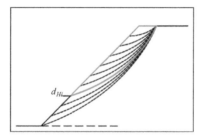

Fig. 2. Nested Newmark model

Integrating the relative velocity, the horizontal displacement of a given block could be obtained:

$$d_i(t) = \int_{t_0+\Delta t}^{t} v_i(t)dt \tag{3}$$

The total horizontal displacement of each nested block is then integrated for a given time increment, starting with the basal region and sequentially proceeding towards the crest (see Fig. 2).

$$d_H = \int_{0}^{H} d_i(t)dH \tag{4}$$

Furthermore, the yield accelerations of the upper nested blocks increase with the height increase. Therefore, the upper regions may not encounter yield and their displacement is solely a result of displacement of underlying failures.

3   Results and Discussions

An earth dam example which is located in a gentle canyon is introduced here, as shown in Fig. 3, to assess the site amplification effects of the canyon on the post-earthquake profile of dam slope. The width and depth of this canyon is 340 m, 25 m, respectively. The shear wave velocity of this canyon is 300 m/s. For a specific seismic wave time history such as the Maoxian wave (Wenchuan 2008) which has been widely used in the academic study and engineering design, it is more than sufficient to compute the transfer functions in its major frequency band within 25 Hz. The incident Maoxian wave (Wenchuan 2008) has a PGA of 0.288 g, as shown in Fig. 4. Using the analytical approach of Zhang et al. [4] can obtain the time domain results of the seismic responses along the height, as shown in Fig. 4. A toe failure which begins from the height 12.76 m is considered in this study.

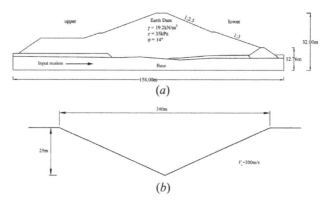

Fig. 3. (*a*) Example for an earth dam subjected to Wenchuan earthquake; (*b*) Profile of the located canyon

From Fig. 5a, it indicates that the yield acceleration increases with the dam slope height, which means difficult for top region of the dam to encounter yield subjected to the same input motion. The peak acceleration obtained by this study demonstrates obvious amplification with the height. Compared to the constant input motion in the origin NNM, the acceleration profile of this study indicates that larger region would encounter yield in Fig. 5b. Due to lager failure zone and integrated time, the estimation obtained from this study is larger than those by NNM and rigid block theorem. From this result, neglecting the amplification effects also yield the displacements for one-block and nested Newmark models, respectively. The site amplification effects can induce a much larger value of the displacement.

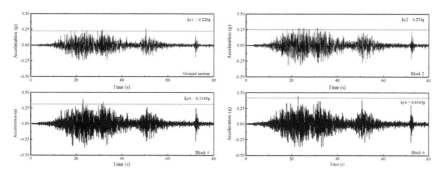

Fig. 4. The ground motion and the motions for the seismic responses with the dam height.

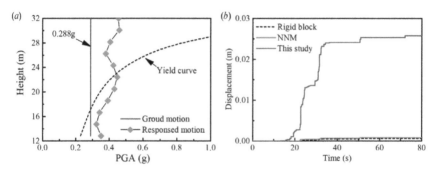

Fig. 5. Calculated results: (a) seismic yield acceleration and responses; (b) displacements derived from Newmark methods.

4 Conclusions

This study presents a modified approach to determine the seismic displacement of slope of earth dam built in a canyon. The site amplification effects of the canyon are involved here using the analytical results calculated by Zhang et al. [4]. Conducting the NNM can obtain the post-earthquake profile of the earth dam slope. When ignoring the site amplification effects could underestimate the seismic displacement significantly. The presented approach to assess permanent displacement of slope is very useful for anti-earthquake design and reinforcement of earth dam in practice. In addition, the soil amplification is not considered here but involving it into the presented analysis is straightforward.

References

1. Gazetas, G.: Seismic response of earth dams: some recent developments. Soil Dyn. Earthq. Eng. **6**, 2–47 (1987)
2. Gazetas, G., Dakoulas, P.: Seismic analysis and design of rockfill dams: state-of-the-art. Soil Dyn. Earthq. Eng. **11**, 27–61 (1992)
3. Newmark, N.M.: Effects of earthquakes on dams and embankments. Géotechnique **15**(2), 139–60 (1965)
4. Zhang, N., Gao, Y., Li, D., Wu, Y., Zhang, F.: Scattering of SH waves induced by a symmetrical V-shaped canyon: a unified analytical solution. Earthq. Eng. Eng. Vibra. **11**(4), 445–460 (2012)
5. Leshchinsky, B.A.: Nested Newmark model to calculate the post-earthquake profile of slopes. Eng. Geol. **233**, 139–45 (2018)

Dynamic Response of a Central Clay Core Dam Under Two-Component Seismic Loading

Z. Zhu[1(✉)], M. Kham[2], V. Alves Fernandes[2,3,5], and F. Lopez-Caballero[4,5]

[1] Ecole des Ponts ParisTech, 6-8 av. Blaise Pascal, Cité Descartes, Champs-Sur-Marne, 77455 Marne la Vallée cedex 2, France
zhehao.zhu@enpc.fr
[2] Electricité de France/R&D, 91120 Palaiseau, France
[3] IMSIA, UMR EDF-CNRS-CEA-ENSTA 9219, 828 Boulevard des Maréchaux, 91762 Palaiseau Cedex, France
[4] Laboratoire MSS-Mat CNRS UMR 8579, CentraleSupélec, Université Paris-Saclay, 91190 Gif-sur-Yvette, France
[5] SEISM Institute, 91190 Saint-Aubin, France
https://institut-seism.fr

Abstract. The seismic response of Aratozawa rockfill dam (Japan) is simulated in the present work by using ECP constitutive model and Aster FE code [1]. Static calculation is first performed for computing initial stress and strain fields. Then, the numerical convergence problem in dynamic step under two-component seismic loadings is discussed. It is found that the numerical damping induced by time-integration algorithms does not help solving the numerical convergence problem. However, the augmentation of subdivision level of calculation step is very efficient. The total duration of calculation is extended but practically acceptable. Finally, the dynamic response of Aratozwa dam is compared with real measurements presented in the literature with three sets of model parameters. All results show great agreement with the measurements and also be reconfirmed by Anderson's criteria.

Keywords: Rockfill dam · Seismic loading · Computational convergence

1 Introduction

With the development of calculation power, methodologies of numerical simulation on clay core rockfill dams have been developed [2]. Most of the severe damage to high embankment dams and geotechnical structures during earthquakes can be attributed to shear wave propagation, which causes destructive horizontal movements. Thus, the effect of compression wave is frequently neglected and most numerical studies have extensively focused on dynamic non-linear response induced uniquely by shear wave. In addition, the literature contains very few investigations of the combined influence of shear and compression waves on embankment dams. Whether the addition of compression wave into a finite element code could largely affect the dynamic response and cause special computational problem is still unclear. Moreover, more than 50 large dams of Class A

© Springer Nature Switzerland AG 2020
J.-M. Zhang et al. (Eds.): ICED 2020, SSGG, pp. 231–236, 2020.
https://doi.org/10.1007/978-3-030-46351-9_23

($H \geq 20$ m) in France need to be reevaluated under complete strong seismic loading in order to respond to changes in safety regulations. As a consequence, the emphasis of this paper is on presenting the numerical simulation of the seismic response of a Japanese rockfill dam, Aratozawa. The static and dynamic calculation procedures are then respectively detailed with special attention paid to the numerical convergence problem. Finally, the numerical results are compared with real in-situ measurements.

Aratozawa dam is located in Miyagi Prefecture, Japan. It is a 74.4 m high rockfill dam with a central clay core. The dam was built in 1991 and activated in 1998 for irrigation and flood control purposes. The dam consists of five zones, and three sets of 3-component strong motion seismometers were installed at three locations, i.e., the dam crest, the mid-core and the bottom gallery in the largest cross-section of the central part of the dam (see Fig. 1). The Iwate-Miyagi Nairiku earthquake ($M_J = 7.2$) [3] occurred on June 14, 2008, with its epicenter 8 km below southwestern Iwate Prefecture. During the main shock, a peak acceleration of 10.24 m/s^2 was measured by strong motion accelerometers at the bottom gallery. However, the dam remained in a stable working condition and the reservoir functions were not clearly affected by the main shock.

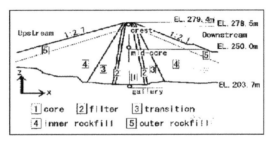

Fig. 1. Cross section of the Aratozawa dam with seismometers located at the crest, mid-core and bottom gallery (from [3]).

2 Soil Constitutive Model

Experimental studies indicate that soil behaves in a non-linear way. Thus; in this paper, an elastoplastic non-linear constitutive model ECP [4] is applied, which was developed at MSSMat laboratory of Ecole Centrale Paris. The model considers one isotropic and three deviatoric mechanisms, which are coupled by the volumetric plastic strain in order to correctly simulate soil behavior in a large range of loading paths. Both coarse and fine granular soils can be simulated since the model's yield surface varies from Mohr-Coulomb criteria to Cam-clay surface according to a model parameter b. The behavior is decomposed in three functional regimes by an internal parameter quantifying the degree of mobilized friction: pseudo-elastic, hysteretic and mobilized regimes. A calibration procedure of ECP model parameters was proposed by Lopez-Cabellero ([5] and [6]).

2.1 Mechanical Properties of Aratozawa Dam

In order to maintain the precision of input signals, the rock foundation of Aratozawa dam is assumed to be rigid, described by a simple elastic linear equation with the following parameters: $E = 143$ GPa, $v = 0.3$ and $\rho = 2200$ kg/m³.

Filters and rockfills are assumed to be drained during Miyagi earthquake. However, central clay is simulated under undrained condition by ECP model. Due to the lack of feasible deep detection methods, the knowledge about the mechanical behavior of central clay is incomplete. Thus, based on the known plasticity index of central clay, three sets of ECP model parameters with different fine fractions: Jeu1, Jeu2 and Jeu3 are respectively proposed by Kham [7].

3 Simulation of Aratozawa Dam

3.1 Finite Element Mesh

The adopted mesh in this study is bi-dimensional in plane strain deformation. In order to accurately capture the wave propagation, the size of the used finite element mesh needs to be inferior to earthquake predominant wavelength. Note that Eq. (1) is pertinent uniquely for linear elastic media. Thus, a security factor must be taken into account for soil non-linear degradation (i.e. shear wave velocity).

$$\Delta z < \frac{\lambda}{N} = \frac{V_s}{N * f_{max}} \tag{1}$$

where λ, N, V_s and f_{max} are respectively wavelength, security factor, shear wave velocity and maximum frequency. Finally, $\Delta z < 1$ m is chosen and the finite element mesh is presented in Fig. 2 which contains 3462 quadratic elements and 7564 nodes.

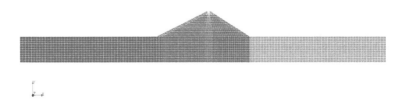

Fig. 2. 2D mesh of Aratozawa dam (visualized in Salome-Meca).

3.2 Static Calculation

The objective of static calculation, including construction by layers and the rise of water level in dam reservoir, is to estimate the non-linear initial state of the dam before dynamic calculation, by assessing the initial stress, strain and internal variables fields. In this step, the displacement of the dam foundation is blocked. The construction by layer is performed in order to reproduce the real construction over 3 years and each layer is approximately equal to 5 m. A consolidation phase of 10 years is then simulated, which represents the time interval between the end of construction and the occurrence of the Miyagi earthquake. Hydrostatic pressure condition is applied on the upstream of the core for simulating the in-flow water whereas constant zero pressure is applied to simulate free out-flow water on downstream side (see Fig. 3).

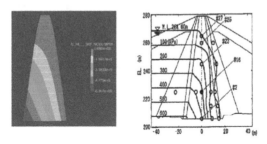

Fig. 3. Comparison of the water pressure distribution of Aratozawa dam (after [3]).

3.3 Dynamic Calculation

The addition of seismic loadings is impossible while the dam foundation is immovable and a transition of boundary conditions is necessary. Thus, the embedding of dam foundation is first replaced by very rigid springs. Nodal forces are then calculated from them and re-applied to the contour of the dam foundation. In addition, damping elements are placed on the periphery of the foundation in order to absorb the waves diffracted by the dam.

Numerical Convergence Studies and Dynamic Response of Aratozawa Dam
In the initial dynamic calculation with two-component seismic loadings, computational convergence problem is revealed by the presence of considerable accelerations (Acc > 100 m/s^2) at the crest of the dam.

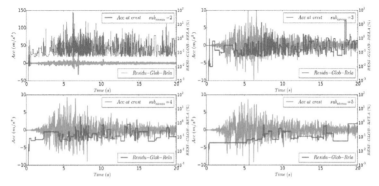

Fig. 4. Comparison of acceleration at the crest of the dam and global relative residue with different subdivision step.

First, attempts have been made by adding numerical damping in time-integration algorithms [8]. However, the results do not appear to be improved. Then, the calculations' information files are verified with special caution and it seems that almost all unreal 'peak' accelerations happen suddenly while the global relative error largely exceeds convergence criterion. Therefore, a quantitative study of subdivision level of calculation step is investigated. The results of the same dynamic calculation with four different subdivision levels are presented in Fig. 4.

Red solid line represents the acceleration at the crest of the dam and blue solid line represents the evolution of the global relative error. It is clear that the global relative error decreases sensitively with the increase in sub_{niveau} making 'peak' accelerations to disappear. Calculation time is also checked: it increases from 37 h when $sub_{niveau} = 2$ to 40 h when $sub_{niveau} = 4$, which is practically acceptable. Then, the calculated horizontal

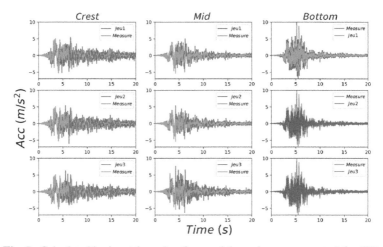

Fig. 5. Calculated horizontal accelerations and the real measurements (after [3]).

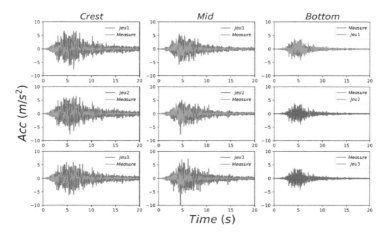

Fig. 6. Calculated vertical accelerations and the real measurements (after [3])

and vertical accelerations of Aratozawa dam are compared with the real measurements presented in the literature.

Figure 5 and 6 show that all three sets of parameters give qualitatively correct results compared with real measurements. Accordingly, 6 of 10 parameters of Anderson's Criteria [9] are applied ($C1$, $C2$, $C3$, $C4$, $C5$ and $C8$) in order to evaluate the quality of a calculated signal with respect to a reference one. Finally, Anderson average scores of three sets of parameters are all greater than 8.5/10, which reconfirms this present work.

4 Conclusion

Seismic response of Aratozawa rockfill dam is simulated in this study. The coherence between numerical results and real in-situ measurement demonstrates that the methodology of Aster FE code and soil constitutive model used in this paper are accurate to reproduce the seismic response of large rockfill dams. In addition, subdivision level needs to be increased to avoid computational convergence problem.

Acknowledgements. The authors acknowledge JCOLD for sharing the data during the CFBR-JCOLD benchmark exercise. The help provided by Miss Li is also gratefully acknowledged.

References

1. Code-Aster: general public licensed structural mechanics finite element software. http://www.code-aster.org
2. Kham, M., Kolmayer, P., Matsumoto, N.: Advanced numerical modeling of Aratozawa dam response under Miyagi 2008 strong earthquake. In: 7th International Conference on Earthquake Geotechnical Engineering, Roma, Italy (2019)
3. Ohmachi, T., Tahara, T.: Nonlinear earthquake response characteristics of a central clay core rockfill dam. Soils Found. **51**(2), 227–238 (2011)
4. Hujeux, J.C.: Une loi de comportement pour le chargement cyclique des sols, Génie Parasismique, V. Davidovici, Presses ENPC; 1985, pp. 278–302 (1985). (in French)
5. Lopez-Caballero, F.: Influence du comportement non linéaire du sol sur les mouvements sismiques induits dans des géo-structures. Doctoral dissertation, Châtenay-Malabry, Ecole centrale de Paris (2003). (In French)
6. Lopez-Caballero, F., Razavi, A.M.F., Modaressi, H.: Nonlinear numerical method for earthquake site response analysis I—elastoplastic cyclic model and parameter identification strategy. Bull. Earthq. Eng. **5**(3), 303 (2007)
7. Kham M.: Modélisation numérique du barrage en remblai d'Aratozawa sous sollicitation sismique. Documentation 6125-1714-2017-03830-FR, EDF R&D (2018). (In French)
8. Hilber, H.M., Hughes, T.J., Taylor, R.L.: Improved numerical dissipation for time integration algorithms in structural dynamics. Earthq. Eng. Struct. Dyn. **5**(3), 283–292 (1977)
9. Anderson, J.G.: Quantitative measure of the goodness-of-fit of synthetic seismograms. In: 13th World Conference on Earthquake Engineering Conference Proceedings, Vancouver, Canada, vol. 243 (2004)

Soil Mechanics for Embankment Dams

Modeling of Soil Migration Phenomena in Embankment Dams

Francesco Federico[(✉)] and Chiara Cesali

Geotechnics, University of Rome Tor Vergata, Rome, Italy
fdrfnc@gmail.com, cesali@ing.uniroma2.it

Abstract. The problems of granulometric stability related to particle migration phenomena *(i)* at the contact between materials affected by different grain size or *(ii)* through widely or gap graded soils (i.e. *suffusion*) are well recognized, as shown by several historical dams incidents. For a complete simulation of these phenomena and their evolution towards possible limit conditions (i.e. *clogging, blinding, complete erosion*), the (space and time) variability of soils granulometric properties (i.e. voids volume, porosity, permeability, flow velocity, local piezometric gradients, flow direction,…), as well as the particles erodibility and deposition, must be taken into account. The available empirical and analytical (derived from "*continuum*" mechanics) methods to analyze particle migration phenomena generally don't considerer the coupled effects of these micro-structural (at the grain scale), meso-structural (porosity, permeability) and hydraulic variables. Thus, a numerical procedure allowing to simulate coupled 1D seepage and particle migration processes, by taking into account the mutual dependency of the above cited governing variables, has been developed and applied to carry out detailed analyses and review of available, selected experimental data.

Keywords: Soil migration processes · Numerical procedure · Review of experimental data

1 Introduction

The granulometric compatibility between materials characterized by different grain size distribution (GSD) plays a key role in safety of embankment dams and levees [1]. The possible migration of fine particles through voids larger than their size, at the contact between different materials due to seepage, may be analyzed through different approaches: empirical criteria (e.g. [2–5]), analytical (e.g. [6–9]) and numerical (e.g. [10, 11]) models.

For a reliable simulation and a better comprehension of these dangerous, often hidden, phenomena, as well as their evolution towards possible granulometric limit states (i.e. clogging, blinding, complete erosion), the (*space and time*) variability of granulometric properties, particularly voids volume, porosity (n), permeability (k), flow velocity and direction, local piezometric gradients, should be considered [12].

© Springer Nature Switzerland AG 2020
J.-M. Zhang et al. (Eds.): ICED 2020, SSGG, pp. 239–252, 2020.
https://doi.org/10.1007/978-3-030-46351-9_24

To this purpose, a numerical procedure that couples the effects of geometrical (at the "grains" micro-scale, e.g. GSD, porosity, volume voids and constriction sizes distributions) and geotechnical (e.g. permeability, piezometric gradients, seepage velocity....) governing variables, has been recently developed and refined [13].

The proposed procedure and a "*traditional*" continuum model have been applied to the interpretation of some experimental measurements in the paper.

Results of simulations show that the procedure allows to highlight/understand the micro-mechanical aspects that govern the growth of the effective filtering zone; the "*continuum*" approach allows only a partial description (particularly at the macro-scale) of the local variability of the granulometric properties.

2 "Continuum" Modeling Approach

By considering a representative elementary volume (R.E.V.) of a granular material composed by the volume of particles in suspension, the fluid phase and the solid phase (stable and deposited particles) volumes (Fig. 1), the particle migration process can be described, along space x and time t, by the following system of three partial differential equations (PDEs) [7, 8]:

- *fluid mass balance equation*

$$\nabla \cdot \left[(1 - c_s) \cdot n \cdot \vec{v}_f \right] = -\frac{\partial[(1 - c_s) \cdot n]}{\partial t} \tag{1}$$

- *solid mass balance equation*

$$\nabla \cdot \left(c_s \cdot n \cdot \vec{v}_{sp} \right) = \frac{\partial[(1 - c_s) \cdot n]}{\partial t} \tag{2}$$

$\vec{v}_f$ being the velocity vector of the fluid phase; c_s, the concentration of particles in suspension; n, volumetric porosity; $\vec{v}_{sp}$, the velocity vector of the suspended and transported particles ($\vec{v}_{sp} = \chi \cdot \vec{v}_f$, with $\chi \in (0; 1]$, [14]);

'*kinetic equation*', describing the deposition and erosion processes of the deposited/accumulated particles within voids; to this purpose, different formulations have been proposed in the past [6–9]. The available relationships in general do not allow to simulate simultaneously coupled erosion and deposition processes.

The (three) unknown variables of the problem are n, c_s, v_f, depending on time (t) and space (x). If the following hypotheses: (*i*) $v_{sp} = v_f$ ($\chi = 1$, [14]) and (*ii*) unidirectional flow (1-D case) are assumed, since $\vec{v}_D = n \cdot \vec{v}_f$ ($\vec{v}_D$, Darcy's velocity), by combining the Eqs. (1) and (2), it is obtained:

$$v_{D,x} \cdot \frac{\partial c_s}{\partial x} = \frac{\partial[(1 - c_s) \cdot n]}{\partial t} \tag{3}$$

Considering the "*kinetic*" equation proposed by Vardoulakis [6], which allows to simulate fluidization and erosion (not deposition) processes [15],

$$\frac{\partial n(x,t)}{\partial t} = \lambda \cdot [1 - n(x,t)] \cdot c_s(x,t) \cdot n(x,t) \cdot v_{sp}(x,t) \tag{4}$$

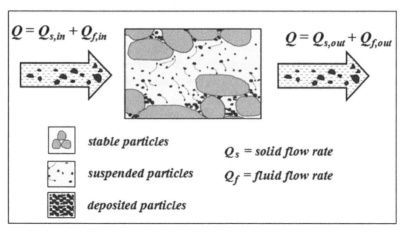

Fig. 1. Physical scheme of a R.E.V. within a granular material subject to particle migration: particularly, suspended (eroded) particles move through voids formed by stable (not eroded) particles; deposited particles remain trapped within these voids.

λ being an experimentally calibrated parameter [6], the above system of three governing PDEs becomes:

$$\begin{cases} v_{D,x} \cdot \dfrac{\partial c_s(x,t)}{\partial x} = \dfrac{\partial [n(x,t) \cdot (1-c_s(x,t))]}{\partial t} \\ \dfrac{\partial n(x,t)}{\partial t} = \lambda \cdot [1 - n(x,t)] \cdot c_s(x,t) v_{D,x} \end{cases} \tag{5}$$

The system is solved by an explicit second order method Finite Difference Method, FDM (Δx, Δt are space and time integration intervals, respectively; $i = i\text{-}th$ element of the system composed by two different contacting materials):

$$\frac{\partial c_s}{\partial x} \cong \frac{1}{2\Delta x}\left(c_{s,i+1}^{t-1} - c_{s,i-1}^{t-1}\right) \tag{6.a}$$

$$\frac{\partial c_s}{\partial t} \cong \frac{1}{\Delta t}\left[c_{s,i}^{t} - \frac{1}{2}\left(c_{s,i+1}^{t-1} + c_{s,i-1}^{t-1}\right)\right] \tag{6.b}$$

$$\frac{\partial n}{\partial t} \cong \frac{1}{\Delta t}\left[n_i^{t} - \frac{1}{2}\left(n_{i+1}^{t-1} + n_{i-1}^{t-1}\right)\right] \tag{6.c}$$

The final explicit resolving equation is thus obtained:

$$c_{s,i}^{t} = \frac{\frac{1}{2}\left(c_{s,i+1}^{t-1} + c_{s,i-1}^{t-1}\right) \cdot \frac{n_i^{t-1}}{\Delta t} - v_{D,i,t-1} \cdot \frac{\left(c_{s,i+1}^{t-1} - c_{s,i-1}^{t-1}\right)}{2\Delta x}}{\left[\frac{n_i^{t-1}}{\Delta t} - \left(1 - c_{s,i}^{t-1}\right) \cdot \lambda_i \cdot \left(1 - n_i^{t-1}\right) \cdot v_{D,i,t-1}\right]} \tag{7}$$

3 Proposed Numerical Procedure

A numerical procedure to simulate coupled particle migration and seepage processes, previously developed by the Authors and based on the model proposed by [10], is herein presented. As the model by Indraratna and Vafai [10], the proposed procedure is founded on the fluid and solid mass balance equations, and on the concept of critical hydraulic gradient derived from limit equilibrium considerations, where the migration of particles is assumed to occur under applied hydraulic gradients exceeding the critical value, this one depending on particle diameter and its confining conditions (plugged or unplugged particles).

With respect to the model by Indraratna and Vafai [10], the proposed procedure allows to considerer the effective "internal geometry" of the porous materials (i.e. Constriction Sizes Distribution, CSD, instead of an average diameter of voids) in the particle migration phenomena simulation. To this purpose, an advanced geometrical and hydraulic characterization of the granular material is worked out.

3.1 Geometrical Characterization

To determine the distribution of the volume of voids (*VVD*) within porous media (and the corresponding Constriction Size Distribution, *CSD*), several methods are available in technical literature (e.g. [16, 17]).

In the proposed procedure, the *'geometric-probabilistic'* model by Musso and Federico [18] is applied, according to which the *cumulative probability function $F(V)$* of the dimensions of voids is expressed as:

$$F(V) = \frac{e^{-\beta V} - e^{-\beta V_{min}}}{e^{-\beta V_{max}} - e^{-\beta V_{min}}} \tag{8}$$

Specifically, for an assigned volume V of a pore, the volume of the largest particle (V_{cs}), able to move through the porous material, satisfies the relation $V_{cs} < V$ (Fig. 2).

By assuming spherical particles (D, diameter), on the basis of geometric observations [14], it is possible to determine the diameters of the smallest ($D_{cs,min}$) and largest ($D_{cs,max}$) particles passing through the smallest ($V_{cs,min}$) and largest ($V_{cs,max}$) pores. In other words, the minimum and maximum constriction sizes (Fig. 2):

for pores formed by *three spherical particles*

$$D_{cs,min} = 2 \cdot \left(3^{0.5}/3 - 1/2\right) \cdot D \rightarrow V_{cs,min} = 1.94 \cdot 10^{-3} \cdot D^3 \tag{9}$$

for pores formed by *four spherical particles*

$$D_{cs,max} = \left(2^{0.5} - 1\right) \cdot D \rightarrow V_{cs,max} = 3.72 \cdot 10^{-2} \cdot D^3 \tag{10}$$

By defining the coefficient $\eta = V_{cs}/V$, it is obtained:

$$\eta_{min} = \frac{V_{cs,min}}{V_{min}} = 1.15 \cdot 10^{-2}; \quad \eta_{max} = \frac{V_{cs,max}}{V_{max}} = 7.81 \cdot 10^{-2} \tag{11, 12}$$

Thus, the volume of constriction sizes (V_{cs}) and the corresponding diameter (D_{cs}) can be generally evaluated as:

$$V_{cs} = \eta(V) \cdot V \tag{13}$$

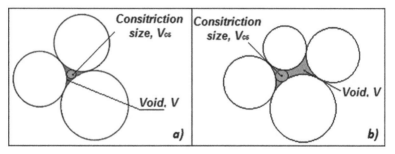

Fig. 2. Constriction size for pores formed by (a) three and (b) four particles

$$D_{cs} = \sqrt[3]{\frac{\eta(V) \cdot V \cdot 6}{\pi}} \qquad (14)$$

The corresponding passing percentage P_{cs} is assumed equal to the passing percentage P_v ($= F(V)$) associated with related void.

If a linear change of η with volume V is simply assumed, it is obtained:

$$\eta(V) = \frac{1}{V_{max} - V_{min}} \cdot [(\eta_{max} - \eta_{min}) \cdot V + (\eta_{min} V_{max} - \eta_{max} V_{min})] \qquad (15)$$

3.2 Hydraulic Characterization

The permeability coefficient (k) represents the fundamental parameter on which the seepage velocity through a porous medium mainly depends. k can be evaluated through the Kozeny-Carman relationship [19], according to grain size properties and porosity, which may vary along space and time due to particle migration phenomena:

$$k = \chi \cdot \frac{\gamma_w}{\mu_w} \cdot \frac{n^3}{(1-n)^2} \cdot D_h^2 \qquad (16)$$

D_h is the equivalent diameter of grains: $D_h = 1/\sum_i \frac{\Delta P_i}{d_i}$; χ is a numerical coefficient [19]; μ_w is the water viscosity.

The soil particles can be scoured if subjected to a seepage velocity greater than a critical value (v_{cr}); the analysis of the actions on a movable particle and the dynamic equilibrium along the flow direction [14] allow to estimate v_{cr}.

Particularly, two types of kinematics of particles can be distinguished: *(a)* frictional and rolling (unconfined particles, $D \leq D_{v,0}$); *(b)* purely frictional (confined particles, $D \cong D_{v,0}$). If the drag force F_h (Stokes law) overcomes the maximum local shear force related to the effective weight of the particle and the acting confining stresses (F_s), the particle can be eroded.

By imposing $F_h = \sum_i F_{s,i}$, for a horizontal flow path under laminar flow conditions, the following general expression for v_{cr} is obtained:

$$v_{cr} = \frac{n}{3\mu_w} \cdot \left[(\gamma_s - \gamma_w)\frac{D^2}{6} + \frac{\lambda D}{2}\left(\sigma_z' + \sigma_y'\right) \right] \tan\varphi \qquad (17)$$

λ is a coefficient allowing to consider the density of the granular matrix ($0 < \lambda \leq 4/\pi$); $\lambda = 4/\pi$ for granular matrix composed by spherical particles arranged in hexagonal configuration, most dense state [16, 19]; γ_s is the volume unit weight of particles; γ_w is the volume unit weight of interstitial fluid; σ'_z is the effective stress along the direction z; σ'_y is the effective stress orthogonally acting to the plane x-z; φ is the internal friction angle.

3.3 Problem's Setting and Governing Equations

Referring to a B-T system (B = fine grained material; T = transition granular material), the heterogeneous porous medium is decomposed into several elements (l_i = length of ith element; N = number of elements), each characterized by initial grain size curve ($D_{i,j,0}$; $P_{i,j,0}$), porosity $n_{i,0}$ and permeability $k_{i,0}$ (Fig. 3); i, j and t define the system element, materials granular fractions and the elapsed time, respectively [13, 14]. Each element is schematically composed by original (not eroded) material ($V_{or,i,t}$), deposited/accumulated particles ($V_{dep,i,t}$) and particles in suspension ($V_{susp,i,t}$) due to migration phenomena, and water saturating the ith element ($V_{w,i,t}$).

The variables ($D_{i,j,t}$; $P_{i,j,t}$) and $n_{i,t}$ (and then $k_{i,t}$ according to Eq. 16) evolve because of erosion-deposition processes, associated with particle migration, causing an unsteady seepage flow.

The unsteady state is simply analyzed by considering a sequence of steady states (time interval, Δt; "successive steady states" method); for each Δt, the piezometric load for the ith element ($\Delta h_{i,t}$) and the corresponding seepage velocity ($v_{i,t} = k_{i,t} \cdot \Delta h_{i,t}/l_i$) are determined through the continuity equation expressed as follows:

$$
\begin{bmatrix}
\frac{k_{1,t}}{l_1} & -\frac{k_{2,t}}{l_2} & 0 & \cdots & \cdots & 0 \\
0 & \frac{k_{2,t}}{l_2} & -\frac{k_{3,t}}{l_3} & 0 & \cdots & 0 \\
\vdots & \vdots & \vdots & \vdots & \vdots & \vdots \\
0 & \cdots & \frac{k_{i,t}}{l_i} & -\frac{k_{i+1,t}}{l_{i+1}} & \cdots & 0 \\
\vdots & \vdots & \vdots & \vdots & \vdots & \vdots \\
0 & \cdots & \cdots & 0 & \frac{k_{N-1,t}}{l_{N-1}} & -\frac{k_{N,t}}{l_N} \\
1 & 1 & \cdots & \cdots & \cdots & 1
\end{bmatrix}
\cdot
\begin{pmatrix}
\Delta h_{1,t} \\
\Delta h_{1,t} \\
\vdots \\
\Delta h_{2,t} \\
\vdots \\
\Delta h_{N,t}
\end{pmatrix}
=
\begin{pmatrix}
0 \\
0 \\
\vdots \\
0 \\
\vdots \\
\Delta H
\end{pmatrix}
\tag{18}
$$

being ΔH the hydraulic load imposed to the system.

Therefore, the suspension rate Q through the elements of the section Ω, and the volume of the suspension $V_{m,t}$, composed by the scoured particles dragged by the seeping fluid, entered and washed out from each element, is the same during each Δt:

$$
Q = \Omega \cdot k_{i,t} \cdot \frac{\Delta h_{i,t}}{l_i}; \quad V_{m,t} = Q \cdot \Delta t
\tag{19}
$$

So, the total volume of each element (V_i), generally composed of the original (not eroded) material ($V_{or,i,t}$), accumulated/deposited material ($V_{dep,i,t}$), material in suspension ($V_{susp,i,t}$) and water saturating the ith element ($V_{w,i,t}$), doesn't vary during Δt ($V_i = V_{or,i,t} + V_{dep,i,t} + V_{susp,i,t} + V_{w,i,t}$ = constant).

To determine the eroded or erodible particles, the hydraulic and geometrical compatibility conditions must be verified. Particularly, the hydraulic compatibility conditions consist in comparing the previous defined critical velocity $v_{cr,i,j,t}$ of the jth granulometric fraction, within the ith element at the instant t, with the effective seepage velocity $(v_{i,t}/n_{i,t})$. Unconfined and confined particles are distinguished according to average diameter of voids $D_{v,0,i,t}$, defined as [10]: $D_{v,0,i,t} = 2.67 \cdot n_{i,t} \cdot D_{h,i,t}/(1 - n_{i,t})$: particularly, *unconfined* if $D_{i,j,t} < D_{v,0,i,t}$; *confined* if $D_{i,j,t} \geq D_{v,0,i,t}$. The geometrical compatibility conditions concern the comparison between the diameter of the jth granulometric fraction $(D_{i,j,t})$ belonging to the ith element with the constriction sizes of the $(i + 1)^{th}$ element and the evaluation of the probability of one *forward step*, $P_{F,i,j,t}$ [11]. For each $D_{i,j,t}$, the percentage $P_{cs,i,j,t}$ of smaller constriction sizes of the $(i + 1)^{th}$ element is defined.

If $D_{i,j,t} < \min(D_{cs,i+1,j,t}; j)$ (smaller diameter of constriction sizes of the $(i + 1)^{th}$ element), $P_{F,i,j,t} = 1$; if $D_{i,j,t} > \max(D_{cs,i+1,j,t}; j)$ (larger diameter of constriction sizes of the $(i + 1)^{th}$ element), $P_{F,i,j,t} = 0$; if $\min(D_{cs,i+1,j,t}; j) < D_{i,j,t} < \max(D_{cs,i+1,j,t}, j)$, $P_{F,i,j,t}$ $(\in (0,1))$ is expressed as follows:

$$P_{F,i,j,t} = \left(1 - P_{cs,i,j,t}\right) + \sum_{w=0}^{3}[1 - \left(P_{cs,i,j,t}\right)^4] \cdot \left(1 - P_{cs,i,j,t}\right) \cdot P_{cs,i,j,t}$$
$$\cdot \left\{\left[1 - \left(P_{cs,i,j,t}\right)^3\right] \cdot \left(P_{cs,i,j,t}\right)\right\}^w \tag{20}$$

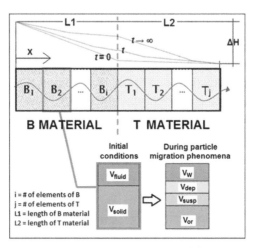

Fig. 3. Problem setting. One-dimensional unsteady seepage flow through a heterogeneous base (B) – transition (T) system; B and T are divided into elements; a constant hydraulic load ΔH is imposed.

$P_{F,i,j,t} \in (0,1)$ means that not all particles belonging to the jth granulometric fraction can be eroded and then pass through the pores network up to the $(i + 1)^{th}$ element; although erodible, the amount $(1 - P_{F,i,j,t})$ is trapped in the pores network and can remain in suspension or deposit [11].

Definitively, the particles with $v_{cr,i,j,t} < v_{i,t}/n_{i,t}$ (hydraulic condition) and $P_{F,i,j,t} > 0$ (geometrical condition) simultaneously, are therefore scoured.

For each eroded particle, the migration path ($L_{mig,i,j,t}$) is determined. $L_{mig,i,j,t}$ depends on the number ($m_{i,j,t}$) of constrictions greater than the particle ($D_{i,j,t}$), along its path, before its arrest [18].

The length of the migration path is compared to the length that the particles can cross during each step Δt; particularly, $L_{mig,i,j,t} = \min(s \cdot m_{i,j,t}; v_{i,t} \cdot \Delta t)$, s being the unit step (average migration path for each constriction).

Therefore, $V_{susp,i,t}$ is composed by particles eroded from the $(i - 1)^{th}$ element ($V_{acc,in}$), particles of the ith element with $P_{F,i,j,t} \in (0,1)$, V_{acc} (i.e. potentially erodible, but trapped in the pores network).

All these particles can deposit; then: $V_{susp,i,t} = V_{acc,i,t} + V_{acc,in,i,t} - V_{dep,i,t-1}$. The deposited particles are determined according to the probability $P_{dep,i,j,t}$ [20]:

$$P_{dep,i,j,t} = 4\left[\left(\frac{\theta_{i,t} \cdot D_{i,j,t}}{D_{v,0,i,t}}\right)^2 - \left(\frac{\theta_{i,t} \cdot D_{i,j,t}}{D_{v,0,i,t}}\right)^3\right] + \left(\frac{\theta_{i,t} \cdot D_{i,j,t}}{D_{v,0,i,t}}\right)^4 \qquad (21)$$

being $\theta_{i,t} = \theta_0 \cdot e^{-[(v_{i,t}/n_{i,t})/v_{clogg}]}$, with θ_0 and v_{clogg} experimental coefficients.

Once re-defined $V_{or,i,t}$ (reduced by eroded particles), $V_{dep,i,t}$ and $V_{susp,i,t}$, it is possible to determined $V_{w,i,t}$ (V_i = constant); therefore, for each element the porosity $n_{i,t}$ (and then the permeability $k_{i,t}$) and the GSD ($D_{i,j,t}$; $P_{i,j,t}$) can be updated.

Thus, the main outputs provided by the proposed procedure at each time step are: grain size distributions, porosity, permeability, eroded and deposited particles volumes, discharge rate and seepage flow velocity.

4   Applications and Comparisons

The *"continuum"* approach (based on Vourdalakis's equation as previously assumed) and the proposed numerical procedure have been applied to the lab experimental results obtained by Locke et al. [11].

Particularly, the effects of a downward seepage flow through a system B-T composed by a well-graded sand (150 mm thickness), as base soil (B), and a well-graded gravel (800 mm thickness), as transition (T), are investigated (Fig. 4).

Transition material (T) is characterized by D_{15} (particles diameter corresponding to the 15% passing) = 10.1 mm; different base soils (B), characterized by d_{85} (particles diameter corresponding to the 85% passing) satisfying the values of the piping ratio $D_{15}/d_{85} = 4$ and 7 [2] are tested. The initial porosity of the base soils (B) was assumed constant (35%); the initial transition (T) porosity was estimated at 40%. The applied hydraulic load (ΔH) corresponds to an initial flow rate of 0.02 l/s/m². In particular, since the permeability coefficients of base soils are $k = 8 \bullet 10^{-5}$ m/s (for $D_{15}/d_{85} = 4$), $k = 4 \bullet 10^{-5}$ m/s (for $D_{15}/d_{85} = 7$), and of the transition $k = 2 \bullet 10^{-2}$ m/s, ΔH is equal to 0.04 m and 0.08 m, respectively. Duration of test is approximately 2 h. To interpret lab measurements, the following input parameters are assigned:

- *"continuum approach"* (*"kinetic equation by Vardoulakis"*): $\Delta x = 0.002$ m; $n = 0.35$ for 0.15 m of base soil and $n = 0.40$ for the remaining 0.80 m granular transition; $\Delta t = 3$ s; duration of simulation = 2 h; the value of λ are parametrically assigned to fit the lab results: values of λ for tested materials are not *"a priori"* available;

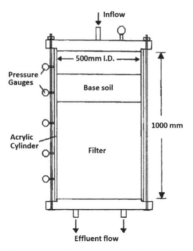

Fig. 4. Experimental apparatus (adapted from [11]).

- *proposed numerical procedure (PNP)*: length of base soil $(B) = 0.15$ m; length of transition $= 0.80$ m; number of elements of base soil $(B) = 3$ $(\Delta x = 0.05\ m)$; number of elements of the transition $(T) = 16$ $(\Delta x = 0.05\ m)$; $n = 0.35$ (base soil); $n = 0.40$ (transition material); $\Delta t = 3$ min; duration of simulation $= 2$ h.

Δx, Δt are related to the computational times. However, by modifying them, the results don't change. The two sets of input parameter are comparable.

The comparison between measured and numerical (according to the both applied approaches) values of the "*mass (M_{PT}) passing through the transition material [g/cm²]*" is carried out (Figs. 5 and 6). M_{PT} is estimated as:

- *proposed procedure (PNP)*

$$M_{PT} = V_{s,out}\ \rho_s / S \qquad (22)$$

$V_{s,out}$ being the volume of eroded particle leaving the base soil and passing through the transition material; ρ_s, the grains density/unit weight; S, area of the sample cross section;

- *continuum approach*

$$M_{PT} = c_s n\ \rho_s V_{tot} / S \qquad (23)$$

V_{tot} $(\Delta x \bullet S)$, the volume of each discretized element.

To apply the "*continuum*" approach (and to fit the lab results), specific values of the a *priori* unknown parameter λ [6] must be assigned. By this way, the measured values of the mass passing through the transition material, i.e. the amount of eroded particles at the interface *B-T*, may be simulated (Fig. 5). Anyway, the "*continuum*" model does not get any information concerning to the progressive generation of the filter thickness, specifically about the reasons for which the new generated zone (the filter) can (or cannot) control further fine particle movements. If the *PNP* is applied, the same appreciable agreement, between the computed values of the mass passing through the transition material and lab measurements, is observed (Fig. 6). In addition to this *"global"* variable, the *PNP* also provides information at the micro (*particle*) scale, particularly through the GSD and CSD curves and their evolution, at the different distances from the interface *B-T* (Figs. 7 and 8). Referring to the base soil (*B*) characterized by $D_{15}/d_{85} = 4$, the constriction sizes of *T* rapidly and considerably reduce (from $D_{cs,50} = 2.2$ mm of the initial *CSD* of *T* up to $D_{cs,50} = 0.8$ mm of the *CSD* of *T* at the end of the test; $D_{cs,50}$ = constriction size diameter corresponding to 50% passing, Figs. 7 and 8). To this purpose, small variations of GSD and CSD of *T* are observed at 0.20 m from the interface *B-T* (Fig. 8); fine particles movements through growing filter are prevented within small distances from the interface *B-T*.

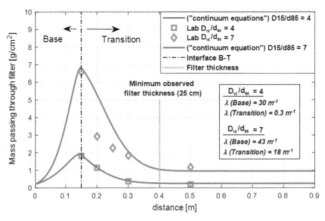

Fig. 5. "Mass passing through transition" vs distance according to the "continuum" approach (kinetic equation by Vardoulakis).

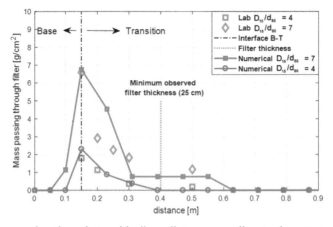

Fig. 6. "Mass passing through transition" vs distance according to the proposed numerical procedure.

Concerning to the base soil (B) characterized by $D_{15}/d_{85} = 7$, the eroded particles of B (finer than those ones of B with $D_{15}/d_{85} = 4$) travel longer distances through the growing filter. Small variations of GSD and CSD of T are observed at 0.05 m from the interface B-T (Fig. 9); thus, the eroded particles of B more easily move through voids larger than their size, beginning to deposit/accumulate at 0.20 m from the interface B-T (Fig. 10).

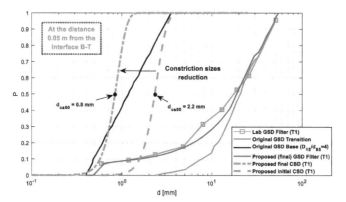

Fig. 7. Proposed procedure: final GSD and CSD curves of transition (T) for soil base (B) with $D_{15}/d_{85} = 4$, at the distance from the interface B-T of 0.05 m.

At this distance, the corresponding constriction sizes of T are thus smaller than those ones of T at 0.05 m from the interface B-T (Fig. 9). The experimental minimum 0.25 m–0.35 m values [11] of the filter thickness (composed by fine particles of B, trapped within the voids network of T) is better interpreted, at the both "*macro*" (in terms of mass passing through T) and "*micro*" (in terms of evolution of GSD and CSD curves of filter material, at different distances from the interface B-T) scales, by applying the proposed procedure.

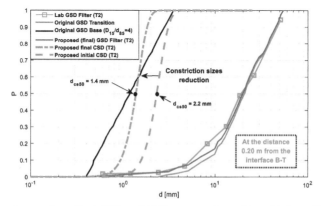

Fig. 8. Proposed procedure: final GSD and CSD curves of transition (T) for soil base (B) with $D_{15}/d_{85} = 4$, at the distance from the interface B-T of 0.20 m.

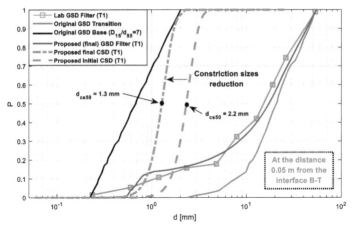

Fig. 9. Proposed procedure: final GSD and CSD curves of transition (T) for soil base (B) with $D_{15}/d_{85} = 7$, at the distance from the interface B-T of 0.05 m.

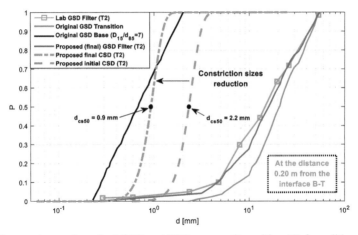

Fig. 10. Proposed procedure: final GSD and CSD curves of transition (T) for soil base (B) with $D_{15}/d_{85} = 7$, at the distance from the interface B-T of 0.20 m.

Therefore, the *PNP* allows to simulate the progressive changes of the transition grain size and the corresponding generation of the filter zone, due to the deposition of fine particles, up to the final, equilibrium condition.

By this way, the efficiency (*macro-behaviour*) of a granular transition to control possible migration phenomena may be checked through the simulation of erosion-deposition phenomena, carried out at the grain scale (*micro-behaviour*).

5 Concluding Remarks

The progressive variations (in the space, 1D, and along time, t) of the physical (porosity n) and hydraulic (permeability k) properties of a granular medium, during the coupled *seepage flow*, particles deposition and scouring phenomena, are simulated through a numerical procedure.

The procedure takes into account voids, constriction sizes and local porosities of the granular materials (geometric-probabilistic model) as well as the rate of the seeping suspension and piezometric gradients.

The main effects of these coupled processes include the amount and the distribution of accumulated material, porosity and permeability changes, as well as the *GSD* and *CSD* variations, and allow to interpret (at the micro "particle" scale) the results of a large scale lab test [11, 21]. The procedure can be furthermore applied to model the filtering action carried out by geotextiles [22].

The comparisons between different numerical results and the corresponding lab measurements allowed to point out the limits of "*continuum*" models, particularly in simulating the micro-scale effects (i.e. variations of GSD and CSD curves, migration path lengths,…) related to the coupled seepage and particle migration processes, which play an essential role in the geotechnical design of granular transitions as protective filters.

References

1. ICOLD: Internal erosion of existing dams, levees and dikes, and their foundations. Jean-Jacques Fry & Rodney Bridle - Bulletin n. 164, vol. 1: "Internal erosion processes and engineering assessment", Vol. 2: Case histories, investigations, testing, remediation and surveillance" (2017)
2. Terzaghi, K.: Theoretical Soil Mechanics. Wiley, Hoboken (1943)
3. Foster, M., Fell, R.: Assessing embankment dam filters that do not satisfy design criteria. J. Geotech. Geoenviron. Eng. **127**(5), 398–407 (2001)
4. Sherard, J.L., Dunnigan, L.P.: Critical filter for impervious soils. J. Geotech. Eng. **115**(7), 546–566 (1989)
5. Wan, C.F., Fell, R.: Investigation of internal erosion by the process of suffusion in embankment dams and their foundations. In: Fell, R., Fry, J.J. (eds.) Internal Erosion of Dams and their Foundations. Taylor & Francis Group, London (2007). ISBN:978-0-415
6. Vardoulakis, I.: Fluidization in artesian flow conditions: hydromechanically unstable granular media. Geotechnique **54**(3), 165–177 (2004)
7. Sakthivadivel, R.: Theory and mechanism of filtration of non-colloidal fines through a porous medium. Tech. rep. HEL 15-5. University of California, Berkele (1966)
8. Litwniszyn, J.: Colmatage-scouring kinetics in the light of stochastic birth-death process. Bull. Acad. Pol. Sci. Ser. Sci. Tech. **14**(9), 561 (1966)
9. Saada, Z., Canou, J., Dormieux, L., Dupla, J.C., Maghous, S.: Modellimg of cement suspension flow in granular porous media. Int. J. Numer. Anal. Methods Geomech. **29**, 691–711 (2005)
10. Indraratna, B., Vafai, F.: Analytical model for particle migration within base soil - filter system. J. Geotech. Geoenviron. Eng. **123**(2), 100–109 (1997)
11. Locke, M.R.: Analytical and laboratory modelling of granular filters for enbankment dams. University of Wollongon, Faculty of Engineering, Research Online (2001)

12. Federico, F., Montanaro, A.: Granulometric stability of moraine embankment dam materials. Theoretical procedure and back-analysys of cases. In: 6th International Conference on Dam Engineering, Lisbon, Portugal, February 15–17, 2011 (2011)
13. Federico, F., Cesali, C.: A numerical procedure to simulate particles migration at the contact between different materials in earthfill dams. In: 26th Annual Meeting of European Working Group on Internal Erosion, Milan, September 2018, LNCE, vol. 17, pp. 124–136. Springer (2018)
14. Federico, F.: Particle migration phenomena related to hydromechanical effects at contact between different materials in embankment dams. Granul. Mater. ISBN 978-953-51-5423-5, INTECH. Sakellariou, M.G. (ed.) NTUA Greece (2017)
15. Bouziane, A., Benamar, A., Tahakourt, A.: Finite element analysis of internal erosion effect on the stability of dykes. In: 26th Annual Meeting of European Working Group on Internal Erosion, Milan, September 2018, EWG-IE 2018, LNCE 17, pp. 113–123. Springer (2018)
16. Silveira, A.: An analysis of the problem of washing through in protective filters. In: Proceedings of 6th International Conference on Soil Mechanics and Foundation Engineering, Montreal, vol. 2, pp. 551–557 (1965)
17. Scheuermann, A., Williams, D.J., To, H.D.: A new simple model for the determination of the pore constriction size distribution. In: 6th International Conference on Scour and Erosion (ICSE-6), Société Hydrotechnique de France, pp. 295–303 (2012)
18. Musso, A., Federico, F.: Geometrical probabilistic approach to the design of filters. Ital. Geotech. J. **XVII**(4), 173–193 (1983)
19. Kovács, G.: Seepage Hydraulics, p. 6519. Elsevier, Amsterdam (1981). ISBN:9780080870014
20. Reddi, L.N., Ming, X., Hajra, M.G., Lee, I.M.: Permeability reduction of soil filters due to physical clogging. J. Geotech. Geoenviron. Eng. **126**, 236–246 (2000)
21. Dos Santos, R.C., Caldeira, L., Das Neves, E.M.: Laboratory tests for evaluation of the actions limiting the progression of internal erosion in zoned dams. In: Proceedings of 25th ICOLD CONGRESS, Stavanger (June 2015)
22. Cazzuffi, D., Moraci, N., Mandaglio, M.C.: Hydraulic properties, behavior, and testing of geotextiles. In: Geotextiles, pp. 151–176 (2016)

An Experimental Study on the Effect of Soil-Structure Interface on the Occurrence of Internal Erosion

Min-koan Kim[1]([✉]), Qian He[2], and Ying Cui[1]

[1] Institute of Urban Innovation, Yokohama National University, Yokohama, Japan
kim-minkoan-zb@ynu.jp

[2] School of Civil Engineering, Dalian University of Technology, Dalian, China

Abstract. Internal erosion affects the stability of embankment dams and other earthen structures, and usually develops at the interface of multiple layer surfaces with different particle sizes. It is characterized by weak areas formed in the embankments around water sluices and pressure conduits. This study presents the phenomenon of internal erosion aided by lab experimentations. Permeability and particle size distribution tests were performed under varying characteristics of the soil-structure interface. Weak zone development in a structure with fine soil particles and vulnerable to displacement was a function of the flow velocity of the water. Surface roughness at the interface adds to the displacement potential of fine particles. The smooth surface interface helps increase the flow velocities, thereby enhancing the rate of internal erosion. Loss of fine particles at the interface helps significantly increase erosion, especially at the boundary, resulting in intense erosive action. A significant difference was observed between the erosion potential of rough and smooth concrete structures and the soil for nearly half of the total area of the sample, which was eroded in both cases. The variation of velocity is mainly seen at the embedded structure's interface in the soil, and remained constant around the other regions of the soil sample, indicating no specific impact. The erosion of the soil particles was mainly dependent on the distance from the soil-structure interface.

Keywords: Contact erosion · Boundary effect · Levee failure · Soil-structure interface

1 Introduction

In general, failures caused by internal erosion of embankment dams and their foundations are classified into three general failure modes. Internal erosion associated with the structure through and through, internal erosion and foundation with internal bank erosion [1]. The term internal erosion here refers to the separation of soil particles from the soil structure due to mechanical or chemical action of the seepage flow, followed by the formation of piping structures. For internal erosion, fine particles are eroded by the seepage flow, which flows between larger particles to leave a coarse skeleton behind.

© Springer Nature Switzerland AG 2020
J.-M. Zhang et al. (Eds.): ICED 2020, SSGG, pp. 253–260, 2020.
https://doi.org/10.1007/978-3-030-46351-9_25

Progressive erosion and soil transportation result in the formation of pipe structures in the soil, which lead to soil surface instability [2].

Internal erosion has been widely investing under set boundary conditions. Increased water permeability around underground structures and the concentration of water ingress can result from sandy ground [3]. Additionally, a critical hydraulic gradient for erosion inside the interface, shear strength of the interface, severity of the soil and the porous soil structure affect its permeability and in turn, the erosion of the interface [4]. Internal erosion is initiated due to the breach of the minimum flow threshold [5]. The repeated water level movements result in an intense erosion of fine particles, and these changes make the soil structure unstable [6–8].

There is a lack of research on the evaluation of the impact of internal erosion on the loss of ground or velocity of seepage flow within a predefined boundary. In this study, a newly design permeability test apparatus that can specify the flow around the soil-structure interface is tested and used to investigate the effect of soil-structure interface on the permeability of the soil around ground structures.

2   Outline of the Experiment

2.1   Experimental Apparatus

An experimental apparatus that can evaluate the impact of the soil-structure boundary on the permeability of the soil has been developed. The experimental apparatus mainly consists of a soil chamber, a vacuum system, a constant-head water tank, and an outflow system with weighing equipment, as shown in Fig. 1.

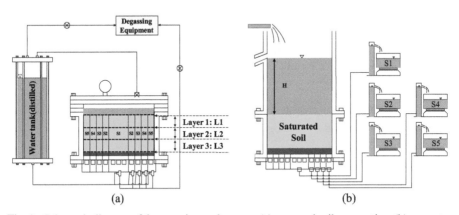

Fig. 1. Schematic diagram of the experimental system; (a) saturated soil preparation, (b) apparatus setup for water permeability test

The schematic diagram of the soil chamber for the preparation of the saturated soil is shown in Fig. 1(a). The bottom of the soil chamber is divided into five sections (S1–S5), and the outflow system is connected to each section. The water first passes through the soil specimen and drains out from the five different sections. The weight of the outflowed water is measured by means of electronic balances (RJ-3200) and the permeability of each section (S1–S5 of Fig. 1(a)) is then evaluated. The specimen is then saturated using the vacuum method. The capped soil chamber and the water tank are connected to a vacuum pump that removes the residual air in the soil specimen. The water tank is then connected to the soil chamber to introduce the water into the soil specimen to ensure complete saturation. The saturated soil is then tested for permeability, where a constant-head water tank is used to apply the water head on the soil chamber, as shown in Fig. 1(b).

2.2 Erodible Soil Materials

Erodible soil material, as referred to in a previous study by Ke and Takahashi [2], is chosen as the soil material to evaluate the potential of internal instability of the soil. The soil sample was prepared by mixing silica sand Nos. 3 and 8 using a weight ratio of 4:1. The particle size distribution of silica sand Nos. 3 and 8 and the mixed soil are shown in Fig. 2. The potential for internal erosion is evaluated by Kenney and Lau's method [9]. The H/F ratio was used to determine the soil stability, where H is the mass fraction between grain size D and 4D, and F represents the mass fraction at any grain size D, as shown in Fig. 2. The erodible particle size in this soil material was found to be 0.425 mm, where H/F < 1.3.

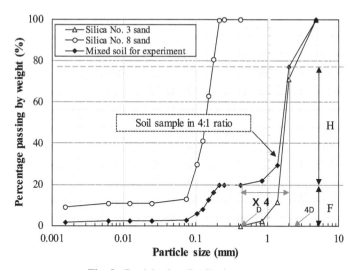

Fig. 2. Particle size distribution curves

2.3 Modeling of the Concrete Structure and Experimental Cases

The impact of the soil-structure boundary and the roughness of the structure surface are discussed in this section. A 270 mm diameter cylinder is used to model the concrete structure buried in the soil, as shown in Fig. 3. Furthermore, the roughness of the surface of the concrete structure is varied by paste-up the sand, with a grain size of 2 mm.

(a) different concrete of roughness   (b) the structure in the soil used in the test

Fig. 3. Testing specimen apparatus for interface roughness

Table 1 shows the details of each experimental case for the soil material with two different compaction degrees (Dc) of 60% and 90%, respectively. In all the cases, the initial water content was set to 3% to ensure uniformity during the mixing and compaction process.

Table 1. Experimental cases

	Soil only	Concrete structure in soil	
		Smooth concrete	Rough concrete (with 2 mm sand)
Dc = 60% (ρ_d = 1.65 g/cm^3)	Soil_60	SC_60	RC_60
Dc = 90% (ρ_d = 1.73 g/cm^3)	Soil_90	SC_90	RC_90

2.4 Experimental Procedure

The experiment process consists of the three parts; soil specimen preparation, saturation, and the constant head permeability test. The detailed procedure is given as follows.

Step 1: Silica sand No. 3, which has a grain size of 20 mm and a specific density, was compacted. This layer behaves as the drainage layer to ensure the eroded particles do not accumulate at the bottom of the soil chamber. For the experiments with the concrete structure, the concrete model was installed before ground compaction.

Step 2: The silica sand and water were mixed well in a specific weight ratio to obtain the pre-designed relative density and molded to a height of 100 mm. The dry density of each of the specimens is shown in Table 1.

Step 3: The soil chamber was capped and connected to the vacuum-pump and water tank, as shown in Fig. 1(a). The specimen was then saturated using the vacuum method. Water was introduced from the bottom to the top, and the saturation was stopped when the water reached the top of the specimen. The cap was then removed after decreasing the vacuum pressure when the specimen was saturated.

Step 4: The constant water level permeability test was then conducted with the Waterhead difference of 200 m (hydraulics gradient i = 2) after installing the acrylic cylinder, which supplied the water at a constant water level, as shown in Fig. 1(b). Simultaneously, the flow from the five sections, as shown in Fig. 1, were measured using the electric balance for the entire testing process. This flow was used to calculate the flow velocity at different sections, to investigate the influence of the soil-structure boundary on the permeability.

Step 5: A sieving test was then conducted to identify the movements of the fine particles. Further, samples were collected from 15 different areas of each of the different sections (S1–S5) and layers (L1–L3), as shown in Fig. 1. The erosion index, calculated from the PSD curve, was then used to discuss the occurrence of internal erosion in each area of the soil.

3 Testing Results

3.1 The Impact of the Soil-Structure Interface on the Permeability

Figures 4 and 5 show the temporal change of the flow velocity, calculated for the five sections at different distances from the soil-structure surface. The velocities are calculated every 1 min. Figure 4 represents the loose sand with a compaction degree of 60%, while Fig. 5 represents dense sand (Dc = 90%).

It can be seen that for all the cases, the soil velocity is almost the same for all the sections for each of the samples tested. Moreover, the flow velocity for the sample Dc 60 is almost twice as that for the sample Dc 90. When the concrete structure is installed in the area S1, the velocity around the soil-structure surface (S1) increases significantly, while there is almost no change in the other areas (S2–S5).

With respect to the roughness of the soil-structure boundary, it can be seen that the velocity for the smooth concrete sample shows a slightly larger value as compared to that of the rough concrete sample. The above results indicate that the velocity is significantly larger near the soil-structure boundary, and is affected by the roughness of the boundary.

3.2 The Impact of the Soil-Structure Interface on the Occurrence of Erosion

A sieving test was conducted on the 15 samples collected from each of the sections and layers mentioned earlier, to investigate the impact of the soil-structure interface on the occurrence of erosion. Figure 6 shows the example of particle size distribution (PSD) curves for the 15 different areas for the case of RC_60. The circular marks (black symbols) show the PSD for the initial soil material before the experiment, and the diamond-shaped marks (red symbols) shows the PSD for the eroded soil after the experiment was conducted. The figure shows that the PSD is significantly affected by seepage in section S1, closest to the soil-structure, as compared to other sections (S2–S5). Additionally, by comparing the PSD for samples at different heights from the bottom, it can be seen that the change in PSD near the bottom (L3) is more significant than the top layer (L1) of the specimen.

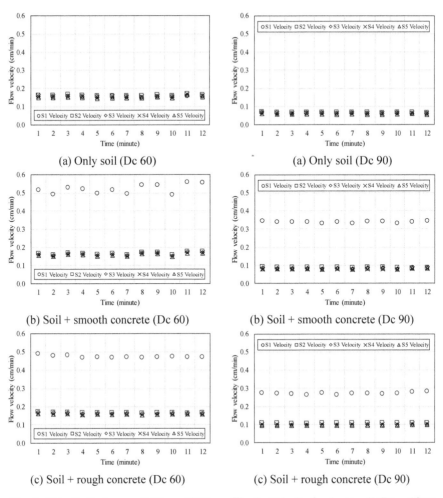

Fig. 4. Velocity for loose soil (Dc = 60%)   **Fig. 5.** Velocity for dense soil (Dc = 90%)

The grading index I_G proposed by Muir Wood [10] was used to evaluate the variation of the PSD and discuss the erosion condition of the samples. I_G represents the changes in particle grading and is defined as the ratio of the area under the current grain size distribution to the area under the initial grain size distribution before the seepage test was conducted. I_G equals one prior to erosion and decreases when erosion occurs. The distribution of the I_G values for the 15 different areas are as shown in Figs. 7 and 8. Figure 7 shows the result for the loose ground condition (Dc = 60%) and Fig. 8 shows the results for the dense ground condition (Dc = 90%).

In the cases without the concrete structure (Fig. 7(a) and Fig. 8(a)), irrespective of the compaction degree, the IG value lies close to 1, i.e., the movement of the fine particles is relatively small. Moreover, the figures suggest that layers L1 and L3 were exposed to a higher degree of erosion than L2. Additionally, in the case where only soft concrete was used, the layer near the structure demonstrated a higher degree of erosion. A similar result was found for the rough concrete cases with additional erosion areas in the upper

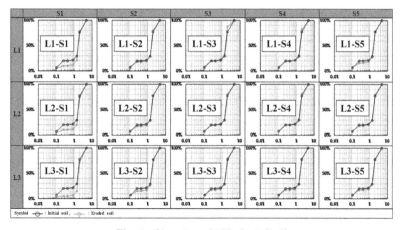

Fig. 6. Changing of PSD for RC_60

	S1	S2	S3	S4	S5
L1	1.00	0.97	0.98	0.98	1.00
L2	1.00	1.00	1.01	1.02	1.03
L3	1.00	0.97	0.97	0.98	1.02

(a)   Only soil (Dc 60)

	S1	S2	S3	S4	S5
L1	0.93	1.00	0.99	1.01	0.98
L2	0.92	0.97	1.00	0.97	0.99
L3	0.91	0.98	1.01	1.02	1.02

(b)   Soil + smooth concrete (Dc 60)

	S1	S2	S3	S4	S5
L1	0.83	0.99	1.06	1.02	0.95
L2	0.83	0.98	1.02	1.03	1.00
L3	0.72	0.93	0.98	0.98	0.98

(c)   Soil + rough concrete (Dc 60)

Fig. 7. IG for each area (Dc = 60%).

	S1	S2	S3	S4	S5
L1	1.02	0.99	0.99	1.00	0.99
L2	1.01	0.98	0.99	1.00	1.03
L3	1.01	0.94	0.96	0.93	1.06

(a) Only soil (Dc 90)

	S1	S2	S3	S4	S5
L1	0.89	0.94	1.01	0.97	1.02
L2	0.94	1.05	1.01	1.00	1.01
L3	0.91	0.97	0.98	0.99	1.05

(b) Soil + smooth concrete (Dc 90)

	S1	S2	S3	S4	S5
L1	0.83	0.98	1.00	1.01	0.98
L2	0.92	1.00	1.02	1.02	1.02
L3	0.78	0.95	0.98	0.98	0.96

(c) Soil + rough concrete (Dc 90)

Fig. 8. IG for each area (Dc = 90%)

and lower areas of the structure. Figure 7 and 8 show the IG values based on the Dc values (60 & 90) for all the samples.

In the absence of the structure, erosion at the upper and lower levels was noted as the leading cause. In the case of the soft structures, erosion was visible near the structure's surface, and was more significant in the case where rough concrete was used. Additionally, although the difference in flow velocity was doubled (as seen in Figs. 4 and 5), there was no significant difference in the IG value between the samples based on their degree of compaction. Studies have shown that the reason for reduced erosion despite the high flow rates is attributed to the roughness of the sample with particle sizes less than 3 mm, which protects the particles against erosion.

4 Conclusions

Based on the results, the weak zone development in the aligning structure, in which the fine soil particles are vulnerable to displacement, is a function of the flow velocity around the interface. The surface roughness at the interface adds to the displacement potential of the fine particles. It was found that the flow velocity through the sample in the case where the Dc of 60% was almost twice the velocity in the case where the Dc of 90%. Consequently, more erosion was seen in the soil with a Dc of 60%. The interface with a smooth surface assists in increasing the flow velocities, thereby enhancing the rate of internal erosion. The loss of fine particles at the interface of the two surfaces helps in increasing the rate of erosion significantly, especially at the boundary. Hence, intense erosive action was seen at the soil-structure interface. There was a significant difference between the erosion potential of the rough and smooth concrete structures and the soil was eroded over nearly half of the total area of the sample in both cases. As the permeability flow is allowed to continue for an extended period, erosion deceases at the lower layer of soil. This phenomenon is due to the accumulation and clogging of fine eroded particles in the rough edges of the soil-structure interface. The variation in velocity is mainly seen in the embedded structure's interface in the soil. The flow velocity remained constant around the other regions of the soil sample indicating no specific impact. The erosion of the soil particles was mainly dependent on the distance from the soil-structure interface and the impacts of the distance on the erosion are being studied.

References

1. Bonelli, S.: Erosion in Geomechanics Applied to Dams and Levees (2013)
2. Ke, L., Takahashi, A.: Strength reduction of cohesionless soil due to internal erosion induced by one-dimensional upward seepage flow. Soils Found. 52(4), 698–711 (2012)
3. Sato, M., Kuwano, R.: Basic study for permeability change around buried structure in sandy ground. In: International Conference of Institute of Industrial Science, Tokyo University, vol. 61, no. 4, pp. 67–70 (2010)
4. Xie, Q., Liu, J., Han, B., Li, H., Li, Y., Li, X.: Critical hydraulic gradient of internal erosion at the soil-structure interface. Processes 6(7), 92 (2018)
5. Jiang, M., Konrad, J., Leroueil, S.: An efficient technique for generating homogeneous specimens for DEM studies. Comput. Geotech. 30(7), 579–597 (2003)
6. Wan, C., Fell, R.L.: Assessing the potential of internal instability and internal erosion in embankment dams and their foundations. J. Geotech. Geoenviron. Eng. 134(3), 401–407 (2008)
7. Gaber, F., Bowman, E.T.: The role of seepage flow rate and deviatoric stress on the onset and progression of internal stability in a gap-graded soil. In: Bonelli, S., Jommi, C., Sterpi, D. (eds.) Internal Erosion in Earthdams, Dikes, and Levees, EWG-IE 2018. Lecture Notes in Civil Engineering, vol. 17, pp. 50–59. Springer, Switzerland (2019)
8. Takizawa, I.: Study on erosion deterioration of soil samples at the boundary between soil and structure (2018)
9. Kenney, T.C., Lau, D.: Internal stability of granular filters. Can. Geotech. J. 22, 215–225 (1985)
10. Muir Wood, D.: The magic of sands, The 20th Bjerrum Lecture presented in Oslo 25 November 2005. Can. Geotech. J. 44, 1329–1350 (2007)

CFD-DEM Simulations of Seepage Induced Erosion

Qiong Xiao$^{(\boxtimes)}$

Southeast University, Nanjing 210096, People's Republic of China
xiaoqiong@seu.edu.cn

Abstract. The increase of seepage force would reduce the effective stress of particles and result in the erosion of particles for the heave failure. The sheet piles/cutoff walls are often employed in the dams to control the seepage. In this paper, a solver of computation fluid dynamics (CFD) involving two-fluid phases is developed and coupled with the discrete element method (DEM) software for mimicking the piping process. The binary-sized particles are selected to study the impact of fine particles on the mechanisms of seepage. The results demonstrated that the developed software could successfully model the seepage process. It adopts the characteristics of particle displacements, drag force and porosity to illustrate the erosion mechanism. The results imply the seepage impact triggers the movement of particles in the downstream side with a looser condition for the skeleton, which resulting in the transportation of particles in the upstream side to achieve a flow channel.

Keywords: Seepage · Hydraulic gradient · Fine particles · Drag force · Porosity

1 Introduction

The erosion phenomenon contains two categories as surface erosion and internal erosion depending on the form of surface flow and internal seepage respectively [1]. It mainly occurs within the water retaining structures, such as the dams and dikes, with a change of permeability, porosity, pore pressure, shear resistance and internal structure for the system. This would affect the internal stability of structures, threaten peoples safety and cause economic loss. Foster et al. [2] found 46% of dam failure was occurred due to piping through the embankment based on the incidents taken from the International Commission on Large Dams (ICOLD). Hence, it is necessary to understand the occurrence and mechanisms of erosion induced by seepage in depth.

Numerous categories of experiment method could be found to study the erosion of soils, as the flume test [3], rotating cylinder test [4, 5], hole erosion test [6] and et al. In literature, the internal erosion is suggested depending on the grain size distribution, stress state and anisotropic condition of the system [7–9]. With the increase of clay content, it could reduce the flow density and increase the internal erosion resistance [8, 9]. Furthermore, a significantly larger local hydraulic gradient is observed around the cutoff walls, which are often employed to control the seepage of dams [10, 11]. With the development of techniques, it draws attention to performing numerical simulations for

© Springer Nature Switzerland AG 2020
J.-M. Zhang et al. (Eds.): ICED 2020, SSGG, pp. 261–266, 2020.
https://doi.org/10.1007/978-3-030-46351-9_26

understanding the mechanisms of internal instability, for instance, the DEM approach without the corporation of the fluid phase [12]. In recent decades, the coupled DEM-CFD technique becomes a popular method, involving the information of continuum-scale and particle-scale, to conduct simulations [13–16]. Tao and Tao [16] employed this technique and found most of the orientation of contact forces concentrated on a more horizontal direction during the erosion process.

It is believed that the factors of particle size distribution/gap-graded particles, the porosity of the system and hydraulic gradient are vital for the process of a piping phenomenon. Works reported in the literature were mainly with one fluid phase circumstance and the hydraulic condition is achieved with the change of fluid velocity, which could not simulate the true excavation process and give rise to this work. A solver of Open-Foam is developed to an open-sourced CFD-DEM software for performing simulations containing two-fluid phases. The binary-sized particles are adopted to study the impact of fine particles on the mechanisms of seepage with results presented in Sect. 3. Finally, it gives a summary of this paper.

2 Methodology

The CFD-DEM approach, coupled with the open-sourced software OpenFOAM and LIGGGHTS, is employed here to investigate the multi-phases properties. The fluid flow is simulated by the OpenFOAM software from solving the Navier-Stokes equations. A solver named as interFoam is developed in this study to handle the two-fluid phases condition and coupled with LIGGGHTS software for performing the simulations. The motion of incompressible flow is governed by the volume-averaged Navier-Stokes equation as,

$$\frac{\partial\left(\alpha_f \rho_f \boldsymbol{u}_f\right)}{\partial t} + \nabla \cdot \left(\alpha_f \rho_f \boldsymbol{u}_f \boldsymbol{u}_f\right) = -\alpha_f \nabla p - \boldsymbol{K}_{sl}\left(\boldsymbol{u}_f - \boldsymbol{u}_p\right) + \nabla \cdot \boldsymbol{\tau} \qquad (1)$$

where $\boldsymbol{u}_f$ and α_f denote the velocity and porosity of fluid flow respectively. $\boldsymbol{u}_p$ refers to the averaged particle velocity. ∇ is the gradient operator. ρ_f is the density of fluid, calculated as $\rho_f = \alpha_1 \rho_1 + \alpha_2 \rho_2$, in which α_1 and α_2 are the phase fraction parameters, and ρ_1 and ρ_2 are the corresponded density of fluid phase. The stress tensor for the fluid phase is calculated as $\boldsymbol{\tau} = \nu_f \nabla \boldsymbol{u}_f$, in which ν_f is the fluid viscosity. $\boldsymbol{K}_{sl}$ denotes the momentum exchange with the particulate phase.

The CFD method divides the momentum exchange into an implicit and an explicit term, stemming from either the hydrostatic forces or the hydrodynamic forces. The typical hydrostatic force is buoyancy force. In this study, the hydrodynamic fluid-interaction force is regarded as the Di Felice drag force. For each cell, the volumetric fluid-particle interaction force ($\boldsymbol{F}_d$) is determined by the total drag force of particles inside the cell.

Regarding the coupling procedures, the DEM solver first gives the positions of particles and calculates the velocities, and then passes the information to the CFD solver for determining the corresponding cells of particles. Based on the particle volume fraction of fluid cells, it estimates the fluid forces acting on the particles and returns the results to the DEM solver for updating the positions and velocities of particles. The simulation is conducted by repeating these procedures.

3 Macro-micro Scale Observations of the Seepage Erosion

The spherical particles are adopted for the simulation using the Hertz contact model, with the properties of materials presented in Table 1. The parameter of stiffness has a negligible impact on erosion behavior [13], and it is thus reduced to 5 MPa for speeding up the simulations. In the simulation process, the steps for data exchange between DEM and CFD is determined as 10. Hence, the timestep of DEM solver and CFD solver are set as 5e-6s and 5e-5s respectively. The binary-sized spheres are selected to study the seepage failure, with the mass ratio set as 0.3 and 0.7 for the fine particles and coarse particles respectively.

Table 1. Properties of particles and boundary

Young's modulus	Radii of particles	Friction coefficient	Poission's ratio	Restitution coefficient	Gravity
5 MPa	$r_{min} = 0.25$ mm $r_{max} = 1$ mm	0.85	0.45	0.1	9.81 N/m^2

The numerical procedures contain two steps. First, it only uses DEM software for randomly generating and depositing particles into the container to prepare the sample under a quasi-static condition. Then, the region of the retaining wall is cleared via deleting particles for importing the boundary. After that, the specimen is submerged with fluid flow to perform simulations.

The literature shows that the piping phenomenon mainly appears around the sheet piles. Hence, it adopts a narrow simulation boundary in this work as shown in Fig. 1(1) for saving the computation effort. To mimic the excavation condition, the downstream side is determined as the left side of the retaining wall with the soil deposit of 0.03 m. Regarding the upstream side, it refers to the right side of the boundary with a deposit height of 0.045 m.

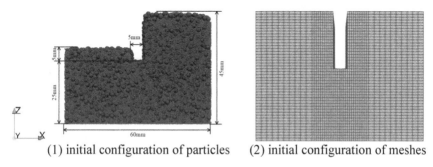

(1) initial configuration of particles   (2) initial configuration of meshes

Fig. 1. The initial configuration for the numerical simulation

Different water tables are set at the upstream side to study the impact of hydraulic gradient on the seepage failure phenomenon, with the water level defined at 0.025 m for the downstream side. Benmebarek et al. [17] found that heave behaviour was difficult to be captured using the coarse meshes. For speeding up the simulations, it adopts fine meshes for the central area. Regarding the other regions, they are constructed with coarse meshes as shown in Fig. 1(2).

The process of seepage failure consists of three main stages as, (1) first visible/initial movement of particles, (2) heave progression to form the seepage path, (3) total heave. The spatial distribution of particle velocity is studied to observe the initial erosion behaviour at $t = 0.1$ s. For each hydraulic condition, the high magnitude appears around the boundary. With the raising of the water table in the upstream side (h), the particle velocity is also enhanced, in particular for the area close to the retaining wall (Fig. 2).

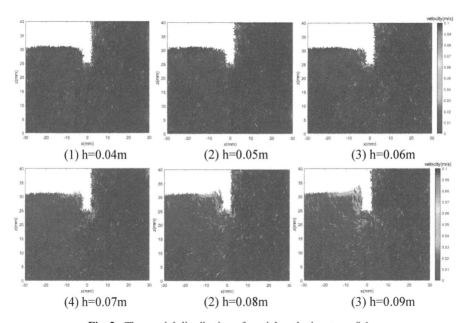

Fig. 2. The spatial distribution of particle velocity at $t = 0.1$ s

The seepage impact on heave behavior is studied in terms of the highest position for particles located on the downstream side. According to Fig. 3(1), the phenomenon is initially occurred at $h = 0.06$ m and becomes more significant with the increase of h value. Taken $h = 0.09$ m as an example, it depicts the behavior of particles displacement in Fig. 3(2). The fluid flow induces higher magnitudes around the retaining wall and lifts some particles out of the soil skeleton. However, regarding the region below the boundary, it is under a relatively static condition. Figure 4 displays the spatial distribution of peak drag force to further understand the seepage failure. The behavior is similar to the observations of the displacements of particles.

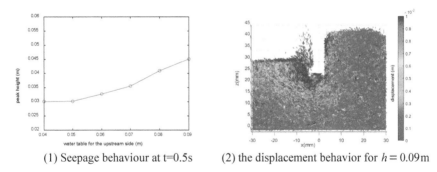

(1) Seepage behaviour at t=0.5s   (2) the displacement behavior for $h = 0.09$m

Fig. 3. Observations of the seepage impact

Fig. 4. Observations of seepage impact at t $= 0.5$ s for h $= 0.09$ m

Figure 5 portrays the temporal-spatial distribution of the porosity for the specimen of $h = 0.09$ m. Depending on the detachment of particles, it produces a higher porosity adjacent to the retaining wall in the downstream side and then regressed to the upstream side during erosion for forming the flow channel. Combined with the results of drag force and displacement, it could be deduced that the occurrence of erosion mainly depending on the drag force for triggering the movement of particles, which would then alter the porosity and internal structure.

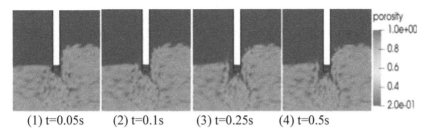

(1) t=0.05s   (2) t=0.1s   (3) t=0.25s   (4) t=0.5s

Fig. 5. The evolution of porosity of specimen during seepage for h $= 0.09$ m

4 Summary

This paper illustrates the CFD-DEM simulation of heave behaviour subjected to the seepage impact. It demonstrates that the developed technique successfully mimicked the piping process. The results show that the seepage erosion depending on the significant large drag force for transporting particles. The migration of particles through soil skeleton would then increase the porosity adjacent to piles/sheet boundary and forms a channel from the downstream side to the upstream side during the seepage process.

References

1. Sterpi, D.: Effects of the erosion and transport of fine particles due to seepage flow. Int. J. Geomech. **3**(1), 111–122 (2003)
2. Foster, M., Fell, R., Spannagle, M.: The statistics of embankment dam failures and accidents. Can. Geotech. J. **37**, 1000–1024 (2000)
3. Arulanandan, K., Perry, E.B.: Erosion in relation to filter design criteria in earth dams. J. Geotech. Eng. **109**(5), 682–698 (1983)
4. Chapuis, R.P., Gatien, T.: An improved rotating cylinder technique for quantitative measurements of the scour resistance of clays. Can. Geotech. J. **23**, 83–87 (1986)
5. Chang, D.S., Zhang, L.M.: Critical hydraulic gradients of internal erosion under complex stress states. J. Geotech. Geoenviron. Eng. **139**(9), 1454–1467 (2013)
6. Reddi, L.N., Lee, I.-M., Bonala, M.: Comparison of internal and surface erosion using flow pump tests on a sand-kaolinite mixture. Geotech. Test. J. **23**(1), 116–122 (2000)
7. Tomlinson, S.S., Vaid, Y.P.: Seepage forces and confining pressure effects on piping erosion. Can. Geotech. J. **37**, 1–13 (2000)
8. Sato, M., Kuwano, R.: Laboratory testing for evaluation of the influence of a small degree of internal erosion on deformation and stiffness. Soils Found. **58**(3), 547–562 (2018)
9. Xie, Q., Liu, J., Han, B., Li, H., Li, Y., Li, X.: Critical hydraulic gradient of internal erosion at the soil–structure interface. Processes **6**(7), 92 (2018)
10. Moffat, R., Fannin, R.J., Garner, S.J.: Spatial and temporal progression of internal erosion in cohesionless soil. Can. Geotech. J. **48**(3), 399–412 (2011)
11. Luo, Y., Nie, M., Xiao, M.: Flume-scale experiments on suffusion at bottom of cutoff wall in sandy gravel alluvium. Can. Geotech. J. **54**(12), 1716–1727 (2017)
12. Shire, T., O'Sullivan, C., Hanley, K.J., Fannin, R.J.: Fabric and effective stress distribution in internally unstable soils. J. Geotech. Geoenviron. Eng. **140**(12), 04014072 (2014)
13. Tang, Y., Chan, D., Zhu, D.: A coupled discrete element model for the simulation of soil and water flow through an orifice. Int. J. Numer. Anal. Meth. Geomech. **41**(10), 1477–1493 (2017)
14. Zou, Y., Chen, C., Zhang, L.: Simulating progression of internal erosion in gap-graded sandy gravels using coupled CFD-DEM. Int. J. Geomech. **20**(1), 0401935 (2019)
15. El Shamy, U., Denissen, C.: Microscale characterization of energy dissipation mechanisms in liquefiable granular soils. Comput. Geotech. **37**(7–8), 846–857 (2010)
16. Tao, H., Tao, J.: Quantitative analysis of piping erosion micro-mechanisms with coupled CFD and DEM method. Acta Geotech. **12**(3), 573–592 (2017)
17. Benmebarek, N., Benmebarek, S., Kastner, R.: Numerical studies of seepage failure of sand within a cofferdam. Comput. Geotech. **32**(4), 264–273 (2005)

Microscopic Aspects of Internal Erosion Processes in Gap-Graded Soils

Yanzhou Yin[1,3], Yifei Cui[2(✉)] [iD], and Yao Jiang[1,3,4] [iD]

[1] Key Laboratory of Mountain Surface Process and Hazards/Institute of Mountain Hazards
and Environment, Chinese Academy of Sciences, Chengdu 610041, China
[2] State Key Laboratory of Hydroscience and Engineering, Tsinghua University,
Beijing 100084, China
yifeicui@mail.tsinghua.edu.cn
[3] University of Chinese Academy of Sciences, Beijing 100049, China
[4] CAS Center for Excellence in Tibetan Plateau Earth Sciences, Chinese Academy of Sciences,
Beijing 100101, China

Abstract. Internal erosion processes in soils play an important role on the instability analyses of hillslopes and embankment dams. Field observations support the assumption that the internal fine particles may migrate among the channels formed by coarser particles under the high hydraulic gradient condition, where the enrichment of fine particles has great potential on the increase of local pore-water pressure due to their low permeability. Although a number of traditional seepage experiments in laboratory have provided data showing the effect of soil properties on the macroscopic permeability, however, much remains unknown particularly for microscopic erosion processes. Therefore, in the current study, a series of one-dimensional soil seepage tests were firstly conducted by controlling the coarse to fine particle size ratio, and then the X-ray tomography tests were carried out at beamline BL13W1 at the Shanghai Synchrotron Radiation Facility (SSRF) to obtain the particle distributions and three-dimensional pore structures. By coupling discrete element method (DEM) with Darcy's law, the internal particle erosion processes were back-analyzed. The results reveal that the preferential erosion can occur in the top and bottom regions of the soil specimen, and the migrated fine particles can be supplied when the pore size is large enough along the seepage path.

Keywords: Internal erosion · Gap graded soil · Seepage test · X-ray tomography · Discrete element method · Fluid-solid coupling

1 Introduction

The migration of fine particles in porous media, also known as internal erosion, has been substantially investigated by researchers from different disciplines. Consequently, the theoretical research has been applied in the fields of piping [1], slope failure initiation [2] and oil extraction [3, 4], and particle migration has become a multi-discipline research topic in past 20 years. Internal erosion is frequently found within dams and is considered

© Springer Nature Switzerland AG 2020
J.-M. Zhang et al. (Eds.): ICED 2020, SSGG, pp. 267–273, 2020.
https://doi.org/10.1007/978-3-030-46351-9_27

as one of the major causes of dam failure initiation. Anderson [5] reported that the Baldwin Hills dam failure was initiated by internal erosion, subsequently killed five people. The analysis of 206 earth dams in the United States found that 29% of dam failures were caused by internal erosion, and 44% and 40% of dams in Japan and Switzerland were due to internal erosion, respectively [6].

Kenney and Lau [7] stated that the soil is composed of soil skeleton and loose fine particles, and proposed a method that could determine the loss of movable fine particles based on the particle gradation curve, which plays an important role in the laboratory study of piping. Ke and Takahashi [8] carried out the one-dimensional upward seepage tests to study the effect of fine particle content, relative density of soil, and hydraulic gradient on the effects of internal erosion and subsequent soil strength. Besides experimental approach, many researchers further analyzed fluid and particle characteristics such as velocity and force during internal erosion by using numerical simulation, such as Computational Fluid Dynamics (CFD) coupled with Discrete Element Method (DEM) [1, 9]. However, both experimental and numerical approaches are still facing challenges. Currently, the laboratory 1D seepage test is not able to directly obtain the variation of migration particles and pore structure during the seepage process, thus not able to provide quantitative feedback for numerical verification. In this paper, imaging technology was introduced into the research method of this problem. The soil samples under the seepage effect were scanned by X-ray, and the three-dimensional structure of soil was reconstructed. The discrete element coupling with Darcy flow was then used to back-analyze the internal erosion process of soil.

2 Methodology

2.1 Experimental Test

The computerized tomography (CT) with high energy is able to visualize the soil sample's micro-pore structure. Furthermore, numerical method can track these fine particles' status including position and velocity over time. Coupling the quasi dynamic CT and the dynamic simulation is an available method to understand the internal erosion problem.

The computerized tomography operation was carried out at the BL13W1 beamline at the Shanghai Synchrotron Radiation Facility. The photon energy was set as 33 keV to permit monochromator work well. The soil samples with fine particle size range from 72 to 100 μm and coarse particle size range from 600 to 800 μm were put into the PVC by compacting, and were firstly experienced the seepage in 1D column test (see Fig. 1a) without confining pressure under the hydraulic gradients set as 21 for 20 h. After the test, samples were frozen and transported to the CT laboratory for scan (see Fig. 1b). The spatial resolution of CT scan was set as 9 μm to make sure enough mesh to reconstruct the fine particles (8–10 mesh element is required for a single particle in 3D reconstruction). These initial obtained images were further processed including bit conversion, phase retrieval, phase projections and 3D reconstruction in the PITER and ImageJ software packages.

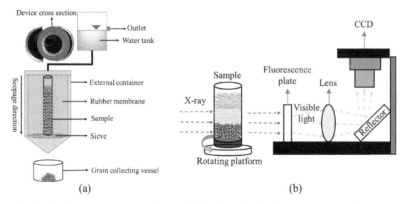

Fig. 1. (a) Device for seepage experiment; (b) Schematic of sample scanning in computerized tomography (CT) laboratory (orange, yellow and gray particles represent the upper, middle and bottom parts of the sample respectively).

2.2 Numerical Simulation

The fluid equation in saturated porous media is applied to solve the fluid action on particles, which follows the Darcy fluid and mass conservation [10]. The equation can be expressed as in Eq. 1:

$$\frac{\partial}{\partial x}(k_x \frac{\partial H}{\partial x}) + \frac{\partial}{\partial y}(k_y \frac{\partial H}{\partial y}) + \frac{\partial}{\partial z}(k_z \frac{\partial H}{\partial z}) + Q = C_w n \gamma_w \frac{\partial H}{\partial t} \tag{1}$$

where H is total fluid head including the pressure head and elevation head; k_x, k_y, and k_z are the coefficient of permeability in three principle directions in the Cartesian coordinate system; Q is the fluid flux; n is the porosity of the porous media with updates from DEM calculation; C_w is the compressibility of water which is normally assumed to be 4.4×10^{-10} Pa^{-1} in porous media; and γ_w is the unit weight of water.

The following head and flow boundary conditions are considered:

Head boundary condition: $H = H_b(t)$

Flow boundary condition: $(k_x \frac{\partial H}{\partial x})l_x + (k_y \frac{\partial H}{\partial y})l_y + (k_z \frac{\partial H}{\partial z})l_z + q(t) = 0 \tag{2}$

where $q(t)$ is the specified flow rate at the flow boundary; l_x, l_y, l_z are direction cosines of the outward unit vector perpendicular to the flow boundary.

The equation is solved by the finite difference method. And the shape functions are used to calculate the fluid velocity at any point based on the assumption of a continuum. The drag force is calculated as follows [11]:

$$\boldsymbol{f}_d = \frac{n}{1-n}\beta V_p \boldsymbol{u}_r \tag{3}$$

where $\boldsymbol{u}_r$ is the relative velocity vector between the particle and the fluid; V_p is the volume of the particle. The parameter β is related to porosity n [12, 13].

Cundall [14] firstly proposed the concept of DEM and used it to study rock slopes with fractured structures. Particle flow discrete element method is able to deal with discontinuous media. It only needs to display iterative analysis of the equations of motion and force-displacement. For the fluid-structure interaction calculation process, the sphere boundary calculation process is also relatively simple by using the discrete element method. In the two-dimensional and three-dimensional models, these elements are respectively represented by disks and spheres. There are mainly two contact relationships between the ball and the ball, and the ball and the wall (see Fig. 2). Both particles and walls follow Newton's second law and force-displacement law [15].

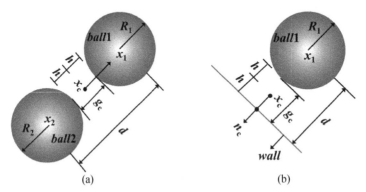

(a)   (b)

Fig. 2. The contact relationships in particle flow discrete element method (a) between particle and particle (b) between particle and wall; (g_c is the contact gap; d is the distance between the centers of the balls or the distance from the center of the ball to the wall; h is the half of contact gap; x_1, x_2 is the position of the ball 1 and the ball 2; R_1, R_2 is the radius of the ball 1 and the ball 2 respectively; n_c is the direction of the wall).

The basic equation of the discrete element method has the following form:

$$mx''(t) + cx'(t) + kx(t) = f(t) \tag{4}$$

where m is the mass of the element; x is the displacement; t is the time; k is the stiffness coefficient; f is the load to the unit. The velocity $x'(t)$ and acceleration $x''(t)$ are solved by the central difference method.

$$x'(t) = [x(t + \Delta t) - x(t - \Delta t)]/(2\Delta t)$$
$$x''(t) = [x(t + \Delta t) - 2x(t) + x(t - \Delta t)]/(\Delta t)^2 \tag{5}$$

Because real particles with irregular shapes increase the difficulty of fine particle migration, in order to better simulate the fine particle migration process, the ratio of coarse to fine particles is set as 6.0. At the same time, considering that the calculation speed of the discrete element is related to the square power of the particle mass, the diameter of the fine particles is set to 0.5 mm. And the model is set to a cube with a side length of 20 mm (see Fig. 3b), which is equivalent to the middle rectangular portion of the circular sample.

3   Results and Discussion

The CT microscopic imaging technology can be used to intuitively obtain the particle distribution, arrangement and combination type of these particles in space (see Fig. 3a). It is observed that coarse particles usually form the skeleton of the soil with the filling of fine particles. Under the action of the fluid (mainly drag force) on particles, fine particles will move along the pore channel in the seepage direction. After a period of time (20 h in the paper), the coarse particles locally displaced and rotated slightly. Part of fine particles were eroded away and even flowed out, which resulted in the formation of the larger holes in the soil sample in Fig. 3a. Furthermore, these holes provide space for the further migration of fine particles.

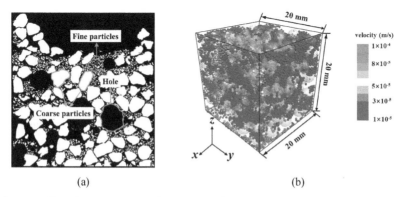

(a)   (b)

Fig. 3. (a) Y_Z section at the center of the sample after seepage test for 20 h; (b) Particles position and velocity profile at 15 s.

The sample is divided equally into three parts in the longitudinal direction: the upper, the middle and the bottom as shown in Fig. 1b. Cross sections (inscribed square of circular sample) were cut at distances from the bottom of the soil sample of 17.4 mm, 10.0 mm, and 2.5 mm respectively to represent the three parts as shown in Fig. 4a–c. Meanwhile, three cross sections were cut at distances from the bottom of the numerical model among [18, 20 mm], [9, 11 mm] and [0, 2 mm] respectively as shown in Fig. 4d–f. Those fine particles located at the upper were locally eroded especially its middle part in Fig. 4a and d, because the relatively less energy of fluid was consumed. The fine particles in the middle of the sample were eroded less as shown in Fig. 4b due to the particle supplies from the upper. But there are relatively less particles in the middle of the numerical result due to the larger space for migration in Fig. 4e. The relatively numerous fine particles located at the bottom were eroded especially in the simulation due to lack of supply of particles from the upper and large space for migration.

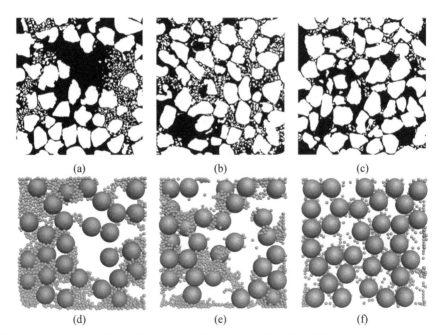

Fig. 4. Comparison of CT and numerical simulation results. (a) and (d) represent the upper part of the sample; (b) and (e) represent the middle part of the sample; (c) and (f) represent the lower of the sample.

4   Conclusions and Outlook

In current study, one-dimensional seepage experiments combined with CT imaging technology were carried out in order to study the characteristics of internal soil erosion. A simplified numerical model coupling DEM and Darcy flow is developed to back analyses the laboratory results. The following conclusions are drawn based on data interpretations:

1. CT imaging technology is an effective method to obtain the characteristic of the eroded soil sample. By comparison of the CT and numerical simulation results, the Darcy flow coupling with DEM is an effective method to calculate the particle migration due to water seepage.

2. Relatively more fine particles located at the upper and bottom of the sample are eroded but less particles at the middle part are eroded due to the particle supplies from the upper side.

Due to the limitation of experimental conditions, time depend continuous seepage scanning experiments could not be carried out, and only samples after the seepage process was scanned. In the future, the seepage experiment device will be improved to conduct simultaneous seepage and scanning experiments on the CT platform. Besides, the effects of shape on the movement of fine particles was not considered in the numerical simulation. Therefore, clump technology will be used to create irregular shaped particles for simulation.

References

1. Ferdos, F.: Internal erosion phenomena in embankment dams: through flow and internal erosion mechanisms, KTH Royal Institute of Technology (2016)
2. Cui, Y.F., Zhou, X., Guo, C.: Experimental study on the moving characteristics of fine grains in wide grading unconsolidated soil under heavy rainfall. J. Mt. Sci. **14**(3), 417–431 (2017)
3. Cui, Y.F., Nouri, A., Chan, D., Rahmati, E.: A new approach to the DEM simulation of sand production. J. Petrol. Sci. Eng. **147**, 56–67 (2016)
4. Civan, F.: Reservoir Formation Damage, 2nd edn. Gulf Professional Publishing, Burlington (2007)
5. Anderson, W. A.: The baldwin hills, california am disaster, vol. 78, no. 5, pp. 84–86 (1964)
6. Zou, S.J.: The research on theory and application of the flow-fitting method for detection of piping and leakage in dykes and dams. Central South University (2009)
7. Kenney, T.C., Lau, D.: Internal stability of granular filters: reply. Can. Geotech. J. **23**, 420–423 (1986)
8. Ke, L., Takahashi, A.: Strength reduction of cohesionless soil due to internal erosion induced by one-dimensional upward seepage flow. Soils Found. **52**(4), 698–711 (2012)
9. Tao, H., Tao, J.: CFD-DEM Modeling of Piping Erosion Considering the Properties of Sands. In: Geo-Chicago 2016, pp. 641–650. ASCE Press, Illinois (2016)
10. Tang, Y., Chan, D.H., Zhu, D.Z.: A coupled discrete element model for the simulation of soil and water flow through an orifice. Int. J. Numer. Anal. Meth. Geomech. **41**(2), 1477–1493 (2017)
11. Tsuji, Y., Kawaguchi, T., Tanaka, T.: Discrete particle simulation of two-dimensional fluidized bed. Powder Technol. **98**(1), 79–87 (1998)
12. Ergun, S.: Fluid flow through packed columns. Chem. Eng. Prog. **48**(2), 89–94 (1952)
13. Wen, C.Y., Yu, Y.H.: Mechanics of Fluidization. In: The Chemical Engineering Progress Symposium Series, vol. 162, pp. 100–111 (1966)
14. Cundall, P.: A computer model for simulating progressive large scale movement in block rock systems. In: Proceedings of International Symposium. Fracture, vol.1, no. ii-b, pp. 11–18. ISRM, Nancy (1971)
15. Potyondy, D.O., Cundall, P.A.: A bonded-particle model for rock. Int. J. Rock Mech. Min. Sci. **41**(8), 1329–1364 (2004)

Numerical Investigation on Sediment Bed Erosion Based on CFD-DEM

Tian-Xiong Zhao and Xia Li[✉]

Southeast University, Nanjing 211189, China
xia.li@seu.edu.cn

Abstract. In this paper, erosion of sediment bed is simulated numerically based on CFD-DEM coupled method. Fluid phase is simulated through computational fluid dynamics (CFD) approach. Such simulation is conducted by solving locally averaged Navier-Stokes equation in CFD solver. The particles are simulated through discrete element method (DEM). The interaction between fluid and particles is considered with an unresolved approach by exchanging the information of interaction forces between the CFD and DEM computation process in CFDEM. Drag force and buoyancy force are considered in this study. The coupling model is first applied to a particle settling test to investigate its performance and the sensitivity of grid size. After benchmarking, the tool is employed to simulate an erosion experiment involving 20000 particles. The results help to describe the erosion pattern of a sediment bed.

Keywords: Erosion · CFD-DEM · Sediment movement

1 Introduction

The movement of sediment-water mixture could be observed in a wide range of engineering cases. Accurate prediction of such phenomenon is vital in engineering applications relevant to granular media, such as local scour, debris flow and erosion of embankment dams. For instance, statistical data showed that from 1950 to 1990, more than 60% of bridges failure are caused by hydraulic accident like scour and debris (Shirole and Holt 1991).

Over the past few decades, a large amount of laboratory experiments have been performed to investigate sediment erosion. Existing laboratory experiments provided considerable experimental data (Smart 1984; Rickenmann 1991; Camenen and Larson 2005). Empirical equations have been established based on these results. Some of them are still widely used today (Bagnold 1973). However, one problem that often arises in experiments is that the tracking of trajectory of individual particles and measurement of velocity profile could be relatively difficult in actual experiments.

Another common approach to study this process is conducting numerical simulation. The direct numerical simulation of sediment-water mixture movement could be conducted by two approaches involving different model types, i.e. continuum model and coupled discrete model. Continuum model is established based on conservation of

© Springer Nature Switzerland AG 2020
J.-M. Zhang et al. (Eds.): ICED 2020, SSGG, pp. 274–279, 2020.
https://doi.org/10.1007/978-3-030-46351-9_28

mass and momentum. Such approach could provide the velocity and concentration profile in the process of erosion (Pudasaini 2012; Domnik and Pudasaini 2012; Domnik et al. 2013). However, continuum model lacks of micro scale description of particle motion (Zhao and Shan 2013). To obtain such information, CFD-DEM coupled model is applied. Papista et al. (2010) applied CFD-DEM model to analyse the initial stage of sediment motion. Zheng et al. (2018) implement a 2-D CFD-DEM simulation to simulate the erosion characteristics of the sediment bed in a micro and macro perspectives. The CFD-DEM model has been already proved to be reliable and efficient in many chemical engineering cases. However, it is relatively rare to apply 3-D CFD-DEM model to study erosion and scour. In this study, a 3-D CFD-DEM model is established to simulate the erosion of 20000 particles under the action of fluid. The theoretical basis of applied model is firstly introduced. After that, the simulated condition is presented. In the end, the performance of numerical model and the characteristics of erosion process are discussed.

2 Methodology of the Erosion Simulation

2.1 Theoretical Basis for CFD-DEM Model

The open-source LAMMPS-based DEM code LIGGGHTS and CFD package Open-Foam are employed in this study. The coupling framework is built based on the CFDEM project (Goniva et al. 2010). The coupling of particles and fluid is considered by the exchange of the interaction forces between two phases, including drag force and buoyancy force. In the particle-fluid interaction process, the particles involved are governed by the Newton's law. Pore fluid is assumed to be continuous in this study, which could be numerically analyzed by solving locally averaged Navier-Stokes equation (Anderson and Jackson 1967). The detailed discussion of governing equations for particles and fluid is shown below.

Governing Equations for Particles. The translational and rotational motions of an individual particle is treated following the equations shown below in LIGGGHTS (Kloss et al. 2012; Zhao and Shan 2013):

$$m_i \frac{dU_i^p}{dt} = \sum_{j=1}^{n_i^c} F_{ij}^c + F_i^f + F_i^g \tag{1}$$

$$I_i \frac{d\omega_i}{dt} = \sum_{j=1}^{n_i^c} M_i^j \tag{2}$$

where m_i and I_i denote the mass and moment of inertia of particle i. U_i^p and ω_i are the translational and rotational angular velocities. F_i and M_i are forces and torque action on particle i. Specifically, F^f denotes the interaction forces, which including buoyancy force and drag force in this study. F^c denotes contact forces between two particles. F^g denotes the force of gravity acting on particle i. In LIGGGHTS, Hertzian contact law and Coulomb's friction criteria are applied.

Governing Equations for Fluid. The equations shown below are calculated by the CFD solver:

$$\frac{\partial(\varepsilon\rho)}{\partial t} + \nabla \cdot (\varepsilon\rho U^f) = 0 \tag{3}$$

$$\frac{\partial(\varepsilon\rho U^f)}{\partial t} + \nabla \cdot (\varepsilon\rho U^f U^f) - \varepsilon\nabla \cdot \left(\mu\rho U^f\right) = -\nabla p - f^p + \varepsilon\rho g \tag{4}$$

Equation (3) and (4) are respectively the continuity equation and locally averaged Navier-Stokes equation solved by CFD module, where U^f denotes the average velocity in a CFD cell. ε denotes the volume fraction of fluid in a cell. ρ is the averaged density calculated by the density of fluid and particle and their volume fraction. P denotes the fluid pressure. f^p denotes the interaction forces applied on fluid by particle.

Particle-Fluid Interaction Forces. The key to accurately simulate the movement of sediment-water movement is the reasonable consideration of interaction forces between two phases. In this study, drag force and buoyancy force are considered. The drag forces is calculated by the equation used by Di Felice (1994):

$$F^d = \frac{1}{8}C_d\rho\pi d_p^2(U^f - U^p)\left|U^f - U^p\right|\varepsilon^{1-\chi} \tag{5}$$

where C_d denotes the particle-fluid drag coefficient. ε^χ is applied to consider the influence of other particles in the system. Both C_d and χ could be calculated by Renolds number. According to previous research, Di Felice equation works well under low Renolds numbers condition (Kafui et al. 2002; Zhao and Shan 2013). The buoyancy force of a spherical particle is calculated following the equation shown below:

$$F^b = \frac{1}{6}\pi\rho d_p^3 g \tag{6}$$

2.2 Interaction Procedures

In this study, the CFD time step is 100 times longer than the DEM time step. The CFD and DEM modules couple once whenever DEM runs 100 time steps. It is should be noted that the interaction approach in this study is one-way interaction. The interaction forces applied on fluid is based on the particle information of the last time step, rather than an iterative result.

2.3 Benchmark Cases

To test if the coupled CFD-DEM model could predict the interaction between the sediment particle and fluid accurately, a single particle settling case is designed to test the performance of the model.

A spherical particle with a diameter 0.1 mm and density of 3000 kg/m^3 is released from rest in water. The size of the fluid container is 0.05 m $\times$ 0.05 m $\times$ 0.1 m. Three different mesh sizes have been applied to investigate the performance of coupling model

Table 1. Settling test result with different mesh sizes

Grid size (mm)	CFD cells number	Error in the particle final velocity	Calculation time (s)
2.5	320	6.093%	1
0.5	40000	2.557%	28
0.25	320000	1.614%	514

and the effect of CFD mesh size on the result error. Obtained simulation results are compared with analytical solution through Stokes equation. The results are shown in Table 1.

When the ration of CFD mesh size to the particle diameter is around 5:1, both the error and the calculation time are acceptable. This became the basis for choosing the mesh size in our subsequent simulation.

3 Erosion Simulation

The erosion simulation is conducted with 20000 spherical particles in a region with a size of $0.05 \, \text{m} \times 0.2 \, \text{m} \times 0.1 \, \text{m}$. Periodical boundary is applied in this model, so that the departing particles return to the calculation region from the opposite side with the same velocity. The velocity profile of fluid at the inlet is linearly distributed. The maximum velocity is at the top and the bottom velocity is 0. The detailed simulated condition is shown in Table 2.

Table 2. Erosion simulation condition

Parameters	Value
Particle number	20000
Density of particles (kg/m^3)	3000
Diameter of particles (mm)	0.15
Top velocity (m/s)	2
CFD time step (s)	1e–3
DEM time step (s)	1e–5

The snapshots of the sediment bed under erosion in different time are shown in Fig. 1. The different colors represent the velocity of the particles.

The erosion form could be observed from the simulation results. At the initial stage of erosion, all simulated particles moved with the fluid for a short time (a). This phenomenon is probably due to the initial velocity of fluid field. After that, the velocity of fluid decreases due to the blocking of particles. That leads to an overall decrease in the velocity of all particles. Most of the particles stop moving except those near the entrance. These

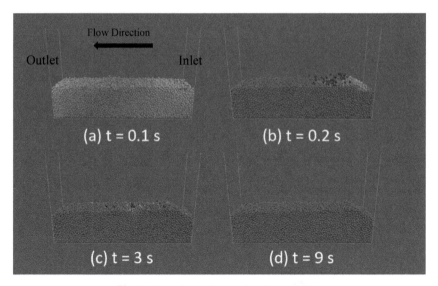

Fig. 1. Snapshots of eroded sediment bed

moving particles are transported through the direction of the current, which forms an erosion pit near the entrance (b). As the pit grows larger, the rolling particles become less. Some particles at the top layer leap, suspend and crash the bed (c). When the number of the particles entering and leaving the erosion pit reaches a balance, the size of the pit gradually stops changing (d).

4   Results and Discussion

A 3-D coupled CFD-DEM model is applied to simulate the erosion of 20000 particles by water. The main conclusion could be summarized as follows: By conducting a spherical particle settling test, the error of simulation result of particle-fluid interaction is analyzed. It is found that when the ratio of CFD mesh size to the particle dimension is around 5:1, both of the accuracy and computational time of calculating the interaction force are acceptable. The subsequent simulation result proved the reliability of applied coupling model.

The erosion pattern is also analyzed. The erosion pit appears near the entrance because of the difference in particle velocity at different locations. As the shape of the erosion pit changes, the number of particles entering and leaving the pit reaches balance, which forms a erosion pit with an equilibrium depth. It can also be observed that the erosion rate is relatively high at the initial stage of erosion, and gradually decreases until the equilibrium depth is reached.

References

Anderson, T.B., Jackson, R.: Fluid mechanical description of fluidized beds. Equations of motion. Ind. Eng. Chem. Fundam. **6**(4), 527–539 (1967)

Bagnold, R.A.: The nature of saltation and of 'bed-load' transport in water. Proc. R. Soc. Lond. A Math. Phys. Sci. **332**(1591), 473–504 (1973)

Camenen, B., Larson, M.: A general formula for non-cohesive bed load sediment transport. Estuar. Coast. Shelf Sci. **63**(1–2), 249–260 (2005)

Di Felice, R.: The voidage function for fluid-particle interaction systems. Int. J. Multiph. Flow **20**(1), 153–159 (1994)

Domnik, B., Pudasaini, S.P.: Full two-dimensional rapid chute flows of simple viscoplastic granular materials with a pressure-dependent dynamic slip-velocity and their numerical simulations. J. Nonnewton. Fluid Mech. **173**, 72–86 (2012)

Domnik, B., Pudasaini, S.P., Katzenbach, R., Miller, S.A.: Coupling of full two-dimensional and depth-averaged models for granular flows. J. Nonnewton. Fluid Mech. **201**, 56–68 (2013)

Goniva, C., Kloss, C., Hager, A., Pirker, S.: An open source CFD-DEM perspective. In: Proceedings of OpenFOAM Workshop, Göteborg, pp. 22–24, June 2010

Kafui, K.D., Thornton, C., Adams, M.J.: Discrete particle-continuum fluid modelling of gas–solid fluidised beds. Chem. Eng. Sci. **57**(13), 2395–2410 (2002)

Kloss, C., Goniva, C., Hager, A., Amberger, S., Pirker, S.: Models, algorithms and validation for open source DEM and CFD–DEM. Prog. Comput. Fluid Dyn. Int. J. **12**(2–3), 140–152 (2012)

Papista, E., Dimitrakis, D., Yiantsios, S.G.: Direct numerical simulation of incipient sediment motion and hydraulic conveying. Ind. Eng. Chem. Res. **50**(2), 630–638 (2010)

Pudasaini, S.P.: A general two-phase debris flow model. J. Geophys. Res. Earth Surf. **117**(F3), 1–28 (2012)

Rickenmann, D.: Hyperconcentrated flow and sediment transport at steep slopes. J. Hydraul. Eng. **117**(11), 1419–1439 (1991)

Shirole, A.M., Holt, R.C.: Planning for a comprehensive bridge safety assurance program. Transp. Res. Rec. **1290**, 39–50 (1991)

Smart, G.M.: Sediment transport formula for steep channels. J. Hydraul. Eng. **110**(3), 267–276 (1984)

Zhao, J., Shan, T.: Coupled CFD–DEM simulation of fluid–particle interaction in geomechanics. Powder Technol. **239**, 248–258 (2013)

Zheng, H.C., Shi, Z.M., Peng, M., Yu, S.B.: Coupled CFD-DEM model for the direct numerical simulation of sediment bed erosion by viscous shear flow. Eng. Geol. **245**, 309–321 (2018)

Risk Assessment and Management

Reliability Analysis for the Surface Sliding Failure of Gravity Dam

Na Hao and Xu Li$^{(\boxtimes)}$

School of Civil Engineering, Beijing Jiaotong University, Beijing 100044,
People's Republic of China
XuLi@bjtu.edu.cn

Abstract. Allowance safety factor design (AFSD) as a traditional method of anti-sliding stability analysis has many limitations. For instance, it cannot consider the uncertainty of variables and make risk assessment. To make up for these shortcomings, this paper provides a simple method to investigate the real failure probability of allowable safety factor K_a through the reliability theory and gives the recommended value of the K_a. The real safety level of the K_a can be represented by its corresponding reliability index distribution, which is composed of the reliability indexes calculated by a group of gravity dams with the same safety factor. Using 90% to ensure the conservative evaluation of the target reliability index β_t of such distribution, we can get the β_t corresponding to the K_a, which is able to roughly assess the risk degree of the K_a. Moreover, the K_a can be calibrated according to the β_t provided by the engineering specification.

Keywords: Gravity dam · Safety factor · Reliability analysis · Anti-sliding stability

1 Introduction

Anti-sliding stability analysis is one of the most concerned problems in gravity dam design. Nowadays, the domestic standards mainly use the allowance safety factor design (AFSD) to analyze this problem.

The current code (SL319-2018) [1] provides the allowable safety factor K_a of gravity dams, which is taken as 3 under normal conditions. Since this K_a is usually determined by engineering experience, it lacks physical significance to assess the risk of damage. In addition, it is unreasonable that the K_a of gravity dams are not graded according to the geometric parameters and geographical locations, which results in the same safety standard of dam with different safety degree.

In order to make up for weaknesses for AFSD, this paper proposes a theory for the reliability-based calibration of the safety factors. In this way, we can evaluate the real safety level of K_a, and the recommended safety factors of different levels are provided, which is likely to bring convenient for practical engineering applications, and makes the selection of safety factor more reasonable.

© Springer Nature Switzerland AG 2020
J.-M. Zhang et al. (Eds.): ICED 2020, SSGG, pp. 283–288, 2020.
https://doi.org/10.1007/978-3-030-46351-9_29

2 Reliability-Based Calibration of Safety Factor Design

Structures that share the same safety factor K can have different β because these two coefficients have different theoretical basis. When a series of gravity dams with the same safety factor can be selected to calculate their reliability indexes, a special statistical distribution can be formed to represent the real safety level of the K.

2.1 RBD

A representative gravity dam is illustrated in Fig. 1, which the uplift pressure at the dam foundation is triangular distribution. The K can be showed as follows:

$$K = \frac{m\rho g h^2 - m\rho_w g h \alpha H}{\rho_w g H^2} f + \frac{2mh}{\rho_w g H^2} c = Af + Bc \qquad (1)$$

Fig. 1. A model of simple gravity dam.

From Eq. 1, we can see that the variables include the friction coefficient f, the cohesion force c, dam height H, the water head height of dam heel h.

In the analysis of anti-slip stability of gravity dams, the performance function needs to express the safety factor as the function of input parameters, and update it as:

$$Z = g(x_1, x_2, \cdots x_m) = K - 1 = Af + Bc - 1 = 0 \qquad (2)$$

where $x_1, x_2, \cdots x_m$ are all the parameters that affect the safety of the gravity dam.

Since c is a positive number, it is more practical to assume that c obeys the log-normal distribution. This means that the reliability index β needs to be obtained for the numerical solution by other methods. This paper chooses the first and second moment method which is one of the most widely used methods at present.

2.2 Reliability-Based Calibration of the Safety Factor for Gravity Dam

Clearly, the value of K_a is based on human experience, which lacks physical significance. When we calibrate K_a based on the reliability theory, it simplifies the application of the reliability-based design (RBD) in practice and makes the AFSD more reasonable. The process of calibration is as follows:

When an allowable safety factor K_a is given, its actual safety level can be represented by the corresponding distribution of reliability index, as follows:

$$K(x_d) = K_a \Rightarrow (\mu_\beta, \sigma_\beta) \qquad (3)$$

where μ_β and σ_β are respectively the mean and standard deviation of the statistical distribution of reliability index.

Based on the actual engineering situation, the assurance rate of 90% points is selected from the statistical distribution of reliability index to estimate the target reliability index β_T, which can be defined as $\beta_T = \mu_\beta - 1.28\sigma_\beta$. That's to say, by calculating the β_T for some samples satisfying $K = K_a$, the following relation is valid:

$$K = K_a \Rightarrow \beta_T = \mu_\beta - 1.28\sigma_\beta \qquad (4)$$

For each structure that satisfies the condition $K = K_a$, we can also define the difference between the real reliability index β and the target value β_T as follows:

$$\varepsilon_i = \beta_i|(K = K_a) - \beta_T \tag{5}$$

where β_i is the real reliability index of case i, and the meaning of "|" is similar to "|" in conditional probability. The total difference degree can evaluate though the mean square error σ_ε, i.e.

$$\sigma_\varepsilon = \sqrt{\int \varepsilon^2 dp} \tag{6}$$

Substituting Eqs. 4, 5 into Eq. 6, the expression of σ_ε will be upgraded to:

$$\sigma_\varepsilon = \sqrt{1 + 1.28^2}\sigma_\beta = 1.62\sigma_\beta \tag{7}$$

Overall, σ_ε and σ_β can measure the degree of dispersion of the reliability index distribution.

3   Main Random Variables of Gravity Dams

In above paragraphs, variables of gravity dams can be divided into shear strength parameters (friction coefficient f, cohesion force c) and geometric parameters (dam height H, the water head height of dam heel h, etc.). It is obvious that f and c, compared with geometric parameters, need to deal with COVs varying from a wide range. Because rock and soil are natural materials, c and f are regarded as random variables [2]. Hence, the five variables need to be considered as:

$$\mu_c, \delta_c, \mu_f, \delta_f, H \tag{8}$$

Table 1 lists the value ranges and distribution types of the above five variables in the specifications and literature [1–5]. According to this, Box-Behnken design method which is a relatively simple and efficient experimental design method is adopted to sample the main variables, as shown in Table 1.

Table 1. Sampling the parameters involved in the stability analysis of gravity dam.

Variable	Range	Type	Complete samples
μ_f (°)	0.70–1.50	Normal	{0.94, 1.2, 1.46}
δ_f	0.15–0.2	Normal	{0.16, 0.18, 0.2}
μ_c (MPa)	0.9–1.50	Log-normal	{0.83, 1.1, 1.41}
δ_c	0.25–0.3	Normal	{0.3, 0.33, 0.35}
H(m)	30–200	Normal	{50, 100, 150}

4   Calibration of Safety Factor in Gravity Dam Engineering

4.1   Calibration Process

The calibration process requires a series of gravity dams with the same safety factors, which are used to calculate the reliability index and establish the reliability index distribution corresponding to the safety factor. Table 2 shows that these dams can be obtained from the following two groups of samples:

Table 2. Gravity dam cases used in this study.

Group name	Parameters for sampling	Count of cases	Constraint condition	Results
B-1	$\mu_c, \delta_c, \delta_f, H$	81	$K = K_a$	μ_f, β
B-2	$\mu_f, \delta_c, \delta_f, H$	81	$K = K_a$	μ_c, β

When the allowable safety factor is determined K_a, the gravity dam that meets the condition of $K = K_a$ can be obtained by using B-1 and B-2. The steps of calibration are these:

(1)   Obtain 162 special gravity dams with the allowable safety factor $K_a = 2.5$ though the groups of B-1 and B-2. We can calculate the reliability index β of this group of gravity dams through the first order second moment method.
(2)   With gaining many reliability indexes, we establish the reliability index distribution, and then quantify the μ_β and σ_β of such distribution;
(3)   Calculate the target reliability index β_t according to Eq. 4.
(4)   Replace the allowance safety factor $K_a = 2.5, 3, 3.5, 4, 4.5$, then repeat steps 1–4 to obtain the corresponding calculated reliability index distribution under different K_a (See Fig. 2) and the target reliability index β_t.
(5)   Determine a nonlinear fitting between K_a and β_t, which this fitting expression is shown in Eq. 9.

4.2   Calibration Results

According to Sect. 3, the results are drawn in Fig. 2 and Fig. 3. The following conclusions can be obtained:

(1) The statistical distribution of the reliability index is approximate to the normal distribution, so that the target reliability of β_t applying Eq. 4 has a reasonable basis.

(2) When reliability indexes β have the same safety factor K_a, they are likely to be very discrete, which can be reflected by σ_β. Consequently, the safety factor K is a rough estimate of the stability in gravity dam projects.

(3) It is clearly that the variation σ_β increase rapidly when K_a increases, which indicates that the safety factor performance becomes worse. That is to say, for important a gravity dam with high safety factor K, RBD is still required.

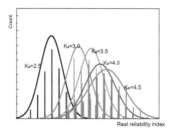

Fig. 2. The distribution of $\beta|(K = K_a)$.

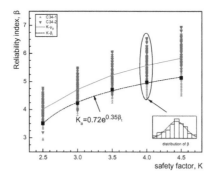

Fig. 3. Real reliability index caused by ASFD.

We can also get the fitting expression between K_a and β_t from Fig. 3:

$$K_a = 0.72e^{0.35\beta_t} \tag{9}$$

In the limit state, the common target reliability indexes β_t of gravity dams are 4.2, 3.7, 3.2 [6], which can modify K_a by the Eq. 9. Table 3 lists the recommended values of K_a corresponding to different structural safety degrees based on reliability theory. Therefore, according to the structural safety degree provided by the geological exploration, the allowable safety factor K_a can be reasonably selected in the future.

Table 3. The recommended design chart for ASFD.

Safety degree	I	II	III
Target reliability indexes β_t	4.2	3.7	3.2
Allowance safety factor K_a	3.1	2.7	2.4

5   Conclusion

In this paper, we put forward a simple way to calibrate the K_a based on the reliability analysis to compensate for the over-empirically selected K_a in the code. This method is to select a group of special gravity dam projects with the same K, and then use the statistical distribution of those dams calculated β to represent the real safety degree of the K_a. In addition, we can utilize σ_β to measure the discrete degree of the reliability index distribution, which are able to evaluate the performance of the calibrated safety factor K_a.

When the calibration process is carried out based on the reliability theory according to different safety degrees, we list the suggested values of K_a under this method. These recommendations are based on careful estimates of the 90% guarantee.

As we all know, there are three kinds of failure surfaces in the gravity dam projects, such as foundation surface, deep sliding surface and concrete surface [7]. Each surface can calibrate the safety factor and sub-coefficient by the simple method introduced in this study. It is a pity that we only explore the stability against sliding along the foundation base of gravity dam, and the other two modes will continue in the future.

Acknowledgement. This study is supported by the National Natural Science Foundation of China (No. 51979002).

References

1. SL 319-2018: Design specification for concrete gravity dams. China Ministry of Water Resources (2018). (in Chinese)
2. Phoon, K.K.: Role of reliability calculations in geotechnical design. Georisk Assess. Manage. Risk Eng. Syst. Geohazards **11**(1), 4–21 (2017)
3. NB/T 35026-2014: Design code for concrete gravity dams. China Energy Industry, Beijing (2014). (in Chinese)
4. Chen, Z.Y., Xu, J.C., Sun, P.: Reliability analysis on sliding stability of gravity dams: part II, determination of shear strength parameters and partial factors. J. Hydroelectr. Eng. **31**(3), 28 (2012)
5. Chen, W.: Study on Partial Coefficient of Anti-sliding Stability Limit State of Gravity Dam. Tsinghua University, Beijing (2010)
6. GB 50199-2013: Unified Standard for Reliability Design of Hydraulic Engineering Structures. China Architecture & Building Press, Beijing (2013). (in Chinese)
7. Ran, H.Y., Li, Q., Du, H.D.: Research on applicability and safety control standards of shear-resistant formula in stability analysis of gravity dam foundation. J. Water Resour. Archit. Eng. **09**(5) (2011)
8. Duncan, J.M.: Factors of safety and reliability in geotechnical engineering. J. Geotech. Environ. Eng. **126**(4), 307–316 (2000)
9. Chen, Z.Y., Xu, J.C., Sun, P.: Reliability analysis on sliding stability of gravity dams: part I, an approach using criterion of safety margin ratio. J. Hydroelectr. Eng. **31**(3), 27 (2012)
10. Wang, D., Chen, J.K.: Reliability research on sensitivity of concrete gravity dam to random variables. J. Sichuan Univ. **33**(4), 1–5 (2001)
11. Ching, J.Y., Phoon, K.K.: Quantile value method versus design value method for calibration of reliability-based geotechnical codes. Struct. Saf. **44**, 47–58 (2013)

Quantification of Uncertainties for Risk Management of Landslide Dam Break Emergency

Jian He[1], Limin Zhang[1,2,3(✉)], Te Xiao[1], and Chen Chen[2]

[1] Department of Civil and Environmental Engineering, The Hong Kong University of Science and Technology, Hong Kong, China
cezhangl@ust.hk
[2] State Key Laboratory of Hydraulics and Mountain River Engineering, College of Water Resource and Hydropower, Sichuan University, Chengdu, China
[3] HKUST Shenzhen Research Institute, Shenzhen, China

Abstract. The break of large landslide dam will trigger catastrophic flood hazard to the downstream area. To manage the flood risk, multiple mitigation measures are required, such as evacuation and removal of obstacles in the river channel. Design of these mitigation measures relies on the estimation of critical flood parameters, namely, peak discharges and stages along the river. However, uncertainties in outburst flood prediction and flood routing analysis have significant influence on the estimation of these flood parameters. Ignoring these uncertainties will undermine the reliability of the estimated flood parameters, which might lead to insufficient design of risk mitigation measures. To enhance the reliability of landslide dam break risk management, we quantify the influence of uncertainties in both landslide dam material erodibility and river channel roughness on the estimation of flood parameters. Successive Baige landslide dams on the Jinsha River in 2018 are investigated to show how the uncertainty quantification facilitates robust decision making for risk management.

Keywords: Landslide dam · Dam break · Flood · Risk management · Uncertainty quantification

1 Introduction

The Jinsha River, the upper reach of the Yangtze River, was blocked twice by landslide mass in 2018. The first landslide dam was about 61 m high, corresponding to a barrier lake of 249 million m^3. While the second dam was 35 m higher than the first one, forming a lake of 757 million m^3. Large amount of water will be released once a landslide dam as large as these two dams breaks. Risk management for the downstream flood risk introduced by the break of a landslide dam is of the utmost importance. Generally, risk management involves an iteration of several steps: design of risk mitigation measures, prediction of outburst flood, flood routing analysis, evaluation and modification of risk mitigation measures. Common risk mitigation measures include engineering measures

© Springer Nature Switzerland AG 2020
J.-M. Zhang et al. (Eds.): ICED 2020, SSGG, pp. 289–294, 2020.
https://doi.org/10.1007/978-3-030-46351-9_30

(e.g., excavating a diversion channel on the dam crest and removing obstacles in the river) and non-engineering measures (e.g., evacuation and emptying downstream reservoirs). Dam breaching flood can be estimated using empirical equations [1] or numerical simulation [2–4], and then flood routing analysis is carried out to estimate the flood and water level at downstream areas [2, 3]. Based on results from dam breaching and flood routing analyses, risk mitigation measures to be implemented in downstream areas, such as removal of obstacles in the river channel and evacuation, can be designed. However, there are various uncertainties underlying the analyses, namely, uncertainties in erodibility of dam materials and riverbed roughness. These uncertainties have great impact on the estimation of flood parameters. Ignoring these uncertainties might lead to insufficient design of risk mitigation measures. Previous study considered the uncertainty in erodibility of dam materials to facilitate decision-making for dam break emergency management [5]. Further analysis is desired to take the uncertainty in riverbed roughness into consideration.

In order to enhance the reliability of landslide dam break risk management, this study proposed an uncertainty quantification method, which quantifies the impact of both uncertainties in erodibility of dam materials and riverbed roughness on the estimation of flood parameters. Firstly, peak outflow rate and breach time are predicted by empirical equations considering probabilities of different erodibility. Then, the breaching hydrograph is reconstructed based on a typical landslide dam breaching hydrograph. Flood routing analysis is carried out based on the reconstructed hydrograph considering different Manning's coefficient. Finally, the distribution of peak discharge and water surface elevation can be obtained to serve as a basis for robust decision making. The proposed method is illustrated by an application to Baige landslide dams in 2018.

2 Methodology

2.1 Prediction of Outburst Flood

The outburst flood can be predicted by empirical methods [1, 6, 7] or numerical methods (e.g., BREACH [8], DABA [9], DB-IWHR [10]). Empirical equations proposed by [1] are used in this study as an example:

$$Q_p = g^{0.5} H_d^{-0.407} V_l^{0.512} e^a \tag{1}$$

$$T_b = H_d^{-0.43} V_l^{0.241} e^b \tag{2}$$

where Q_p is the peak outflow rate (m/s); g is the gravitational acceleration (m/s^2); H_d is the dam height (m); V_l is the lake volume (m^3); a is an erodibility-related coefficient, which equals 1.276, −0.336, −1.532 for high, medium and low erodibilities, respectively; T_b is the breach time (h); b is also an erodibility-related coefficient which equals −0.805 and −0.674 for high and medium erodibilities, respectively.

Based on the estimated peak outflow rate and breach time, a breaching hydrograph can be reconstructed based on a previous landslide dam breaching hydrograph. Outflow rate is scaled by the ratio of the predicted peak outflow rate to the measured peak outflow rate in the previous case. Time is scaled by the ratio of the predicted breach time to the observed breach time in the previous case. In this study, Tangjiashan landslide dam breaching hydrograph is used as an example. The peak outflow rate and breach time are 6500 m^3/s and 14 h [1], respectively, in Tangjiashan case.

The erodibility of dam materials can be assessed through in-situ tests or engineering judgement. In most cases, landslide dam sites are hard to access due to transportation difficulties and limited time. Therefore, engineering judgement plays a more important role in the risk management stage. The erodibility can be assessed based on local geology, landslide travel distance, unmanned aerial vehicle photos, and site investigation on previous landslide dams. However, the reliability of engineering judgement is based on individual experience. Thus, uncertainty in this kind of judgement is inevitable, and its influence on risk management should be investigated. In this study, each category of erodibility is given a probability reflecting the confidence in engineering judgement.

2.2 Flood Routing Analysis

One-dimensional hydraulic model developed by U.S. Army Corps of Engineers, HEC-RAS is used to perform the flood routing analysis. In HEC-RAS, the roughness of riverbed is characterized by Manning's coefficient, of which the range can be determined according to the literature [11]. Repeated flood routing analysis is required for different combinations of erodibility and Manning's n. Given the breaching hydrograph, the peak discharge and water surface level are assumed to have a linear relationship with Manning's n in a limited range of Manning's n. Under this assumption, flood routing analysis can be performed for only the upper and lower bounds of Manning's n.

3   Case Study

Successive landslide dams near Baige Village blocked the Jinsha River twice in 2018. The height and lake volume of Baige landslide dams are given in Table 1. Zhang et al. [4] rated the erodibilities of the first and the second landslide dams as "medium" and "medium-high", respectively. To illustrate the uncertainty in this judgement, different probabilities are given to different categories of erodibility, as shown in Table 1. Notice that the erodibility is classified into three categories for empirical methods used in this study. The outburst flood in "low-medium" and "medium-high" cases is approximated by averaging results of "low" and "medium" erodibilities, and "medium" and "high" erodibilities, respectively. The predicted breaching parameters are summarized in Table 1. The reconstructed and observed hydrographs are shown in Fig. 1. For the first landslide dam, the breach time is underestimated in all erodibility cases, while for the second landslide dam, the reconstructed breaching hydrograph in "medium-high" erodibility case agrees well with the observed one.

The range of Manning's coefficient for the Jinsha River is assumed to be from 0.04 to 0.06. Considering the uncertainty of Manning's n, the range of peak discharge given a breaching hydrograph can be obtained. Figure 2 shows the distribution of predicted peak discharge at Suwalong Dam site (about 220 km downstream of the landslide dam). The width of bars represents the uncertainty introduced by uncertain Manning's n value, while the difference between bars represents the uncertainty introduced by uncertain

Table 1. Summary of landslide dam information and predicted breaching parameters.

Parameter	First landslide dam			Second landslide dam		
Dam height (m)	61			81[a]		
Lake volume (m³)	249×10^6			494×10^6		
Erodibility[b]	L-M	M	M-H	M	M-H	H
Probability	0.1	0.7	0.2	0.1	0.6	0.3
Peak outflow rate (m³/s)	5160	7995	24116	10116	30515	50914
Breach time (h)	9.8[c]	9.2	8.6	9.6	9.0	8.4

Note: [a]the 15-m depth diversion channel on the dam crest is considered [4]; [b]L = Low, M = Medium, H = High; [c]the breach time for L-M is assumed since the empirical methods cannot predict the breach time for low-erodibility dam.

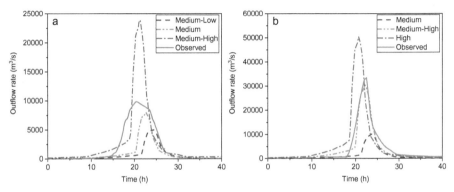

Fig. 1. Reconstructed and observed breaching hydrographs: (a) first landslide dam; (b) second landslide dam.

erodibility of dam materials. The influence of uncertainty in Manning's n on flood parameters increases as the peak outflow rate increases. Therefore, the uncertainties in both erodibility of dam materials and Manning's n should be considered in the risk management for landslide dam break.

The distribution of peak discharge has practical values for decision making. At the time when Baige landslide dams occurred, Suwalong Dam was still under construction. The cofferdams for Suwalong Dam were likely to be overtopped by the flood, which might lead to flood amplification. Whether to remove the cofferdams to prevent escalation of flood was a critical question at that time. The flood-proof capacity of the cofferdams is 6180 m³/s. Based on Fig. 1, the probabilities that peak discharge exceeds the flood-proof capacity of the cofferdams in the first and the second dam break events are 0.2 and 0.9, respectively. Therefore, the cofferdams are very likely to remain intact in the first dam break event. However, they have little chance to survive in the second dam break event, and flood escalation might occur. This indicates that to avoid the cascading failure of dams, the cofferdams should be removed in advance.

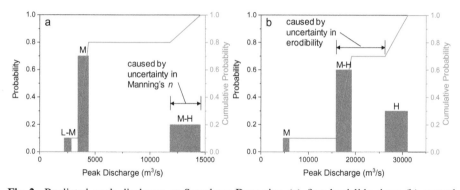

Fig. 2. Predicted peak discharge at Suwalong Dam site: (a) first landslide dam; (b) second landslide dam.

Peak water surface elevation is another important flood parameter. Evacuation plan is designed based on the peak water surface elevation. Figure 3 shows the distribution of peak water surface elevation at Benzilan Town (about 380 km downstream of the landslide dam). For the first dam break, residents located in region where elevation is below 2009.7 m have a probability of 0.8 to be affected by the flood. While for the second dam break, residents living in land below 2022.9 m elevation have a probability of 0.7 to be affected by the flood. Therefore, the evacuation elevation should be at least 2009.7 m and 2022.9 m for the first and the second landslide dam break, respectively.

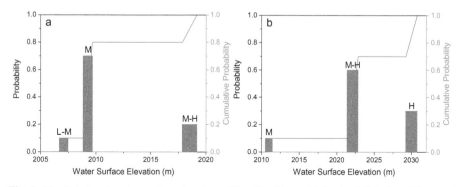

Fig. 3. Predicted peak water surface elevation at Benzilan Town: (a) first landslide dam; (b) second landslide dam.

4 Summary and Conclusions

This study proposed an uncertainty quantification method for risk management of landslide dam break emergency. The uncertainties in erodibility of dam materials and roughness of riverbed are considered, and their influence on the estimation of peak discharge and water surface elevation at downstream areas is quantified. Based on the distributions of downstream flood parameters, risk mitigation measures can be designed in a rational and robust manner. The proposed method was applied to the successive Baige landslide dams in 2018. Results show that uncertainties in both erodibility of dam materials and Manning's n have great influence on the prediction of flood parameters, and should be considered in the risk management for landslide dam break.

Acknowledgement. Financial support was from the National Key Research and Development Program of the Ministry of Science and Technology of China (Project No. 2018YFC1508600).

References

1. Peng, M., Zhang, L.M.: Breaching parameters of landslide dams. Landslides **9**(1), 13–31 (2012)
2. Fan, X., Tang, C.X., Van Westen, C.J., Alkema, D.: Simulating dam-breach flood scenarios of the Tangjiashan landslide dam induced by the Wenchuan Earthquake. Nat. Hazards Earth Syst. Sci. **12**(10), 3031 (2012)
3. Peng, M., Zhang, L.M.: Analysis of human risks due to dam break floods—part 2: application to Tangjiashan landslide dam failure. Nat. Hazards **64**(2), 1899–1923 (2012)
4. Zhang, L., Xiao, T., He, J., Chen, C.: Erosion-based analysis of breaching of Baige landslide dams on the Jinsha River, China, in 2018. Landslides **16**(10), 1965–1979 (2019)
5. Peng, M., Zhang, L.M.: Dynamic decision making for dam-break emergency management–part 2: application to Tangjiashan landslide dam failure. Nat. Hazards Earth Syst. Sci. **13**(2), 439–454 (2013)
6. Costa, J.E.: Floods from dam failures, US Geological Survey, Open-File Report, No. 85-560, Denver, Colorado (1985)
7. Walder, J.S., O'Connor, J.E.: Methods for predicting peak discharge of floods caused by failure of natural and constructed earthen dams. Water Resour. Res. **33**, 2337–2348 (1997)
8. Fread, D.L.: BREACH: an erosion model for earth dam failures, National Weather Service (NWS) Report, NOAA, Silver Spring, Maryland, USA (1988)
9. Chang, D.S., Zhang, L.M.: Simulation of the erosion process of landslide dams due to overtopping considering variations in soil erodibility along depth. Nat. Hazards Earth Syst. Sci. **10**(4), 933–946 (2010)
10. Chen, Z., Ping, Z., Wang, N., Yu, S., Chen, S.: An approach to quick and easy evaluation of the dam breach flood. Sci. China Technol. Sci. **62**, 1–10 (2019)
11. Arcement, G.J., Schneider, V.R.: Guide for selecting Manning's roughness coefficients for natural channels and flood plains. U.S. Geological Survey Water-Supply Paper, vol. 2339, U.S. Geological Survey, Washington, DC (1989)

Dike-Break Induced Flood Simulation and Consequences Assessment in Flood Detention Basin

Shui-Hua Jiang[1(✉)], Zhong-Fa Huang[1], and Jinsong Huang[2]

[1] School of Civil Engineering and Architecture, Nanchang University, 999 Xuefu Road,
Nanchang 330031, People's Republic of China
sjiangaa@ncu.edu.cn
[2] Discipline of Civil, Surveying and Environmental Engineering, Faculty of Engineering
and Built Environment, The University of Newcastle, Callaghan, NSW 2308, Australia

Abstract. To mitigate the flood risk in flood detention basin, it is of great significance to accurately simulate the flood movement process and estimate the potential flood consequences induced by dike-break. In this paper, a MIKE21-based numerical approach for modeling of flood movement in the flood detention basin is developed. The approaches for estimating the consequences (e.g., life losses, economic losses and environment losses) caused by dike-break are also presented. The flood movement in the Kangshan flood detention basin in Poyang Lake district, China subjected to the historical highest water level is simulated using a MIKE21 Flow Mode (FM) model. The dike-break induced flood consequences are estimated based on the flood simulation results (including inundation area, water depth, flow velocity and flood peak appeared time). Then, the flood disaster zones are defined and the emergency evacuation planning against dike-break flood is made. The research results can provide important references for quantitatively evaluating the flood risk and formulating flood control and rescue decisions.

Keywords: Dike breach · Flood detention basin · Flood simulation · Inundation pattern · Flood consequences assessment

1 Introduction

Dikes may break during extreme floods due to various breach mechanisms, which can cause considerable damage to downstream communities. To reduce the dike-break flood risk, detention basins that are surrounded by dikes and will be filled during flood events are often additionally constructed (e.g., [1, 2]). Nevertheless, the flood detention basins are not strictly controlled for use so that the population is growing fast and the industries and agricultural crops develop rapidly in the detention basins. The water conservation engineering measures are not put into effect to mitigate the dike-break flood risk in the detention basins. As a result, a large of flood consequences (including life, economic and

© Springer Nature Switzerland AG 2020
J.-M. Zhang et al. (Eds.): ICED 2020, SSGG, pp. 295–310, 2020.
https://doi.org/10.1007/978-3-030-46351-9_31

environment losses) may be induced once the dike surfers from extreme floods (e.g., [3, 4]). To prevent flood disaster and guide emergent flood control and rescue, it is of great necessity to realistically simulate the dike-break flood movement in the flood detention basin.

Flood simulation in the flood detention basin is a crucial step for assessment of flood consequences, which has received increasing attention (e.g., [1, 5, 6]). For example, Tucciarelli et al. [7] employed a finite element method to simulate flood movement in the flood detention basin. Liu et al. [8] presented a coupled hydrodynamic model linking the channel and detention basin for flood simulation under complex topography and irregular boundary conditions. It is noted that the flood will propagate in all directions in the case of dike breach. One-dimensional models cannot well represent the flow conditions, two-dimensional models are required [6]. With the development of high performance computer and geographic information system, a two-dimensional hydrological method (MIKE21) was developed by Danish Hydraulic Institute (DHI) [9] and has been widely applied to the modelling of flood evolution for dam breaks, rivers and lakes (e.g., [5, 6, 10, 11]). However, it is rarely applied to the flood simulations in the flood detention basins. Based on the two-dimensional flood simulation results (e.g., inundation area, water depth and flow velocity distributions), dike-break induced flood consequences and risk can be evaluated [12], and relevant early-warning and emergency response measures and decision-making can be timely formulated to guide flood control and rescue works.

In this study, a MIKE21-based numerical approach for modeling the flood movement in the flood detention basin is developed. The approaches for estimating the dike-break induced flood consequences (i.e., life, economic and ecological environment costs) are presented. The Kangshan flood detention basin in Poyang Lake area in China is taken as a typical case to illustrate the effectiveness of the developed approach. Finally, the flood disaster zones are defined and the emergency evacuation planning against the dike-break flood is made based on the obtained flood simulation results.

2 Mike21-Based Flood Simulation Approach

MIKE21 is a software for simulating the two-dimensional unsteady free surface flows, which is originally developed for simulation of flow in seas, estuaries and coastal areas (e.g., [5, 6, 9]). It comprises of four modules: coastal hydrology and oceanography, environmental hydrology, sediment transport process and wave. In MIKE21, the alternating direction implicit finite difference method is adopted to solve the governing equations on a staggered grid. The two-dimensional hydrodynamic method is utilized to simulate the water flow movement, from which the water level and flow velocity distributions for different inundation durations can be obtained. Additionally, the MIKE21 can support soft start and predefine a variety of hydraulic structures, dry and wet nodes and elements [9]. The MIKE21 is extended in this study for modeling flood movement in flood detention basin.

2.1 Governing Equations

In the MIKE21, the river flow is treated as an incompressible Newtonian fluid, and the law of river flow motion is described using the Navier-Stokes equations [9]. To account for the frictional and turbulence effects of the river bed, the eddy viscosity is introduced to establish a two-dimensional hydrodynamic model. Neglecting Coriolis and wind force, the governing equations can be expressed as continuity and momentum equations in x and y directions. The two-dimensional flow continuity equation is given by (e.g., [6, 9])

$$\frac{\partial \eta}{\partial t} + h\frac{\partial u}{\partial x} + h\frac{\partial v}{\partial y} = 0 \tag{1}$$

where t is the time coordinate; x and y are the Catesian coordinates; u and v are the flow velocities in the x and y directions, respectively; h is the water depth. The momentum equations in x and y directions are given by

$$\frac{\partial u}{\partial t} + u\frac{\partial u}{\partial x} + v\frac{\partial u}{\partial y} + g\frac{\partial \eta}{\partial x} + \frac{gu\sqrt{u^2+v^2}}{c^2 h} = v_t\left(2\frac{\partial^2 u}{\partial x^2} + \frac{\partial^2 u}{\partial y^2} + \frac{\partial^2 v}{\partial x \partial y}\right) \tag{2}$$

$$\frac{\partial v}{\partial t} + u\frac{\partial v}{\partial x} + v\frac{\partial v}{\partial y} + g\frac{\partial \eta}{\partial y} + \frac{gv\sqrt{u^2+v^2}}{c^2 h} = v_t\left(\frac{\partial^2 v}{\partial x^2} + 2\frac{\partial^2 v}{\partial y^2} + \frac{\partial^2 u}{\partial x \partial y}\right) \tag{3}$$

where η is the water surface elevation; g is the gravity acceleration; v_t is the eddy viscosity; c is the Chezy friction coefficient, $c = \sqrt[6]{h}\big/n$, in which n is the roughness coefficient.

2.2 Boundary Conditions

The boundary conditions for the MIKE21 model include the upper and lower boundary conditions, special boundary condition and dynamic boundary condition. With respect to the flood detention basin, the historical highest, design or real-time measured discharges and water levels are generally taken as the upper boundary conditions. The lower boundary condition is not required if the flood detention basin is a closed region. Otherwise, the discharges or water levels at the boundaries of the detention basin can be taken as the lower boundary conditions. The drainage structures (e.g., culverts, culvert pipe) and the impeding water structures (e.g., roads, dikes) are often taken as special boundary conditions. The dynamic boundary condition is defined where the water depths for the dry and wet boundaries are 0.005 and 0.1 m, respectively (e.g., [2]). It means that the dynamic boundary will be a dry boundary and do not participate in the calculation if the water depth is less than 0.005 m, whereas it will be a wet boundary and participate in the calculation if the water depth is larger than 0.1 m.

2.3 Model Parameters

In the MIKE21, the model parameters including numerical parameters and physical parameters should be determined. The numerical parameters mainly include calculation step size and number of steps which typically take the default values. The physical parameters include the roughness coefficient of river bed, eddy viscosity and dynamic boundary parameters. The roughness coefficient reflects the resistance of the water flow, which can be determined according to the flood risk maps or through calibrating with the condition of the river bed. The eddy viscosity is often determined using Smagorinsky equation based on the flow velocity [13].

2.4 Flood Diversion

The flood diversion port gate is reserved at the dike during the dike construction stage for future flood control and rescue during extreme floods. The natural, gate or blasting flood diversion measures can be taken to open the reserved flood diversion port gate once the water level of the outer river reaches the prescribed water level. Then, a dike beach will be generated and the flood begins to enter the detention basin through the breach. The discharge through the breach can be calculated using the drawn weir formula (e.g., [1]):

$$Q_b = m\sigma B\sqrt{2g}(Z - Z_b)^{1.5} \tag{4}$$

where Q_b is the dike breach discharge (m^3/s); m and σ are the discharge coefficient and inundation coefficient, respectively; Z is the water level of the river at the breach (m); Z_b is the elevation of the weir crest (m); B is the width of the dike breach (m).

3   Estimation of Flood Consequences

Once the dike-break flood with huge energy enters the detention basin, the flood consequences including the life, economic and environment losses will be caused. The life losses represent the number of deaths or disasters. Economic losses include direct and indirect economic losses. The direct economic losses mainly include, but not limited to, the damages to residential buildings, agricultures (e.g., crop, animal husbandry, fishery and forestry) and personal assets in the detention basin. The indirect economic losses comprise of contingency costs, losses caused by reduced agricultural crops and increased costs of socioeconomic operation. The environment losses represent the damages of natural and social resources and the damages caused to the ecology environment, water environment, soil environment and living environment [14]. The economic and environment losses are generally counted in monetary terms.

3.1 Estimate of Life Losses

Following Piers et al. [15] and Graham et al. [16], the flood detention basin can be divided into several sub-regions according to the population concentration and administrative division. The life losses (*LOL*) are then estimated using the risk population integration algorithm as follows:

$$LOL = \sum_{i=1}^{a}\sum_{j=1}^{b} PAR_{ij} I R_{ij} \tag{5}$$

where a is the total number of risk regions; b is the number of risk population groups; PAR_{ij} is the number of risk population in the j-th group of the i-th risk region; IR_{ij} is the life loss rate of the individual in the j-th group of the i-th risk region, which can be evaluated as [17]

$$IR_{ij} = \sum_{k=1}^{l} P_k f_k \qquad (6)$$

where l is the level of flood; P_k and f_k are the probability of dike break and the mortality rate of the individual suffering from the k-th level of flood, respectively. f_k can be estimated as [18]

$$f_k = f_0 + q m_1 + (1 - q) m_2 + \beta \qquad (7)$$

where f_0 is the mortality rate of the individual at risk in China; m_1 is the disaster severity factor of direct influence, $m_1 \leq 1$; m_2 is the disaster severity factor of indirect influence, $m_2 \leq 1$. The suggested values for m_1 and m_2 are given in Table 1. q is the weight that can be determined using analytic hierarchy process method, $0.5 < q < 1$ is usually chosen; β is the correction coefficient, $\beta = 1.4$ is selected because the population density around the rivers or lakes in China is typically high [19].

Table 1. Suggested values for m_1 and m_2.

Cases	m_1	Cases	m_2
Slight	0–0.2	Extremely beneficial	0–0.2
General	0.2–0.4	Favorable	0.2–0.4
Medium	0.4–0.6	General	0.4–0.6
Serious	0.6–0.8	Unfavorable	0.6–0.8
Extremely serious	0.8–1.0	Extremely disadvantageous	0.8–1.0

3.2 Estimate of Economic Losses

The loss rate method that is particularly applicable for estimating the losses of various types of circulating assets and fixed assets is adopted to evaluate the direct economic losses S_D as follows (e.g., [17]):

$$S_D = \sum_{i=1}^{e} \sum_{j=1}^{s} \sum_{k=1}^{m} \beta_{ijk} W_{ijk} \qquad (8)$$

where β_{ijk} and W_{ijk} are the loss rate and the value of the j-th asset in the i-th category suffering from the k-th level of flood inundation, respectively; e is the number of classifications of assets; s is the number of assets in the i-th category; m is the level of the inundation water depth. Based on the field surveys and analyses of a large amounts of historical data, the indirect economic losses S_I can be estimated as (e.g., [14])

$$S_I = kS_D + d \qquad (9)$$

where k and d are the loss coefficients. $k = 0.63$ and $d = 0$ are chosen after synthesizing the loss relations of many agricultural crops and industries [14].

3.3 Estimate of Environment Losses

Few attempts have been made to investigate the environment losses caused by the dike-break flood. This lies in the fact many complicated factors should be taken into account, and there are not unified standards and methods for estimating the environment losses. To this end, a contingent valuation method that is commonly used to assess the nonuse value of the environment is adopted to estimate the environment losses (e.g., [20]). To determine the values of the environment resources to the respondents, the willingness of the population to pay in the flood disaster zone are evaluated through questionnaires (e.g., [21]). The average willingness to pay for each person per year (E) can be estimated as

$$E = \sum_{i=1}^{u} D_i A_i \qquad (10)$$

where u is the number of pay; A_i is the money that each person is willing to pay; D_i is the probability for each person who pays the money A_i. The environmental losses (F) can be estimated as

$$F = ENn_p \qquad (11)$$

where N is the durable years; n_p is the number of risk population. Note that the determinations of E and F also rely on the flood inundation data.

4 Engineering Application

4.1 Study Site

The Kangshan dike built in 1966 is one of the key dikes in Poyang Lake district, which is located in Yugan County of Jiangxi province, China. It is on the south bank of Poyang Lake and has a total length of 36.25 km. The elevation of the dike top is 24.55 m. The inner and outer slopes are both 1:3. The design flood water level is 22.5 m. There are five hydraulic structures along the dike: medium size Meixi water lock with a design flow of 103 m³/s, large (2) size Lugushan pump station with a design flow of 73.8 m³/s, medium size Dahukou water lock with a design flow of 160.5 m³/s, medium size Luojiaohu pump station with a drainage flow of 15.5 m³/s and small size Lixi water lock with a drainage flow of 7.1 m³/s. The Kangshan dike undertakes the flood control task of the protected area of 343.4 km², 22,560 households and 103,437 persons [22].

The Kangshan flood detention basin was approved by China State Council in 1985 and constructed in 1986, which is the largest flood detention basin in Jiangxi Province, China. The effective flood storage capacity is about 16.58×10^8 m^3 with a rainfall collection area and a flood storage area being 450.31 and 312.37 km^2, respectively. Geographic information data and administrative division data of the Kangshan flood detention basin are collected on the scales of 1:10000 and 1:50000, respectively. The entire Kangshan flood detention basin is taken as the study area. Figure 1 presents the division of study area. There are Shikou town, Santang town, Datang town, Ruihong town, and Kangshan town, Kanshan reclamation and a modern agricultural demonstration in the detention basin. There are 29,895 persons, a farmland area of 117.21 km^2 and an aquaculture area of 139.01 km^2. The registered residential assets in the flood detention basin include a contracted land area of 262.49 km^2, an agricultural crops area of 117.21 km^2, an aquatic products area of 139.01 km^2, and an economic forest area of 6.27 km^2. Besides, 11,919 domestic animals, 19,384 domestic birds, 19,432 houses, 7,244 agricultural machines, 1,395 serving livestock products and 17,250 durable consumer products are also registered [22]. The north flood diversion port at the Kangshan dike is treated as the flood-in and flood-fall boundaries. The north Kangshan dike and the south dry lands are treated as the physical boundaries.

4.2   Dike Breach Discharge

A flood diversion port gate is reserved at the 20k + 045–20k + 470 of the Kangshan Dike, as shown in Fig. 1. Deep cement mixing piles are installed at both sides of the port gate to prevent it from unlimited expansion. According to the flood diversion planning of Yangtze River, the blasting method is adopted to open the reserved port gate when the

Fig. 1. Division of study area and layout of monitoring points in the Kangshan flood detention basin.

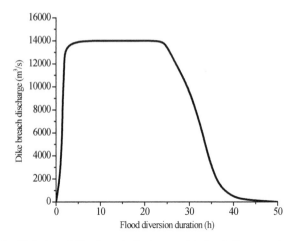

Fig. 2. Variation of dike breach discharge with flood diversion duration.

water level of Poyang Lake monitored at Kangshan gauging station reaches 20.68 m. Then, the dike-break flood enters the Kangshan flood detention basin through the breach mouth with an initial bottom width of 12.5 m. The dike is breached with a bottom width of 300 m and a bottom elevation of 15.93 m is allowed to develop gradually due to flood scouring within two hours [22].

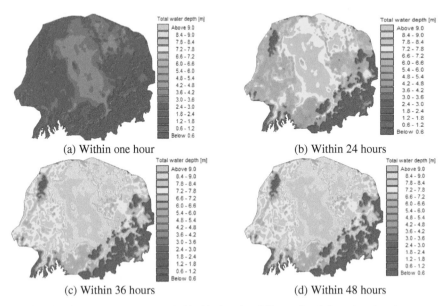

Fig. 3. Contours of water depth distribution for different flood diversion durations.

Figure 2 presents the variation of the dike breach discharge with the flood diversion duration. It can be observed that the discharge rapidly reaches the maximum value of 14,000 m³/s within two hours after the dike is breached, it then almost keeps unchanged. It begins to decrease since the 24th hour of dike-break. The entire flood-in process is completed in 48 h wherein the discharge is close to zero and the water level of the flood detention basin reaches 20.68 m which is exactly equal to the water level of the Poyang Lake.

4.3 Dike-Break Flood Simulation

To simulate the flood routing process in the Kangshan flood detention basin, a two-dimensional non-constant mathematical model of water flow is constructed. The finite volume method in the MIKE21 is utilized for solving the governing equations in Eqs. (1)–(3). The steady-state initial conditions are considered with the dike breach discharge in Fig. 2 as input. The calculation step size and the number of steps are set as 120 s and 2160, respectively. In view of the complexity of the terrain of the study area, the MIKE21 FM unstructured grids with a maximum area less than 0.01 km² are adopted. The denser grids with the smaller sizes are adopted to model the resident zone, compact building zone and planting zone. The study area is discretized into a total of 63,880 grids and 58095 nodes. The maximum and minimum distances between the grid midpoints are 300 m and 26 m, respectively. The historical highest water level of 22.68 m that was monitored at the Kangshan gauging station is treated as the upper boundary condition, and the land border is treated as the lower boundary condition. The water levels for the dry and wet boundaries are 0.005 and 0.1 m, respectively. The roughness coefficients for different underlying surfaces are summarized in Table 2.

Table 2. Roughness coefficients for different underlying surfaces.

Underlying surfaces	Villages	Bushes	Dry farmlands	Paddy fields	Roads	Vacant places	Rivers
n	0.07	0.065	0.06	0.05	0.035	0.035	0.025–0.035

To facilitate the understanding of the flood movement, 23 monitoring points are arranged in the Kangshan town, Ruihong town, Datang town, Santang town, Shikou town and Kangshan reclamation areas, as shown in Fig. 1. Figures 3(a)–(d) present the contours of water depth distribution underlying four representative flood diversion durations. It can be observed that the flood reaches the Kangshan reclamation area and Dahu administration after one hour of dike-break. The Kangshan reclamation area and the lakeside area of Kangshan town are greatly affected by the flood scouring because they are located in the vicinity of the breach and the maximum flow velocities at these sites exceed 1.0 m/s. The flood advances to the Kangshan town, the Ruihong town and Shikou town areas within 24 h, to the Datang town area until the 36th hour after the dike is breached. This is due to the fact that the Datang town area has a relatively high terrain and is far away from the breach mouth. The southeast Santang town area is slightly affected by the dike-break flood even when the entire flood diversion process is finished. As seen from Fig. 3(d), the average water depth of the flood detention basin is 6.95 m, while the water depths of the Kangshan reclamation and Shikou town areas reach 8 m at the end of the flood diversion.

4.4 Flood Inundation Pattern

Before the dike is breached, the water level and the area filled with water in the detention basin are 15.10 m and about 69 km², respectively. The total water flow inside the detention basin and the inundation area are up to 1.658 billion m³ and 288 km², respectively, at the end of the flood diversion. Based on the inundation area and water depth, the flood detention basin is subdivided into light, moderate, severe and dangerous disaster zones, as listed in Table 3. Note that the light disaster zone has a water depth less than 0.5 m, which can be treated as a personnel or asset resettlement zone. The percentages of light, moderate and severe disaster zones are 0.66%, 0.82% and 1.77%, respectively, while that of dangerous disaster zone is the largest, up to 96.75%, and the corresponding inundation area is 278.63 km². It implies that the vast majority of the Kangshan flood detention basin is significantly affected by the flood. The area that the flood does not reach is only 31.3 km². These zones can also be treated as the personnel or asset resettlement zones.

Table 3. Classification of flood disaster zones in the Kangshan flood detention basin.

Flood disaster zones	Water depth h (m)	Inundation area (km²)	Percentage (%)
Light disaster zone	$h \leq 0.5$	1.92	0.66
Moderate disaster zone	$0.5 < h \leq 1.0$	2.36	0.82
Severe disaster zone	$1.0 < h \leq 2.0$	5.09	1.77
Dangerous disaster zone	$h > 2.0$	278.63	96.75

The variations of water depths and flow velocities at the 23 monitoring points in the flood detention basin are also investigated in this section. Based on the water depth and flow velocity distributions for these six key areas, the time for the flood reaching the specified monitoring points, the maximum water depth and peak flow velocity of each monitoring point, and the flood peak appeared time can be deduced.

As observed in Fig. 4, the flood reaches the Fuqian village of Kangshan town area after 9.67 h of dike-break although it is very close to the breach mouth. This is because the terrain of the Kangshan town is high as a whole. Both the maximum water depth and peak velocity appear in the Fuqian village, which are 2.65 m and 0.013 m/s, respectively. The flood peak appeared time is 30 h. The Jinsan and Dashan villages are almost not affected by the dike-break flood. Similarly, it can be found that the flood reaches the Xigang village of Ruihong town after 9.17 h of dike-break. The maximum water depth and peak flow velocity appear in the Xigang village and Luojia village, respectively, which are 4.55 m and 0.054 m/s. The flood peak appeared time is 24.5 h. Among the

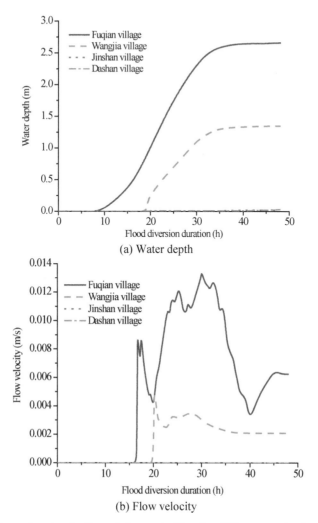

(a) Water depth

(b) Flow velocity

Fig. 4. Variations of water depths and flow velocities for different monitoring points in the Kangshan town area.

Ruihong town area, only the Baxia village is slightly affected by the dike-break flood because it has a relatively higher terrain. The flood reaches the Datang town after 50 min of dike-break. Both the maximum water depth and peak flow velocity appear in the Dahu village, which are 7.65 m and 0.0731 m/s, respectively. The flood peak appeared time is 2.67 h. Among the Datang town area, the Nanlong and Chenjiatang villages are also slightly affected by the dike-break flood because their terrains are relatively high. The flood reaches the Mashan village of the Santang town area until 23.67 h of dike-break. This is due to the fact the Santang town is located in the southern mountainous area of the Kangshan flood detention basin that is far away from the breach mouth, and has the highest terrain in the study area. The maximum water depth and peak flow velocity are 3.67 m and 0.2635 m/s, respectively. The flood peak appeared time is 29 h. The other zones in the Santang town area including Weijia and Yujiaqiao villages are not flooded, which can also be taken as the personnel or asset resettlement zones. Similar to the Datang area, the flood advances to the Shikou town after 50 min of dike-break. The maximum water depth and peak flow velocity appear in the Guzhu and Xi villages, respectively, which are 6.24 m and 0.027 m/s, respectively. The flood peak appeared time is 28.83 h. Among the Shikou town area, the Hubin and Liubu villages are not flooded, which can also be taken as the personnel or asset resettlement zones. The flood advances to the Lixi and Ganquanzhou reclamations of the Kangshan reclamation area after 40 min of dike-break. It is obvious that the Kangshan reclamation area is the earliest affected by the flood. Both the maximum water depth and the peak flow velocity appear in the Lixi reclamation, which are 8.32 m and 0.074 m/s, respectively. The flood peak appeared time is 2.5 h.

Moreover, the emergency evacuation planning against the dike-break flood can be made based on the flood simulation results (i.e., inundation area, water depth, flood peak appeared time) [4]. As seen from Fig. 3, the Kangshan reclamation area is flooded with an average water depth exceeding 2.0 m. It indicates that the people and residential assets in this area should be quickly evacuated to the neighboring Shikou town and Datang town areas within the warning time. The personnel or asset resettlement buildings can be constructed in the Liubu and Hubin villages of the Shikou town area and the Nanlong and Chenjiatang villages of the Datang down area. The people and residential assets in the Xigang and Luojia villages of the Ruihong town area should be quickly evacuated to the neighboring Baxia village which is slightly affected by the flood. The people and residential assets in the Kangshan town should be evacuated to the Dashan and Jinshan villages that are not affected by the flood. Finally, only the Mashan village in the Santang town area is affected by the flood. In general, the people and residential assets in the Santang town area will be the safest.

4.5 Dike-Break Induced Flood Consequences

The dike-break induced flood consequences including the life, economic and ecological environment losses are estimated based on the flood simulation results using the methods in Sect. 3. Figure 5 shows the variations of the life losses for different warning time with the flood diversion duration. It is clear that the warning time can greatly affect the life losses. Figure 6 presents the variations of the direct economic losses with the flood diversion duration. It can be observed from Fig. 6(b) that the component with the largest proportion among the direct economic losses is the agricultural crops losses. The total direct and indirect economic losses induced by the Kangshan dike-break are estimated as ¥CNY 953 million and 600 million, respectively. The environmental losses are estimated by means of questionnaires. The average willingness to pay underlying the key dike remediation projects in the Poyang Lake district is calculated using Eq. (10). It is ¥CNY 776.8–780 for each person per year. As mentioned earlier, currently there are 29,895 residents in the Kangshan flood detention basin. Figure 7 shows the variation of the environment losses with the flood diversion duration. It can be found that the environment losses induced by the dike-break will be ¥CNY 464 million for the coming 20 years by taking the average willingness to pay as ¥CNY 776.8 for each person per year.

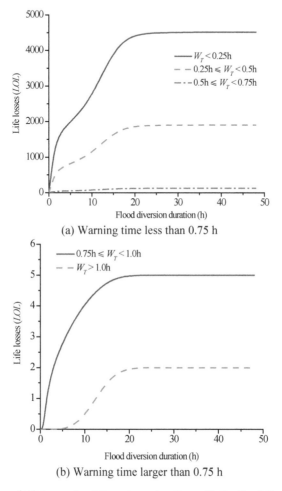

(a) Warning time less than 0.75 h

(b) Warning time larger than 0.75 h

Fig. 5. Variation of life losses for different warning time with the flood diversion duration.

The total dike-break induced flood consequences are approximatively equal to the sum of the life, economic and environment losses. After the flood diversion is stopped, the dike-break induced economic and environment losses can be up to ¥CNY 20.2 billion. Therefore, the early warning and emergency response measures should be worked out to reduce such huge flood consequences. Some engineering or non-engineering measures should be taken to mitigate the dike-break induced flood risk.

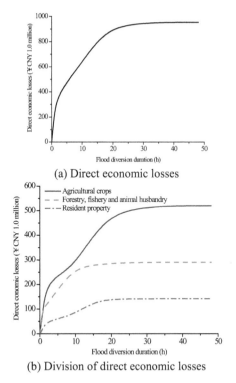

(a) Direct economic losses

(b) Division of direct economic losses

Fig. 6. Variation of direct economic losses with the flood diversion duration.

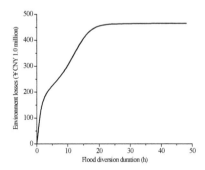

Fig. 7. Variation of environment losses with the flood diversion duration.

5 Conclusion

In this paper, a MIKE21-based numerical method is developed for modeling the flood routing process in the flood detention basin. The approaches for estimation of dike-break induced flood consequences including the life, economic and environment losses are presented. The proposed approach has been applied to flood simulation and flood consequences and corresponding risk assessment in the Kangshan flood detention basin in the Poyang Lake district of Jiangxi Province, China. The inundation area, water depth and flow velocity distributions, the time for the flood reaching the specified locations and the flood peak appeared time are readily estimated based on the flood simulation results. Furthermore, the emergency evacuation planning against the dike-break flood is made and the life, economic and environment losses induced by dike-break are estimated. It can be found that the vast majority of the Kangshan flood detention basin is significantly affected by the flood. The percentage of the dangerous disaster zone is the largest, up to 96.75%, and the corresponding inundation area is 278.63 km^2. The area that the flood does not reach is only 31.3 km^2. The dike-break will cause serious consequences. The economic and environment losses can be up to ¥CNY 20.2 billion. Further research needs to be conducted to validate the effectiveness of the proposed approach in modeling the flood routing process and evaluating the dike-break flood risk quantitatively, and apply it to other flood detention basins.

Acknowledgments. This work was supported by Jiangxi Provincial Natural Science Foundation (Project Nos. 2018ACB21017, 20181ACB20008 and 20192BBG70078).

References

1. Vorogushyn, S., Lindenschmidt, K.E., Kreibich, H., Apel, H., Merz, B.: Analysis of a detention basin impact on dike failure probabilities and flood risk for a channel-dike-floodplain system along the river Elbe, Germany. J. Hydrol. **436**, 120–131 (2012)
2. Guo, F.: Research and Application of Risk Analysis and Assessment of Flood Disasters in Flood Detention Basin, pp. 1–23. Science Press, Beijing (2016)
3. Danka, J., Zhang, L.M.: Dike failure mechanisms and breaching parameters. J. Geotech. Geoenviron. Eng. **141**(9), 04015039 (2015)
4. Zhang, W., Zhou, J., Liu, Y., Chen, X., Wang, C.: Emergency evacuation planning against dike-break flood: a GIS-based DSS for flood detention basin of Jingjiang in central China. Nat. Hazards **81**(2), 1283–1301 (2016)
5. Zolghadr, M., Hashemi, M.R., Hosseinipour, E.Z.: Modeling of flood wave propagation through levee breach using MIKE21, a case study in Helleh River, Iran. In: World Environmental and Water Resources Congress 2010: Challenges of Change, pp. 2683–2693 (2010)
6. Zolghadr, M., Hashemi, M.R., Zomorodian, S.M.A.: Assessment of MIKE21 model in dam and dike-break simulation. Iran. J. Sci. Technol. Trans. Civil Eng. **35**(C2), 247–262 (2011)
7. Tucciarelli, T., Termini, D.: Finte-element modeling of floodplain flow. J. Hydraul. Eng. **126**(6), 416–424 (2000)
8. Liu, Q., Qin, Y., Zhang, Y., Li, Z.: A coupled 1D-2D hydrodynamic model for flood simulation in flood detention basin. Nat. Hazards **75**(2), 1303–1325 (2015)
9. Danish Hydraulic Institute (DHI): MIKE21: A modeling system for rivers and channels reference manual. DHI Water and Environment, Denmark (2007)
10. Bladé, E., Gómez-Valentín, M., Dolz, J., Aragón-Hernández, J.L., Corestein, G., Sánchez-Juny, M.: Integration of 1D and 2D finite volume schemes for computations of water flow in natural channels. Adv. Water Resour. **42**, 17–29 (2012)
11. Rahdarian, A., Niksokhan, M.H.: Numerical modeling of storm surge attenuation by mangroves in protected area of mangroves of Qheshm Island. Ocean Eng. **145**, 304–315 (2017)

12. Yan, B., Li, S., Wang, J., Ge, Z., Zhang, L.: Socio-economic vulnerability of the megacity of Shanghai (China) to sea-level rise and associated storm surges. Reg. Environ. Change **16**(5), 1443–1456 (2016)
13. Zhu, S., Yu, Y., Yu, F., Liu, J., You, Z.: Flood routing simulation of Dongting lakeshore plain city based on MIKE21 FM model. J. Water Resour. Water Eng. **29**(2), 132–138 (2018). (in Chinese)
14. Zhang, S., Tan, Y.: Risk assessment of earth dam overtopping and its application research. Nat. Hazards **74**(2), 717–736 (2014)
15. Piers, M.: Methods and models for the assessment of third party risk due to aircraft accidents in the vicinity of airports and their implications for societal risk. In: Jorissen, R.E., Stallen, P.J.M. (eds.) Quantified Societal Risk and Policy Making. Kluwer Academic Publishers, Dordrecht (1998)
16. Graham, W.J.: A procedure for estimating loss of life caused by dam failure. US Department of the Interior, Bureau of Reclamation (1999)
17. Song, J., He, X.: Discussion on analysis method for risk of life loss caused by dam failure in China. J. Hohai Univ. (Nat. Sci.) **36**(5), 628–633 (2008). (in Chinese)
18. Li, L., Zhou, K.: Methods for evaluation of life loss induced by dam failure. Adv. Sci. Technol. Water Resour. **26**(2), 76–80 (2006). (in Chinese)
19. Brown, C.A., Graham, W.J.: Assessing the threat to life from dam failure. JAWRA J. Am. Water Resour. Assoc. **24**(6), 1303–1309 (1988)
20. Hanemann, A., Loomis, J., Kanninen, B.: Statistical efficiency of double-bounded dichotomous choice contingent valuation. Am. J. Agric. Econ. **73**(6), 1225–1263 (1991)
21. Kristrm, B.: Spike models in contingent valuation. Am. J. Agric. Econ. **79**(3), 1013–1023 (1997)
22. Jiangxi Provincial Water Conservancy Planning and Design Institute: Feasibility study report on safety construction project of Poyang Lake flood detention basin in Jiangxi province. Jiangxi Provincial Water Resources Department (2014)

Flood Consequences Under Extreme Storms over Hong Kong Island

T. Abimbola Owolabi and Limin Zhang[(⊠)]

Department of Civil and Environmental Engineering,
Hong Kong University of Science and Technology, Clear Water Bay, Hong Kong
cezhangl@ust.hk

Abstract. Flood damage has been extremely severe in recent decades and is responsible for a greater number of damaging events than any other type of natural hazard. Floods are anticipated to happen more frequently in the future because of climate change, unplanned rapid urbanization, change in land use pattern, poor watershed management and declining recharge of groundwater by extension of impermeable surfaces in urban areas. Therefore, assessment of flood consequence under large storms is an important issue. This paper quantifies the elements at risk in Hong Kong under large storm scenarios with rain intensities of 29 and 85% of the 24-h Probable Maximum Precipitation (PMP). The buildings affected were obtained by overlaying flood maps and building maps in a GIS environment. The numbers of buildings of all types and exposed individuals affected by the flood under the two storm scenarios were also quantified. The results indicate that residential buildings appear to be the most vulnerable among all structures and facilities in Hong Kong and thereby may lead to the highest affected population under the two rainfall scenarios. The western part of Hong Kong Island is more susceptible to flooding as a result of its steeper slope terrain and densely populated infrastructures. Enhanced flood risk assessment methods and improved understanding about flood risk will support decision makers in decreasing damage and fatalities.

Keywords: Flooding · Risk analysis · Risk map · Element at risk

1 Introduction

Flooding is a leading cause of losses from natural phenomena and responsible for a greater number of damaging events than any other type of natural hazard. Flood damage has been extremely severe in recent decades. However, the effect of flooding on buildings and population has received less attention. Flooding acts as a temporary covering of land by water as a result of surface water escaping from its normal confines or as a result of heavy precipitation (IWRA 2005). There are three main types of flood and several special cases. The main types are storm surge, river flood, and flash flood; special cases include tsunami, waterlogging and backwater. Hong Kong was struck by a major storm on 7 June 2008, causing at least 622 floods and 16-h closure of the North Lantau Airport Highway (HKSAR Government Press Releases 2008).

© Springer Nature Switzerland AG 2020
J.-M. Zhang et al. (Eds.): ICED 2020, SSGG, pp. 311–317, 2020.
https://doi.org/10.1007/978-3-030-46351-9_32

Floods are anticipated to happen more often in the future because of climate change, unplanned rapid urbanization, change in land use pattern, poor watershed management and declining recharge of ground water by extension of impermeable surfaces in urban areas. This means that many urban areas across the globe are likely to be under serious threat of floods; the adverse impacts of which are already believed only next to that of earthquakes (Nasiri et al. 2016). Managing hazards with the aim of safety and wellbeing of people and their environment is important (Li et al. 2012, Glade and Crozier 2005; Zhang et al. 2012; Owolabi and Zhang 2019a; Lee and Jones 2004; Nadim 2016; van Westen et al. 2008; Vranken et al. 2015; Chen and Zhang 2015; Fell et al. 2005; Lacasse and Nadim 2009; Owolabi et al. 2018). Floods are part of the hydrological cycle, but due to dispute natural function of river flood plains in transporting water and sediment as a result of human land use, risk has increased (Schanze et al. 2007). Indeed, urban flood vulnerability varies time to time and in diverse places because of environmental conditions, human activities, and the culture of society in face of the threats (Ahmad and Simonovic 2013). Improved assessment methods and understanding about flood risk vulnerability can support decision makers in decreasing damage and mortalities. Many buildings and streets are affected by larger and faster moving flood. Gao et al. (2019) developed a model for simulating overland flow and underground drainage hydraulics simultaneously based on northern Hong Kong Island and introduced flood control measures to address a major societal need to mitigate flooding problems, but observed that the densely urbanized region may still be affected by flooding when severe events occur.

2   Study Area

The study area of this research is Hong Kong Island, which covers an area of 78.59 km^2 with a population of 1,289,500 and a population density of 16,390/km^2. The steep terrain and high frequency of tropical rainstorms lead to frequent hazards on the natural terrain (Fuchu and Chack 2002). The urban area of Hong Kong surrounded by mountains, when a major rainstorm occurs, is vulnerable to both flash flooding in mid-mountain areas and local flooding in low-lying areas. Conditions can worsen when the urban drainage systems are overloaded. The tropical weather makes the vulnerability to flooding worse because more than 80% of the annual total precipitation occurs in the wet season (Gao et al. 2019).

3   Rainfall-Induced Runoff and Flooding

The governing equations adopted in the simulation of flooding in this research are shallow water equations which are frequently used to describe unsteady surface flow (e.g. Gao et al. 2019; Chen and Zhang 2015; Gao et al. 2016; FLO-2D Software Inc. 2015):

$$\frac{\partial h}{\partial t} + \frac{\partial (h v_x)}{\partial x} + \frac{\partial (h v_y)}{\partial y} = i - q \tag{1}$$

$$S_{fx} = S_o - \frac{\partial h}{\partial x} - \frac{v_x}{g}\frac{\partial(v_x)}{\partial x} - \frac{1}{g}\frac{\partial(v_x)}{\partial t} \tag{2}$$

$$S_{fy} = S_o - \frac{\partial h}{\partial y} - \frac{v_y}{g}\frac{\partial(v_y)}{\partial y} - \frac{1}{g}\frac{\partial(v_y)}{\partial t} \tag{3}$$

where h is the flow depth; t is time; v_x and v_y are the depth-averaged flow velocities in the x and y directions, respectively; i is the excess rainfall intensity on the flow surface; q is the flow discharge from the ground surface to the drainage works per unit surface area, which is determined by the total water head values in the pipe and on the ground, and may be negative when pipes or conduits effuse in the reverse direction; S_{fx} and S_{fy} are the friction slope components in the x and y directions, respectively, which are based on Manning's equation; S_0 is a bed slope term; g is the gravitational acceleration.

Historical records have depicted that extreme rainfall events are often expected to be more recurrent and intense in Hong Kong as a result of climate changes (Ho et al. 2016). Probable flood scenarios of 29% and 85% of the 24-h probable maximum precipitation (PMP) reported by Zhang et al. (2017) in which a physically based model (EDDA 1.0) was adopted to predict likely flood occurrence in large scale were used in this study for the flood risk assessment. The result of the floods under 29%PMP and 85PMP rainfall scenarios is shown in Fig. 1. Both flooding scenarios cover a larger part of Hong Kong Island especially the north-western region where urban infrastructures are densely developed.

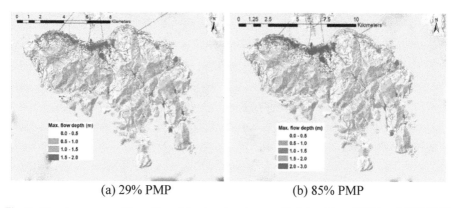

(a) 29% PMP   (b) 85% PMP

Fig. 1. Flood maps over Hong Kong Island under storms with intensities of 29% and 85% PMP.

4 Consequence Assessment

In analysing the flood consequence, the effect of flooding on the numbers of buildings and population are determined on GIS by integrating both flood simulation result (29% and 85% of the 24-h probable maximum precipitation (PMP)) and buildings layer. The number of buildings of all types and exposed individuals affected by the floods under the two scenarios were quantified. The number of populations at risk inside each building was based on evaluation of the number of floors, number of flats and the average number of people living in a flat in Hong Kong during the time of the event. The distribution of population at risk by flood is shown in Fig. 2, and the number of buildings affected and population at risk are shown in Fig. 3. The results indicate that residential buildings appear to be the most vulnerable compared with all structures and facilities in Hong Kong while temple buildings are the least vulnerable. The population at risk follows the same trend.

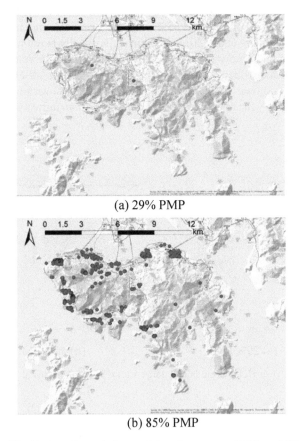

(a) 29% PMP

(b) 85% PMP

Fig. 2. Distribution of population at risk under storms of 85% and 29% PMP.

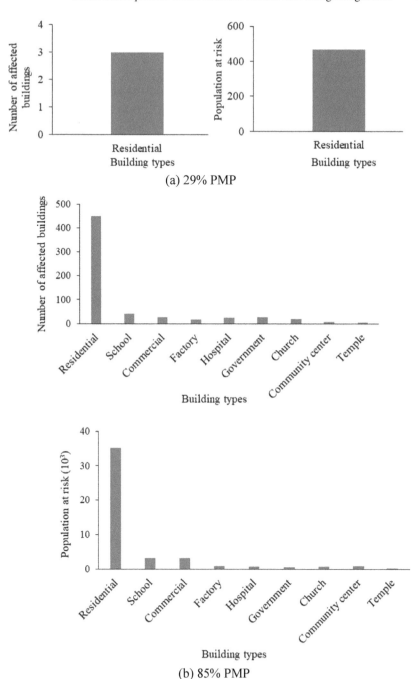

(a) 29% PMP

(b) 85% PMP

Fig. 3. Number of buildings and population at risk under storms of 29% and 85% PMP.

5   Conclusions

This paper assesses consequences of flooding on Hong Kong Island under two storm scenarios with rainfall intensities of 29% and 85% of the 24-h probable maximum precipitation (PMP). The storm of 29%PMP leads to fewer number of affected building and population under flooding while 85%PMP gives higher numbers. This shows that the risk intensifies as the storm becomes more intense. The study has provided information for establishing the need for any mitigation measures to be put in place to prevent the recurrence of severe flooding on Hong Kong Island and deal with the relevant flood events in the future. It provides an organised and clear framework to aid decision-making. The uncertainties involved in the analyses should be addressed in the future.

Acknowledgment. The authors wish to acknowledge the support from the Research Grants Council of the Hong Kong SAR (No. C6012-15G and No. 16206217).

References

Ahmad, S.S., Simonovic, S.P.: Spatial and temporal analysis of urban flood risk assessment. Urban Water J. **10**(1), 26–49 (2013)

Chen, H.X., Zhang, L.M.: EDDA 1.0: integrated simulation of debris flow erosion, deposition and property changes. Geosci. Model Dev. **8**(3), 829–844 (2015)

Chen, H.X., Zhang, L.M.: A physically based distributed cell model for predicting regional rainfall-induced shallow slope failures. Eng. Geol. **176**, 79–92 (2014)

Fell, R., Ho, K.K.S., Lacasse, S., Leroi, E.: A framework for landslide risk assessment and management. In: Proceedings of the International Conference on Landslide Risk Management, Vancouver, Canada, pp. 3–25 (2005)

Fuchu, D., Chack, F.L.: Landslides on natural terrain physical characteristics and susceptibility mapping in Hong Kong. Mt. Res. Dev. **22**(1) (2002)

Gao, L., Zhang, L.M., Chen, H.X., Shen, P.: Simulating debris flow mobility in urban settings. Eng. Geol. **214**, 67–78 (2016)

Gao, L., Zhang, L.M., Li, X., Zhou, S.: Evaluating metropolitan flood coping capabilities under heavy storms. J. Hydrol. Eng. **24**(6), 1–13 (2019)

Glade, T., Crozier, M.J.: The nature of landslide hazard impact. In: Glade, T., Anderson, M.G., Crozier, M.J. (eds.) Landslide Hazard and Risk, pp. 43–74. John Wiley, Chichester (2005)

HKSAR Government Press Releases: Flooding problem in the central and western district. Hong Kong Government (2008)

HK GOV, Hong Kong Population Projections: Demographic Statistics Section, Census and Statistics Department, The Government of Hong Kong Special Administrative Region (2012)

Ho, K.K.S., Sun, H.W., Wong, A.C.W., Yam, C.F., Lee, S.M.: Enhancing slope safety preparedness for extreme rainfall and potential climate change impacts in Hong Kong. Region Report for Hong Kong. In: Ho, K.K.S., Lacasse, S., Picarelli, L. (eds.) Preparedness for Climate Change Impact on Slope Safety, pp. 104–146. Taylor & Francis, London (2016)

International Water Resources Association: Water International, vol. 30, no. 1, pp. 58–68 (2005)

Johnson, K., Depietri, Y., Breil, M.: Multi-hazard risk assessment of two Hong Kong districts. Int. J. Disaster Risk Reduct. **19**, 311–323 (2016)

Lacasse, S., Nadim, F.: Landslide risk assessment and mitigation strategy. In: Sassa, K., Canuti, P. (eds.) Landslide Disaster Risk Reduction, pp. 31–61. Springer, Heidenberg (2009)

Lee, E.M., Jones, D.K.: Landslide Risk Assessment, p. 454. Thomas Tilford Publishing, London (2004)

Li, C., Ma, T., Sun, L., Li, W., Zheng, A.: Application and verification of a fractal approach to landslide susceptibility mapping. Nat. Hazards **61**, 169–185 (2012)

Nasiri, H., Johari, M., Yusof, M., Ahmad, T., Ali, M.: An overview to flood vulnerability assessment methods. Sustain. Water Resour. Manag. **2**(3), 331–336 (2016)

Nadim, F.: Challenges in managing the risk posed by extreme events. In: 6th Asian-Pacific Symposium on Structural Reliability and its Applications (APSSRA 6), Shanghai, China, pp. 62–70 (2016)

Owolabi, T.A., Zhang, L.M., Lacasse, S.: Landslide risk in Hong Kong under extreme storms. In: 6th International Symposium on Reliability Engineering and Risk Management (6ISRERM), pp. 885–891 (2018)

Owolabi, T.A., Zhang, L.M.: Assessment of consequences of debris flows over Hong Kong Island under extreme rainfall. In: Proceedings of the 7th International Symposium on Geotechnical Safety and Risk (ISGSR), pp. 978–981 (2019a). https://doi.org/10.3850/978-981-11-2725-0

Owolabi, T.A., Zhang, L.M.: Natural terrain landslide risk to population under extreme storms. In: International Symposium on Reliability of Multi-disciplinary Engineering Systems under Uncertainty (ISRMES) (2019b)

Schanze, J., Zeman, E., Marsalek, J.: Flood Risk Management: Hazards, Vulnerability and Mitigation Measures, vol. 67. Springer (2007)

Van Westen, C.J., Castellanos, E., Kuriakose, S.L.: Spatial data for landslide susceptibility, hazard, and vulnerability assessment: an overview. Eng. Geol. **102**(3), 112–131 (2008)

Vranken, L., Vantilt, G., Eeckhaut, M.V., Vandekerckhove, L., Poesen, J.: Landslide risk assessment in a densely populated hilly area, PP. 787–798 (2015)

Yen, Y.C., Hung, E.C., Keh, C.Y.: Investigation of the influence of rainfall runoff on shallow landslides in unsaturated soil using a mathematical model. Water **11**, 1178 (2019)

Zhang, S., Zhang, L.M., Peng, M., Zhang, L.L., Zhao, H.F., Chen, H.X.: Assessment of risks of loose landslide deposits formed by the 2008 Wenchuan earthquake. Nat. Hazards Earth Syst. Sci. **12**(5), 1381–1392 (2012)

Zhang, L.L., Zhang, J., Zhang, L.M., Tang, W.H.: Stability analysis of rainfall-induced slope failure: a review. Proc. ICE-Geotech. Eng. **164**(5), 299 (2011)

Zhang, L.M., Gao, L., Zhou, S.Y., Cheung, R.W., Lacasse, S.: Stress testing framework for managing landslide risks under extreme storms. In: Mikos, M., et al. (eds.) Advancing Culture of Living with Landslides, pp. 17–32 . Springer International Publishing AG (2017)

Monitoring, Early Warning and Emergency Response

Analysis of the Effects of Permeability Defects on Seepage Flow and Heat Transport in Embankment Dams

Chiara Cesali[1]([✉]), Walter Cardaci[2], and Francesco Federico[1]

[1] University of Rome Tor Vergata, Rome, Italy
cesali@ing.uniroma2.it, fdrfnc@gmail.com
[2] ERG Hydro S.r.l., Terni, Italy
wcardaci@erg.eu

Abstract. Permeability defects are practically unavoidable in earthen structures due to heterogeneity of the grain size of quarried materials, inappropriate compaction, discontinuities of displacements, dynamic effects, internal erosion processes and animal actions. For their detection, thermal monitoring systems may allow today an advanced surveillance of embankment dams, through the recent advancements in distributed sensing technologies based on optical fibers sensors. By exploiting the dependency of the thermal properties of the soil (i.e. thermal conductivity and heat capacity) on its water content, in saturated conditions, and their possible variations under seepage flow regimes, temperature measurements may provide information about the development and evolution of zones affected by permeability higher than the original one. The coupling between the seepage and heat transport processes, and particularly the effects of permeability defects on piezometric head and temperature distributions, has been parametrically investigated, through two Finite Element Analysis (FEA) codes (i.e. SEEP/W and TEMP/W), in order to evaluate the sensibility of thermal sensors in detecting these undesired phenomena. Finally, an experimental documented case has been back analyzed and re-interpreted.

Keywords: Permeability defects · Monitoring · Numerical analyses

1 Introduction

The need to understand how permeability defects (p.d.), often hidden, modify the seepage flow characteristics (e.g. discharge rate, interstitial pressure distribution and free surface profile) through embankment dams, causing exceedance of serviceability (change of discharge, turbid water) or ultimate limit states (local or global instabilities, piping, structural collapses), is well recognized [1, 2, 3, 4, 5].

p.d. may induce undesirable phenomena, such as leakages, redistribution of interstitial pressures, soil shear strength reduction, instability phenomena and local increases of the hydraulic gradients, whose related drag forces may trigger internal erosion and particles migration that might evolve up to embankment collapses [6].

© Springer Nature Switzerland AG 2020
J.-M. Zhang et al. (Eds.): ICED 2020, SSGG, pp. 321–332, 2020.
https://doi.org/10.1007/978-3-030-46351-9_33

Appropriate monitoring of the seepage flow within earthen dams/levees may allow to detect possible internal erosion phenomena before the safety of these structures is totally compromised. In addition to "conventional" instruments (e.g. piezometers, discharge measurements), among the available methods for monitoring seepage and erosion processes, thermal sensors can contribute to detect these phenomena [6, 7].

Temperature can be used as an indirect measure of the soil saturation degree, exploiting the dependency of the thermal properties of the soil (conductivity and heat capacity) on its water content. In saturated conditions, temperature measurements can provide information on seepage flow velocities, allowing identification of the development and evolution of erosion channels (pipes) and zones of higher permeability, caused by particles migration phenomena. However, the correct interpretation of temperature measurements, coupled to piezometric heads (p.h.) readings, must considerer the mutual dependency of thermal and hydraulic properties of soils.

To this purpose, numerical analyses of the coupled seepage flow process and heat transport mechanism through earth structures have been carried out to evaluate the effects of p.d. (particularly, of their geometrical sizes) on the main hydraulic variables (e.g. piezometric heads, interstitial pressure) and temperature.

2   Coupled Seepage and Heat Transfer Processes Through Embankment Dams and Modelling of Permeability Defects

To evidence the capability of thermal monitoring methods in the detection of possible, practically unavoidable, permeability defects in earthen structures, the coupling between the seepage and heat transfer processes has been firstly investigated through two FEA commercial codes, i.e. *SEEP/W* and *TEMP/W*, belonging to the Geostudio (Geoslope) Package [8].

The code *SEEP/W* solves two-dimensional (2D) seepage flow problems for steady, unsteady, saturated and unsaturated conditions. By assuming the validity of Darcy's law and by considering that the variations of volumetric water content (θ_w) are related to soil properties and stress state (in turn dependent on the suction), the continuity equation of water flow can be expressed as follows:

$$\frac{\partial}{\partial x}\left(k_x\frac{\partial h}{\partial x}\right) + \frac{\partial}{\partial y}\left(k_y\frac{\partial h}{\partial y}\right) + Q = m_w \cdot \gamma_w\frac{\partial h}{\partial t} \tag{1}$$

Where h is the piezometric head; k_x and k_y, the hydraulic conductibility in the x and y directions, respectively; Q, the applied boundary water flow; m_w, the slope of water retention curve; γ_w, the unit weight of water.

The code *TEMP/W* simulates the thermal variations in the ground due to environmental changes, or due to the construction of facilities, such as buildings or pipelines. The principal mechanism for heat flow in soils in most engineering applications is conduction (Fourier). The governing differential equation expressing the distribution of T, running in the TEMP/W code, is:

$$\frac{\partial}{\partial x}\left(k_{T,x}\frac{\partial T}{\partial x}\right) + \frac{\partial}{\partial y}\left(k_{T,y}\frac{\partial T}{\partial y}\right) + Q_T = \lambda\frac{\partial T}{\partial t} \tag{2}$$

being $k_{T,x}$ and $k_{T,y}$ the thermal conductivity in the x and y directions, respectively; Q_T, the applied boundary thermal flux; λ, the capacity for heat storage. λ depends on the volumetric heat capacity (c) of the material and the latent heat (L) associated with the phase change:

$$\lambda = c + L \frac{\partial w_u}{\partial T} \tag{3}$$

w_u being the total unfrozen (liquid) volumetric water content, defined as: $w_u = W_u \cdot w$, with $W_u =$ unfrozen (liquid) water content (ϵ [0,1]) and $w =$ volumetric water content of soil.

TEMP/W can be coupled with *SEEP/W* to model the effects of moving water on heat transfer process and *viceversa*. This coupling requires: *(i)* volumetric water content expressed in function of temperature; *(ii)* thermal conductivity dependent on the volumetric water content (according to [8]). Thus, the partial differential equation for heat flow (Eq. (2)) is modified as follows:

$$\frac{\partial}{\partial x}\left(k_{T,x}\frac{\partial T}{\partial x}\right) + \frac{\partial}{\partial y}\left(k_{T,y}\frac{\partial T}{\partial y}\right) + \theta_w \rho_w c_{pw}\left(\frac{\partial(q_w T)}{\partial x} + \frac{\partial(q_w T)}{\partial y}\right) + Q_T$$
$$= \left(\rho_s c_{ps} + L \cdot \theta_w \frac{\partial w_u}{\partial T}\right)\frac{\partial T}{\partial t} \tag{4}$$

with $\rho_s \cdot c_{ps} =$ volumetric heat capacity of soil; $\rho_w \cdot c_{pw} =$ volumetric heat capacity of water; $q_w =$ the specific discharge of water (Darcy velocity).

The effects of the coupling between seepage and heat flows can be immediately highlighted (Fig. 1) through the analysis of a simplified scheme (1D case), by comparing the temperature distributions obtained through *(a)* uncoupled and *(b)* coupled simulations, under the following hypotheses by way of example: one-dimensional and steady state conditions, upstream temperature (T_{up}) = 90 °C, downstream temperature (T_{down}) = 30 °C, upstream p.h. (H_{up}) = 60 m; downstream p.h. (H_{down}) = 40 m [7].

Fig. 1. Case 1D: temperature distributions obtained through (a) uncoupled and (b) coupled analyses ($T_{up} = 90°$; $T_{down} = 30°$; $H_{up} = 60$ m; $H_{down} = 40$ m)

Referring to a seepage flow through an earthen structure (2D case), the effects of this coupling (between seepage and heat transport) play a key role on the temperature distribution. The numerical (*SEEP/W & TEMP/W*) results obtained by assuming the following boundary conditions by way of example:

Hydraulic conditions: Constant applied hydraulic load ($H = 45\ m$) at the upstream shell; $Q = 0$ and interstitial pressure $p_w = 0$ at the downstream shell; interstitial pressure $p_w = 0$ on the countryside; $Q = 0$ at the bottom of the foundation soil,

Thermal conditions: Constant temperature, $T = 15\ °C$, at the upstream shell; constant temperature, $T = 20\ °C$, at the downstream shell; at the upstream shell, $T = 15.6\ °C$, at the bottom of the foundation soil, are shown in Fig. 2.

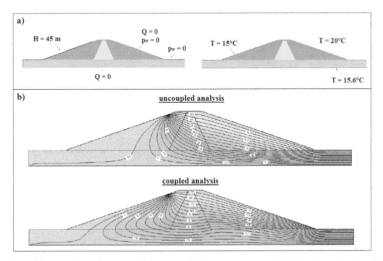

Fig. 2. Case 2D: (a) boundary conditions and (b) temperature distributions obtained through uncoupled and coupled analyses

Secondly, *SEEP/W* and *TEMP/W* codes have been couply applied to the analysis of virtual examples of embankment dams affected by permeability defects aiming to identify the relationship between their characteristics (position and dimensions) and the variation of (*internal*) temperature distribution [6, 7].

Particularly, the effects of permeability defects (p.d.) induced by backward (piping) erosion in foundation soils of homogeneous embankment dams on the temperature are herein investigated and described.

The following (by way of example) earthen structure is considered in the numerical simulations: height $H = 10\ m$; the cross section is trapezoidal; width at the top (L_{crest}) $= 4\ m$; thickness T_f of the foundation soil $= 6\ m$ (Fig. 3a). The corresponding hydraulic (permeability and volume water content) and thermal (conductivity and heat capacity) properties have been simply taken by [8], according to typical construction materials. The assigned (by way of example) hydraulic ($h_w =$ initial upstream hydraulic head, p_w $=$ pore pressure; $Q =$ total water flux) and thermal ($T_1 =$ upstream temperature; $T_2 =$ downstream temperature) boundary conditions are shown in Fig. 3b. Specifically, it is assumed *(i)* steady state seepage flow with constant upstream hydraulic head ($h_w = 8\ m$); *(ii)* unsteady state heat transport process (elapsed time $\cong 12\,h = t_{er}$); *(iii)* tubular p.d. (due to *piping* erosion, 1 m diameter, simply schematized in the proposed 2D hydrothermal model by a horizontal layer of variable length representing the erosion pipe) in pervious foundation soils of a homogeneous levee (permeability of the defect, $k_d = 10^{-2}$ m/s). A linear (uniform) increase of the length (L_d) of the p.d. or pipe during the simulation (up to the maximum value $L_{d,max}$, Fig. 3a) is modelled; particularly, for the examined case, the piping progression lasts 12 h, approximately. Each numerical simulation has been thus divided in 15 steps of 3000 s, for a total duration of 45000 s. The progression of piping has been simply simulated by increasing the length of pipe of 3 m, at each step.

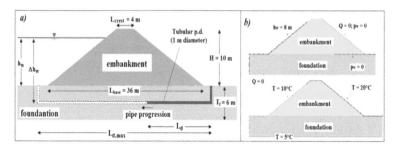

Fig. 3. (a) Considered homogeneous earthen structure; (b) assigned boundary conditions

The results of numerical simulations are shown in the following figure.

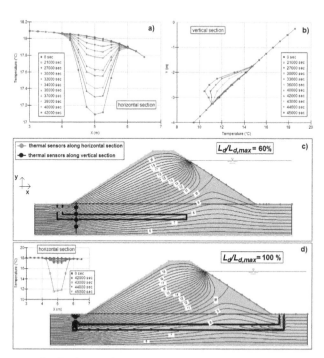

Fig. 4. Temperature distributions vs time (a) along a horizontal section; (b) along a vertical section; (c), (d) within the earthen structure for $L_d/L_{d,max} = 0.60$, 1.0, respectively

Measurable variations ($\cong 0.1$ °C) in temperature (T) distribution along a horizontal section, at the downstream toe, are observed during the progression of piping from $L_d/L_{d,max} \cong 0.60$ (Fig. 4 a, c). By increasing $L_d/L_{d,max}$, the changes in T become more appreciable, up to $\cong 1$ °C for $L_d/L_{d,max} = 0.90$ (Fig. 4a). As soon as the pipe reaches the upstream side ($L_d/L_{d,max} = 1$), great variations in temperature distribution ($\cong 6$ °C) can be observed/measured (Fig. 4d).

On the contrary, variations in temperature distribution along a vertical section become sensible (up to 1 °C) only for $L_d/L_{d,max} = 1$ (Fig. 4b). Thus, fiber optics or thermal sensors along a horizontal section, spaced every 0.50 m for 3 m width, at the downstream toe of earthen structures, appears more reliable to monitor piping erosion progression, allowing to develop early emergency action plans before breaching. Similar positions of fiber optics thermal sensors have been already installed along the Adige River to monitor its levees [9].

3 Back Analysis and Interpretation of a Large Scale Experiment

Very few experimental tests on the development and evolution of permeability defects in earthen structures are reported and available in technical literature. Among these, the IJkdijk piping test by University of Delft (Netherlands) is very significant because the seepage flow, the piping erosion phenomenon progression (inducing p.d. in foundation soils) and the instability of the downstream shell have been carefully recorded and documented [10, 11].

Numerical simulations (through SEEP/W and TEMP/W codes) of the measured temperature and piezometric heads distributions, reported by Bersan et al. (2015, 2017) [10, 11] have been carried out in the paper. The experimental dyke was 3.5 m high, 19 m long and 15 m wide at its base (Fig. 5). The lower part of the dike was constituted of a 0.7 m clay layer, which separated hydraulically the foundation from the dike body.

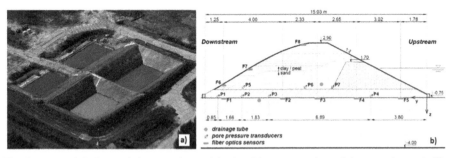

Fig. 5. (a) Aerial photo of the experimental basin; (b) cross section of the experimental dike (adapted from [10, 11])

A small clay dike (1.7 m high) was firstly built at the upstream side; a sand core (behind the small dike) and a cover of organic clay were successively put in place. The test dikes and the clay embankments delimiting the basins enclosed two reservoirs, each one with a volume of approximately 2000 m³, to be filled during the test to apply the desired hydraulic load. Conventional monitoring instruments (Fig. 5), including two liquid level sensors, 4 lines of 17 pore pressure sensors in foundation, 4 lines of 3 pore pressure sensors in the sand core (above the clay layer) were installed to monitor the course of the test.

In addition, distributed fiber optic sensors encased in a geotextile strip were placed within the dyke; particularly, 5 lines of fiber optic sensors in foundation soils, at the base of the dyke, and 3 lines of sensors in the downstream shell.

3.1 Test Procedure

The upstream reservoir was filled according to water level law reported in Fig. 6a. In the downstream basin, the water level was taken constant, at 10 cm above the sand layer, to ensure complete saturation of the foundation soil. The permeability coefficient of the sand layer was estimated to $k = 1.5 \bullet 10^{-4}$ m/s, according to [10, 11].

The first evidence of backward erosion (piping) was detected after two days of testing (elapsed time $\cong 50$ h) at the cross sections corresponding to the (longitudinal) abscissa $x = 4.8$ m, 8.2 m, 11.2 m; as a consequence, the transducers P1 (at 0.85 m from the downstream toe, see Fig. 10a) recorded a slight reduction in interstitial pressures. By increasing the hydraulic load, after 55–60 h of testing, reductions in interstitial pressures were also recorded in transducers P2, indicating the evolution/progression of the piping phenomenon.

Erosion stopped after opening the controllable drainage tube installed in the foundation soil, between the transducers P2 and P3 (see Fig. 5).

Since no appreciable/significant reductions in pore pressure in P3 were recorded, the pipe should have reached a length ranged between 2.5 m–4.3 m, i.e. the distances of the transducers P2 and P3 from the downstream toe, respectively.

However, the failure of the experimental dyke was caused by the instability of the downstream slope, after 5 days of testing [10, 11].

3.2 Temperature Measurements

Figure 6b shows the evolution of the measured temperature at the lines F1–F5, corresponding to the central section of the dam ($x = 10$ m), near the section at the (longitudinal) abscissa $x = 11.2$ m, in which piping phenomena was observed.

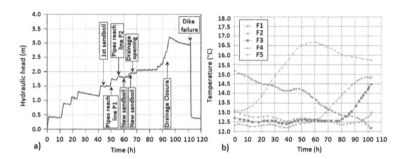

Fig. 6. (a) Applied hydraulic load; (b) Temperature measurements (adapted from [10, 11])

A few hours after the beginning of the test, the temperature recorded in F5 started to increase, suggesting that the water in the upstream reservoir was warmer than the sand layer in the foundation.

Unfortunately, the measurements of reservoir water temperature are not available due to technical problems. At the lines F4 and F3, a temperature increase was recorded after 2 and 3 days, respectively.

In F2, on the contrary, the temperature started to decrease slowly immediately after the beginning of the test and continued to decrease up to 2 days before the end of the test. At the F1 line, the temperature remarkably decreased during the whole duration of the test (Fig. 6b).

3.3 Numerical Simulation

Several *SEEP/W* and *TEMP/W* simulations reproducing the temperature and piezometric head measurements within the experimental dyke have been developed. The schematization of the cross section of the dam is shown in Fig. 7a.

Simulation of initial thermal and hydraulic conditions (uncoupled steady state seepage and heat transport processes) was firstly carried out. To reproduce the initial conditions of temperature (before the start of the experiment), specific *thermal* boundary conditions have been assigned (Fig. 7b) according to [10, 11].

The initial *hydraulic* boundary conditions, H_{up} (upstream load) $= 0.10$ m and H_{dw} (downstream load) $= 0$ m have been imposed. The considered hydraulic and thermal properties of the dyke materials are shown in Table 1 [10, 11].

After the beginning of the test, different boundary conditions must be applied; in particular, by referring to thermal variables defined in Fig. 7b: T_m linearly decreases between 24 °C and 11 °C, $T_{v,BC} = 20$ °C, $T_{v,AB} = 10$ °C, $T_f = 9$ °C, $T_{d1} = 14$ °C and $T_{d2} = 18$ °C (at the opening of drainage); in addition, H_{up} variable with time (t), according to applied hydraulic load (Fig. 6a), and $H_{dw} = 0.1$ m (constant).

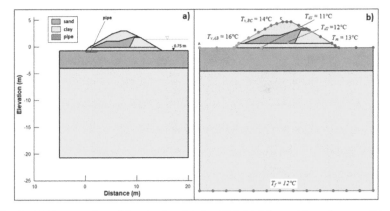

Fig. 7. (a) Cross section of the simulated dyke; (b) Initial thermal boundary conditions

Table 1. Hydraulic and thermal properties of materials (adapted from [10, 11])

Parameter	Sand	Clay	Unit
Volumetric water content	0.4	0.3	–
Permeability	$1.5 \bullet 10^{-4}$	$7.5 \bullet 10^{-6}$	m/s
Thermal conductivity	2.77	1.39	W/(m$\bullet$K)
Volumetric heat capacity	$2.8 \bullet 10^{-6}$	$2.5 \bullet 10^{-6}$	J/(m$^3 \bullet$K)

Therefore, coupled unsteady state seepage and heat transport processes have been simulated following five phases:

$t \in$ [0; 50] hours (pipe reaches P1 sensor);

$t \in$ [50; 65] hours (pipe reaches P2 sensor);

$t \in$ [65; 85] hours (uncompleted opening of the lower drainage tube- possible occlusion of pores is assumed - and total opening of the upper drain);

$t \in$ [85; 94] hours (total opening also for the lower drainage tube);

$t \in$ [94; 110] hours (closure of the both drainage tubes and evolution of hydraulic and thermal processes up to the dyke failure).

The pipe has been simply schematized by a thin layer whose length varies along time (according to experimental observations and measurements), in the foundation sandy soil, just below the basal clay layer. The pipe is affected by the same properties of the material "*sand*" except for the permeability coefficient (k_{pipe}); in particular, k_{pipe} = 0.005 m/s. This value is derived from calibration process of the model aimed to the interpretation of temperature and piezometric head measures.

By referring to the cross section at x = 11.2 m (near the central section, x = 10 m), the obtained results in terms of temperature at the lines F1–F5 and of piezometric head (p.h.) at the sensor P2 are closed to the measured values (Fig. 8).

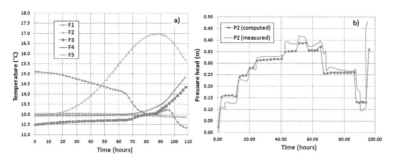

Fig. 8. Numerical results: (a) Temperature at lines F1–F5; (b) Piezometric head at P2

To better evaluate the reliability of the installed monitoring system, and in particular of the (fiber optic) thermal sensors, the developed hydro-mechanical model, almost calibrated on the basis on the fitted piezometric head and temperature measurements (see Figs. 6 and 8), has been re-run by imposing the absence of a growing erosion channel/pipe in foundation soils. For this condition, the comparisons between the p.h. (at transducer P2) and temperature (at line F1),measured and computed, distributions in presence and absence of backward (piping) erosion, are shown in Figs. 9 and 10.

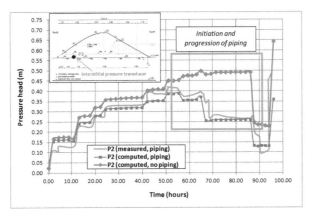

Fig. 9. Comparison between measured data and numerical results, in terms of piezometric head (p.h.), in presence and absence of backward (piping) erosion.

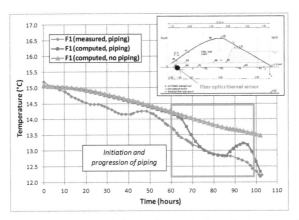

Fig. 10. Comparison between measured data and numerical results, in terms of temperature, in presence and absence of backward (piping) erosion.

Sensible differences between the piezometric head (+0.25 m) and temperature (+1 °C, according to the previous analyses described in the previous chapter) distributions, in presence and absence of piping, can be observed (Figs. 9 and 10).

Furthermore, it is worth observing the appreciable correspondence between the two types of instruments (i.e. transducers and fiber optics thermal sensors) in the detection of the "*initiation phase*"(i.e. 60 h after the beginning of the test, approximately) of the backward (piping) erosion phenomenon within the experimental dike.

4   Concluding Remarks

Coupled implementation of two Finite Element Method (FEM) codes allowed to analyze the seepage flow process (code SEEP/W) and to determine the temperature distributions (code TEMP/W), at specific cross sections, within embankment dams affected by a permeability defect, as well as to evaluate its effects on the possible variations of the main hydrodynamic variables and temperature.

Specifically, results of numerical simulations of virtual examples of embankment dams affected by permeability defects indicate that for a clearer detection of piping phenomena, the temperature sensors should be installed at the downstream toe of earthen structures, preferably along a short horizontal section, spaced every 0.50 m, as also shown by a recently installed, documented, thermal monitoring system.

The interpretation of coupled hydraulic and thermal measures as well as of piping mechanisms, which affected the large scale IJkdijk piping test, highlighted *(i)* the applicability of *SEEP/W* and *TEMP/W* codes to a reliable analysis of the examined coupled hydrodynamic and thermal problem; *(ii)* that the temperature sensors inside the embankment dams (or levees) can contribute, coupled to conventional instruments (i.e. piezometric head transducer), to a more reliable detection of seepage anomalies (e.g. permeability defects). Thus, their installation in existing or new earthen structures, expecially during the construction phase, should be recommended.

References

1. Foster, M., Spannagle, M., Fell, R.: Report on the analysis of embankment dam incidents. UNICIV Report No. R374, University of New South Wales (1998)
2. ICOLD: Internal erosion of existing dams, levees and dikes, and their foundations. Jean-Jacques Fry & Rodney Bridle–Bulletin, no. 164, vol. 1: Internal erosion processes and engineering assessment (2013)
3. Talbot, J.R., Ralston, D.C.: Earth Dam Seepage Control. Symp. On "Seepage and Leakage from Dams and Impoundments", A.S.C.E., Geot. Engineer. Div., vol. 44, issue 6 (1985)
4. Federico, F., Montanaro, A.: Permeability defects in zoned earth structures. Forecasting, F.E. modeling and analyses of effects. In: Pina, C., Portela, E., Gomes, J. (ed.) 6th International Conference on Dam Engineering, Lisbon, Portugal, 15–17 February 2011 (2011)
5. Sjödahl, P., Johansson, S.: Experiences from internal erosion detection and seepage monitoring based on temperature measurements on Swedish embankment dams. ICSE6 Paris, 27–31 August 2012 (2012)
6. Radzicki, K., Bonelli, S.: A possibility to identify piping erosion in earth hydraulic works using thermal monitoring. In: 8th ICOLD European Club Symposium, 22–25 September, Austria, 2010, pp. 618–623 (2010)
7. Cesali, C., Federico, V.: Coupled thermal and piezometric heads monitoring to detect permeability defects within embankment dams and levees. In: Third International DAM WORLD Conference 2018, Foz do Iguassu, Brazil, 17–21 September 2018 (2018)
8. Geostudio Package user's manual by GEO-SLOPE International Ltd. (2012)

9. Bersan, S., Simonini, P., Bossi, G.: Multidisciplinary analysis and modelling of a river embankment affected by piping. 26th Annual Meeting of EWG-IE, Milan, Italy (2018)
10. Bersan, S, Koelewijn, A.R.: Temperature monitoring in piping-prone hydraulic structures. In: Lollino, G., et al. (eds.) Engin. Geology for Society and Territory, vol. 2 (2015)
11. Bersan, S., Koelewijn, A.R., Simonini, P.: Effectiveness of distributed temperature measurements for early detection of piping in river embankments. Hydrol. Earth Syst. Sci. **22**, 1491–1508 (2017)

A Wavelet-Based Fiber Optic Sensors Data Processing Method and Its Application on Embankment Sliding Surface Detection

Guan Chen⬤, Ya-nan Ding, and Yong Liu$^{(\boxtimes)}$ ⬤

State Key Laboratory of Water Resources and Hydropower Engineering Science,
Institute of Engineering Risk and Disaster Prevention, Wuhan University, Wuhan 430072,
People's Republic of China
liuy203@whu.edu.cn

Abstract. As a monitoring sensor, distributed fiber optic sensors (DFOSs) are widely applied in geotechnical engineering. However, existing researches mainly focused on the practice of DFOSs in different engineering fields, few about DFOSs data processing. Under this circumstance, this study proposed a dyadic wavelet-based DFOSs data processing method. To verify the feasibility of the proposed method, a real monitored raw DFOSs signal is tested. In addition, an embankment sliding surface detection scheme by DOFSs using the wavelet-based signal processing method is proposed. An embankment sliding surface detection test based on the finite element method is carried out. The uncertainty of soil properties is taken into account by 3D random fields together with a finite element model. The results show that the dyadic wavelet-based method can effectively avoid the interferences and accurately detect the characteristics of DFOSs data; the proposed scheme can detect the potential embankment sliding surface effectively.

Keywords: Distributed Fiber Optic Sensor (DFOS) · Wavelet transform · Sliding surface

1 Introduction

The Distributed Fiber Optic Sensors (DFOSs) are widely applied in geotechnical engineering for its higher spatial resolutions compared with classical strain gages. Currently, the DFOSs have various categories: mainly including Fiber Bragg Grating (FBG), Optical Time Domain Reflectormeter (OTDR), Optical Frequency Domain Reflectormeter (OFDR), Brillouin Optical Time Domain Reflectometry (BOTDR) and Brillouin Optical Frequency Domain Reflectometry (BOFDR) etc. A brief summary about different DFOSs applied in geotechnical engineering is conducted, shown in Table 1.

Compared with its applications in various engineering fields, researches about DFOSs signals processing are fewer. The wavelet transform is widely applied in data processing as it can well analyze data information from both time and frequency domains [1]. Some researches about DFOS signal processing with wavelet transforms were already carried out. Feng et al. compared the discrete wavelet transform (DWT) and short time

© Springer Nature Switzerland AG 2020
J.-M. Zhang et al. (Eds.): ICED 2020, SSGG, pp. 333–339, 2020.
https://doi.org/10.1007/978-3-030-46351-9_34

Table 1. A brief summary of DFOSs application in geotechnical engineering

Ref.	DFOSs	Objects	Remarks
[4]	FBG	Centrifuges slope model	Proposing a monitoring system combining FBG with Inclinometer pipe and earth pressure gauges that could monitor stress, displacement and the strain
[5]	FBG	Laboratory slope model	Combining FBG inclinometers and particle image velocimetry technique in displacement measurement
[6]	BOTDR	Majiagou landslide	Proposing a BOTDR data based kinematic method to calculate the shear displacement
[7]	BOFDR	Segment joints in shield tunneling	Proposing a segment joints deformation monitoring scheme with BOFDR
[8]	BOTDR	Prestressed concrete bridge	Explaining a practical case for using BOTDR data detecting the constructions damage
[9]	BOTDR	Fiber Reinforced Polymer anchor	Using BOTDR monitoring the tensile strain distribution of the fiber reinforced polymer anchor in the excavation zone
[10]	FBG	Laboratory slope model	Combining FBG and laser displacement transducers monitoring the progressive evolutionary process of the reinforced slope

Fourier Transform in OFDR data processing, and illustrates that the DWT could improve the sensing performance [2]. Feng et al. used stationary wavelet transform to reduce the effects of system noise and Brillouin frequency peak shift distortions and improve the sensing accuracy in BOTDR data processing [3]. Owing to the translation invariance property, the dyadic wavelet transform is adopted to process DFOSs data in this study, named dyadic wavelet-based DFOSs signal processing method. To validate the proposed method, a real monitored DFOSs signal is tested. In addition, an embankment sliding surface detection scheme by DFOSs using the dyadic wavelet-based signal processing method is proposed. To verify the scheme in detecting the sliding surface, a random finite element embankment model with considering the spatial variation of soil property is built.

2   Dyadic Wavelet Transform in DFOSs Data Processing

2.1   Dyadic Wavelet Transform

For signal $f(t)$, its dyadic wavelet transform could be expressed as Eq. (1).

$$Wf(u, 2^j) = \int_0^t f(t)\frac{1}{\sqrt{2^j}}\psi^*(\frac{t-u}{2^j})\mathrm{d}t \tag{1}$$

where u and 2^j is the translation and scale parameter, respectively; ψ^* is the conjugate of the wavelet basis ψ.

Compared with continuous wavelet transform, the dyadic wavelet transform could divide the signal into detail signals and approximate signals effectively as its corresponding à *trous* algorithm [11]. In addition, the dyadic wavelet transform is translation invariance compared with discrete wavelet transform, which means it would keep all time domain information in each decomposition level. Hence, the dyadic wavelet transform is applied in this study to reduce the noises and enhance the DFOSs signal authentic characteristics.

2.2   DFOSs Data Processing

The raw DFOSs data is applied to validate the dyadic wavelet transform corresponding à *trous* algorithm. The raw data are from the open source supplied by Webb et al. [8]. The first column raw data of the file 'beam-1-data' are selected. The raw DFOSs signals are shown in Fig. 1. Due the essence of dyadic wavelet transform, the wavelet coefficients value would increase with the increase of decomposition level. Hence, the data after processing by the à *trous* decomposition algorithm should be scaled down. The first point is applied to calibrate the wavelet coefficients in this study. The processed DFOSs data using the à *trous* decomposition algorithm are shown in Fig. 1.

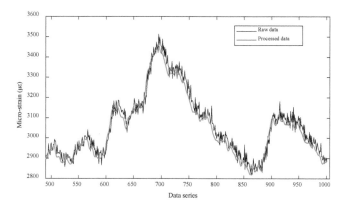

Fig. 1. Raw and processed DFOSs strain signals

From Fig. 1, the raw strain data is contaminated by noise as descripted by Webb et al. [8]. However, after processing with the dyadic wavelet transform corresponding *à trous* decomposition algorithm, the strain signal is smoother and keeps the basic feature of the raw data.

3   Embankment Sliding Surface Detection

3.1   Geometric Model with Random Field

An embankment sliding surface detection scheme by DOFS using wavelet-based processing method is proposed, shown in Fig. 2. To verify the feasibility of the scheme, a finite element embankment model is built. Owing to the symmetry, only the half of embankment is built for convenience. The trapezoidal body parameters and the boundary conditions are shown in Fig. 3. Three paralleled segments of DFOSs are arranged, shown in Fig. 3 as well. In X axis direction, the DFOSs is through the whole model; In Z axis direction, the DFOSs is arranged with interval 2 m. The coordinate system is set in Fig. 3.

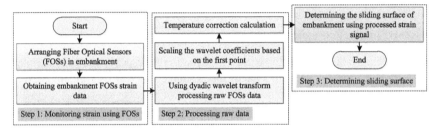

Fig. 2. Embankment sliding surface detection scheme

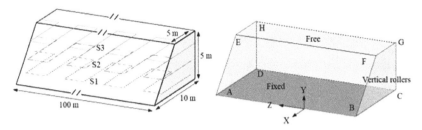

Fig. 3. Sketch of geometric model, boundary conditions and fiber Optic Sensors (S1, S2 and S3) arrangement. Surface ABCD is fixed; Surface BCGF ADHE and CDHG is vertical rollers; Surface ABFE and EFGH is free.

In addition, the uncertainty of soil properties is taken into account in this study. The random filed of undrained shear strength is generated using Liu et al. proposed method [12]. The marginal probability distribution is adopted as the lognormal distribution and the autocorrelation lengths in X, Y and Z axis directions are 2 m, 2 m and 10 m, respectively.

3.2 Strain Signal

Due to the DFOSs detecting the axial direction strain, the numerical model strain data in X axis direction are adopted as DFOSs strain data. The length in Z axis is 100 m and the DFOSs are arranged with interval 2 m, which mean 50 sets of data would be obtained for each segment. The length in X axis direction would decrease with the DFOSs moving up to the top surface. Three segments of DFOSs strain data from the numerical model are show in Fig. 4.

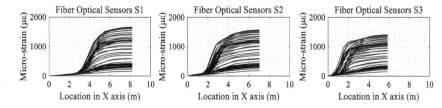

Fig. 4. DFOSs strain signals of S1, S2 and S3 segments

3.3 Embankment Sliding Surface Detection

In this study, the numerical analysis is performed instead of the field test. In addition, since the numerical analysis data are not contaminated, the monitoring step and raw data processing step described in Fig. 2 are not carried out in this part, i.e. the numerical analysis smoothing data are deemed as the processed data to detect the sliding surface of embankment. Based on DFOSs data processing test in Sect. 2, the processed DFOSs data could effectively reduce noises and keep the authentic data characteristics using the dyadic wavelet-based method. Therefore, using the numerical model strains as the processed DFOSs data are reasonable.

The pseudo-color figures of processed DFOSs strain data are plotted in Fig. 5. The location ranges from 40 m to 90 m in Z axis direction and from 2 m to 8 m in X axis direction are the danger zone, especially, locating at around 65 m in Z axis direction. To validate the accuracy of the DFOSs data in detecting the sliding surface, the random finite element results are shown in Fig. 6. In practice, Fig. 6 represents the real sliding; Fig. 5 represents the monitoring data. As the DFOSs detected danger zone are consistent with the finite element results (i.e. the results in Fig. 5 and Fig. 6 are consistent), the feasibility of using DFOS in embankment sliding surface detection is validated. Therefore, the proposed DFOSs-based potential sliding surface detection scheme is effective. In addition, based on the strain data in Fig. 4, the used numerical model strains are lower than 2000 $\mu\varepsilon$ (around 2 cm displacement in this case), which means the proposed scheme has promising potential in forecast the danger zone before destruction.

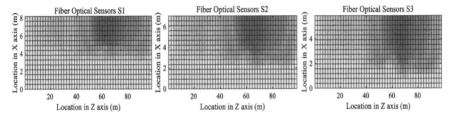

Fig. 5. The potential sliding surface detected by processed DFOSs strain data

Fig. 6. Displacement contour of finite element results

4 Concluding Remarks

This study proposed a dyadic wavelet-based DFOSs data processing method. A real monitored DFOSs raw data are tested to validate the proposed method. The results shown that the dyadic wavelet-based method could effectively reduce the noise and keep the authentic strain data characteristics.

In addition, an embankment sliding surface detection scheme by DFOS using the dyadic wavelet-based signal processing is proposed. A random finite element model is built to verify the scheme. The finite element model strains in X axis direction are deemed as processed DFOSs data due to the essence of DFOSs. The results show that the proposed scheme could detect the sliding surface and has promising potentials in forecasting the danger zone. In further study, an experimental model test using DFOSs in sliding surface detection would be carried out.

Acknowledgements. This research is supported by the National Natural Science foundation of China (Grant No. 51879203).

References

1. Chen, G., Li, Q.Y., Li, D.Q., Wu, Z.Y., Liu, Y.: Main frequency band of blast vibration signal based on wavelet packet transform. Appl. Math. Model. **74**, 569–585 (2019)
2. Feng, K., Cui, J., Dang, H., Jiang, D., Jin, Y., Sun, X., et al.: A OFDR signal processing method based on wavelet transform for improving its sensing performance. IEEE Photonics Technol. Lett. **31**(13), 1108–1111 (2019)
3. Feng, X., Zhang, X., Sun, C., Motamedi, M., Ansari, F.: Stationary Wavelet Transform Method for Distributed Detection of Damage by Fiber-Optic Sensors. J. Eng. Mech. **140**(4) (2014)
4. Zhang, L., Shi, B., Zeni, L., Minardo, A., Zhu, H., Jia, L.: An fiber Bragg grating-based monitoring system for slope deformation studies in geotechnical centrifuges. Sensors (Basel) **19**(7), 1591 (2019)

5. Pei, H., Zhang, S., Borana, L., Zhao, Y., Yin, J.: Slope stability analysis based on real-time displacement measurements. Measurement **131**, 686–693 (2019)
6. Zhang, C.C., Zhu, H.H., Liu, S.P., Shi, B., Zhang, D.: A kinematic method for calculating shear displacements of landslides using distributed fiber optic strain measurements. Eng. Geol. **234**, 83–96 (2018)
7. Wang, X., Shi, B., Wei, G., Chen, S., Zhu, H., Wang, T.: Monitoring the behavior of segment joints in a shield tunnel using distributed fiber optic sensors. Struct. Control Health Monit. **25**(1), e2056 (2018)
8. Webb, G.T., Vardanega, P.J., Hoult, N.A., Fidler, P.R.A., Bennett, P.J., Middleton, C.R.: Analysis of fiber-optic strain-monitoring data from a prestressed concrete bridge. J. Bridge Eng. **22**(5) (2017)
9. Xu, D.S., Yin, J.H.: Analysis of excavation induced stress distributions of GFRP anchors in a soil slope using distributed fiber optic sensors. Eng. Geol. **213**, 55–63 (2016)
10. Zhu, H.H., Shi, B., Yan, J.F., Zhang, J., Wang, J.: Investigation of the evolutionary process of a reinforced model slope using a fiber-optic monitoring network. Eng. Geol. **186**, 34–43 (2015)
11. Shensa, M.J.: The discrete wavelet transform: wedding the a trous and Mallat algorithms. IEEE Trans. Signal Process. **40**(10), 2464–2482 (1992)
12. Liu, Y., Lee, F.H., Quek, S.T., Beer, M.: Modified linear estimation method for generating multi-dimensional multi-variate Gaussian field in modelling material properties. Probab. Eng. Mech. **38**, 42–53 (2014)

Simplified Models for the Interpretation of Total Stress Measurement of Embankment Dams

E. Fontanella[1](✉), L. Pagano[2], and A. Desideri[1]

[1] Department of Structural and Geotechnical Engineering,
Sapienza University of Rome, Rome, Italy
`enzo.fontanella@uniroma1.it`
[2] Department of Civil, Architectural and Environmental Engineering,
University of Naples Federico II, Naples, Italy

Abstract. The paper describes and interprets arching action problems in three zoned earth dams. The principal simplified models available for the interpretation of data records of embankment dams are used. The representativeness of vertical total stress normally monitored in earth dams is discussed. The analysis of measurements permits to investigate how a given measured quantity can be used to interpret the dam's mechanical behavior.

Keywords: Embankment dam · Monitoring data · Representativeness of measurements

1 Introduction

The mechanical behavior and the timely identification of the phenomena that could lead dam to collapse are the main purpose of monitoring and interpretation of measurements.

Slope instability and watertightness problems are the two categories of most feared phenomena. Their triggering can be identified by interpreting the measurements of total stress, pore pressure, seepage flows, and settlements [1–5].

Interpreting a measurement means defining the phenomena that conditioned its evolution and/or distribution and characterize any possible correlations with other known physical quantities.

The reference models can be derived from the knowledge of the solutions of simpler problems known in the literature with strong exemplified hypotheses, such as to ensure solutions in closed form [6, 7].

Simplified models play a key role in any interpretative process. Indicate which are the most effective forms in which to organize and represent the measurements; qualitatively identify many of the factors that have affected the mechanical behavior of dam during construction and working condition; they stimulate doubts and questions about the measurements that present anomalies, suggesting when it is necessary to resort to advanced approaches; they finally guide in the selection of physical quantities that the advanced type model must contain in order to be able to reproduce the observed mechanical behavior.

© Springer Nature Switzerland AG 2020
J.-M. Zhang et al. (Eds.): ICED 2020, SSGG, pp. 340–344, 2020.
https://doi.org/10.1007/978-3-030-46351-9_35

The detailed examination of the measurements of some earth dams has allowed to check the effectiveness of the existing interpretative tools and to develop new simplified models.

The interpretative schemes that seek to establish qualitative and quantitative correlations between "cause" and "effect" magnitudes are therefore associated with representations in which both appear. For these reasons the measurements will be represented in graphs that link them to the following factors: (a) load from embankment dam; (b) dam geometry; (c) material properties; (d) time.

Hydraulic fracturing is a central issue in the evaluation of the safety conditions in earth dams. In general, safety against hydraulic fracturing is checked for by measuring vertical total stress, pore water pressures in the earth dam, rate of flow through the dam; water turbidity is also inspected as a warning sign for erosion phenomena [8].

The analysis of some case histories available in the literature made it possible to identify a crucial mechanism for hydraulic fracturing, namely load transfer phenomena (arching action) due to differences in stiffness between the core and the abutments.

2   Vertical Total Stress and Arching

In a zoned earth dam the different stiffness of materials constituting the core and abutments produces non-homogeneous distributions of vertical total stress. The load transfer leads to a reduction of stress within the dam core and an increase in the abutments.

As an example Fig. 1 shows three distributions of the vertical total stress theoretically obtained along a horizontal section. The distribution obtained for homogeneous dam ($E_{abutment}/E_{core} = 1$) is very similar to the distribution of the overburden pressure. With reference to Beliche Dam geometry, even a modest increase in the stiffness of the abutment ($E_{abutment}/E_{core} = 3$) is able to produce a markedly discontinuous vertical total stress distribution due to load transfer phenomena, and thus increasing arching action.

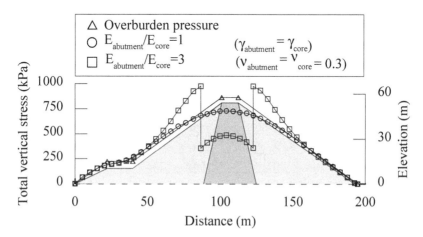

Fig. 1. Beliche Dam: vertical total stress computed along a horizontal plane (dashed line)

A first interpretation of the vertical total stress measurements can be obtained from comparing the measurements along horizontal lines perpendicular to the longitudinal axis of the dam with the theoretical distributions of homogeneous dam or overburden pressure.

Figures 2 and 3 show two examples of interpretative approach above illustrated. They relate to Beliche Dam and Camastra Dam, respectively. In both cases the load transfer phenomena is highlighted with a sudden discontinuity of the values in correspondence with the core-abutment contact. For Beliche dam the comparison between measurements and computed data (model) highlights (filled circles) significant arching action. The vertical total stress measured at Camastra dam, while showing a marked arching action, are characterized by values decidedly superior to the theoretical ones. Beyond the stress concentrations produced by three-dimensional effects (not invocable in this case), the condition of vertical equilibrium in the considered section should produce distributions of vertical total stress, measured and theoretical, which subtend the same area. The marked difference between the areas subtended raise some doubts about the reliability of the measurements.

During construction of earth dam, it is possible to estimate the evolution of the arching action plotting the vertical total stress measured versus overburden pressure. In Fig. 4 this plot is proposed for a point located within the core of the Kastraki Dam. The dotted line corresponds to the condition of equality between the two quantities; the

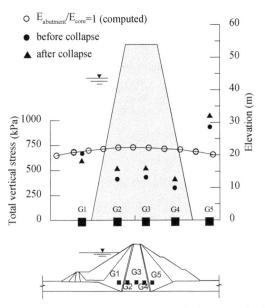

Fig. 2. Beliche Dam: vertical total stress measured at the end of construction before and after the temporary unexpected impounding, compared with a theoretical distribution for a homogeneous dam [Pagano 1998, modified]

greater the deviation from this condition, the greater the arching action, which leads to a load transfer towards the more rigid abutment. This type of representation is particularly effective especially in the presence of "scattered" measurement points.

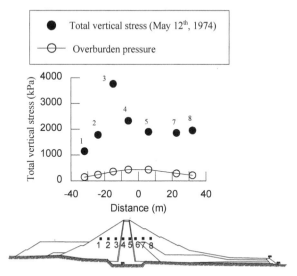

Fig. 3. Camastra Dam: vertical total stress measured at end of construction compared with the overburden pressure

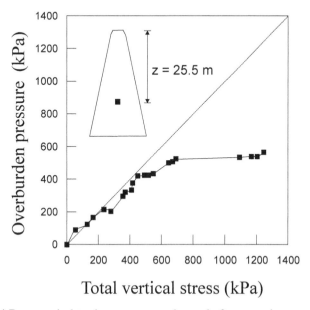

Fig. 4. Kastraki Dam: vertical total stress measured at end of construction compared with the overburden pressure [Pagano et al. 1998]

In the post-construction phases vertical total stress variations can be induced by phenomena that characterize the first impounding and working condition (e.g. saturation collapse, primary consolidation, secondary compression). These are generally phenomena that induce in the generic element within the embankment deformation processes which, if contrasted by the boundary conditions, determine variations in vertical total stress. Figure 2 shows, as an example, the saturation collapse for Beliche Dam during first impounding and the change in total stress produced by the phenomenon. In the upstream side, subject to collapse, there is a reduction of stress, balanced, in the remaining areas of the dam, by an increase in vertical total stress.

3   Conclusion

In this paper the main simplified models available to understand mechanical behaviour of earth dam, were examined. The paper highlighted the main problems associated with the interpretation of the total stress measurements.

A central element of the discussion is the interpretability of measured data, linked to the concepts of reliability of measurement and representativeness of measured physical quantity.

Distribution and evolution of monitored quantities, such as total stress, aid to characterize the overall dam behaviour and detect state condition often associated to crack formation or ti diagnose hydraulic problems. When much lower than overburden pressure, vertical total stress measured in the core of a zoned earth dam indicate core arching and state conditions supposed to enhance crack formation and propagation during impounding stages.

References

1. Pagano, L., Desideri, A., Vinale, F.: Interpreting settlement profiles of earth dams. Journal of Geotechnical And Geoenvironmental Engineering **124**(10), 923–932 (1998)
2. Pagano, L., Sica, S., Desideri, A.: Representativeness of measurements in the interpretation of dam behavior. Can. Geotech. J. **43**(1), 87–99 (2006)
3. Pagano, L., Fontanella, E., Sica, S., Desideri, A.: Pore water pressure measurements in the interpretation of the hydraulic behaviour of two earth dams. Soils and Foundation **50**(2), 295–307 (2010)
4. Desideri, A., Fontanella, E., Pagano, L.: Pore water pressure distribution for use in stability analyses of earth dams. In: Margottini, C., Canuti, P., Sassa, K. (eds.) Risk Assessment, Management and Mitigation 2nd World Landslide Forum, WLF 2011, vol. 6, pp. 149–153. Springer, Heidelberg (2013)
5. Fontanella, E., Pagano, L., Desideri, A.: Actual and nominal pore water pressure distribution in earth dams. Electron. J. Geotech. Eng. **17**(S), 2485–2494 (2012)
6. Marsal, R.J.: Analisis de asentarnientos en la presa Presidente Aleman. Oaxaca, No.5, Instituto de Ingenieria, UNAM, Mexico City, Mexico (1958). (in Spanish)
7. Poulos, H.G., Booker, J.R., Ring, G.J.: Simplified calculation of embankment deformations. Soils Found. **12**(4), 1–17 (1972)
8. Penman, A.D.M.: On the embankment dam. Geotechnique **36**(3), 303–348 (1986)

On the Role of Weak-Motion Earthquakes Recorded on Earth Dams

Stefania Sica[1]([⊠]) and Luca Pagano[2]([⊠])

[1] University of Sannio, Benevento, Italy
stefsica@unisannio.it
[2] University of Naples Federico II, Naples, Italy
lupagano@unina.it

Abstract. Seismic accelerations recorded on earth dams through permanent monitoring systems could be very useful to handle the seismic safety assessment of earth dams. First, they allow to characterize the seismic loads, which in conjunction with the static ones, might induce critical changes in those physical quantities controlling dam safety, i.e. stresses, strains, permanent displacements, pore water pressures, seepage flows. In this way, control and quantification of all loads affecting dam response in its whole lifetime are assured. Second, seismic recordings could be useful to characterize some important features of dam mechanical behavior at relatively none or low cost. Focusing on this latter objective, the paper examines the basic procedures that could be adopted to interpret the seismic signals acquired on a zoned earth dam equipped with a permanent seismic network consisting of five accelerometer stations placed both on rock outcrops at the specific site and on the main cross section of the embankment (top, bank, base). As the dam is placed in a quite seismically active zone of Southern Italy, five years of continuous monitoring have delivered dozens of records associated to weak earthquakes having different source-site distance, magnitude and frequency content. Interpretation of these signals has yielded important aspects on the dynamic behaviour of the sample dam at low strain levels, such as natural frequencies, signal amplification, spatial variability of the motion. The recordings of weak-motion earthquakes, which surely are more frequent than strong motion events during the dam lifetime, may be easily interpreted and could be extremely convenient in completing - at relatively low cost - the huge amount of information needed to carry out the crucial task of seismic safety assessment of dams, especially in case of existing earth dams that have been in operation for several years.

Keywords: Earth dam · Seismic monitoring · Weak-motion events

1 Introduction

Recent guidelines and regulations developed worldwide (AGI 2005; ICOLD 1988–1989; NTD 2014), recommend dams in seismic areas be instrumented to record seismic accelerations. For an earth dam, the earthquake represents a transient change of mechanical boundary conditions and the availability of acceleration recordings on rock outcrops and on the dam body allows both quantifying to what extent the earthquake excited the dam

© Springer Nature Switzerland AG 2020
J.-M. Zhang et al. (Eds.): ICED 2020, SSGG, pp. 345–356, 2020.
https://doi.org/10.1007/978-3-030-46351-9_36

and consequently interpreting the seismic-induced effects (Pagano et al. 2019). Similarly to ordinary loads acting on earth dams, i.e. dead load and impounding pressure, the seismic loading represents the "causative" factor for a number of "effects", which are observed soon after an earthquake: pore water pressure build up, permanent displacements, changes in total stress distribution, and increase in seepage flows. Quantifying the seismic loads through a permanent monitoring network makes, hence, easier the task of interpreting dam response during and after an earthquake and performing reliable assessment of dam safety in post-seismic conditions.

Other than representing the main cause for plenty of effects, the accelerations recorded on the dam body are themselves effects that elucidate how the dam body reacted to an earthquake. Modification of the seismic signal in its upward traveling in the dam embankment is known to primarily depend on (i) source mechanisms (fault geometry, source-site distance, etc.), which are peculiar of the earthquake shaking the dam site, (ii) mechanical properties of the construction soils of the embankment and (iii) the particular interaction between the incoming wave front and dam geometry. The seismic signals acquired at different locations of the dam (embankment, foundation, rock outcrops) could be interpreted to characterize important factors related to the overall mechanical response of the dam, such as its natural frequencies, ground motion asynchronism, stiffness and damping properties of the dam soils. The above quantifications always should be associated to the strain levels induced by the seismic shaking to the dam soils. Although a seismic monitoring system is primarily installed to record strong motion events able to mobilize medium-to-high strain levels in soils, it frequently acquires only weak motion events, which mobilize dam soils in the range of very small strains. However, the characterization of soil behavior at small strains is an inevitable part of the overall mechanical characterization of dam soils, and the availability of plenty of recordings associated to several weak motion events provides this information almost uncostly. It is only needed to create a protocol of the most suitable and effective procedures to interpret these huge amount of data (signals) provided by the accelerometric stations working continuously on the dam. By recalling a previous work of the authors (Sica et al. 2008b), in which a first attempt to establish a suitable protocol was outlined, in this paper emphasis is given to those characteristics that are specific of dam behaviour at the structure scale, which can effectively be identified by interpreting weak-motion earthquake recordings and help engineers in handling the crucial task of seismic safety assessment through advanced dynamic approaches (Sica et al. 2008a and 2019; Elia et al. 2011), as required by the so-called performance-based philosophy and by the more updated technical codes in Italy (NTC 2018 and NTD 2014) and worldwide.

The paper, referring to the case-history of Camastra Dam (Italy) equipped with five accelerometer stations since July 2002, illustrates the most suitable and effective procedures to interpret the recorded weak-motion earthquakes in order to obtain estimation of soil mechanical properties at small strains and other characteristics of dam response at macroscopic scale. Although the proposed procedures were used to interpret weak motion events, the only ones available to date, they are quite general and could also be adopted for accelerometric recordings caused by intense seismic events.

2   The Camastra Dam and the Installed Seismic Monitoring System

The Camastra dam (Fig. 1) is located South of Italy, in the Basilicata Region, 20 km South-East of Potenza city, on the Camastra stream. It is a zoned earth dam with an internal vertical core made of silty sand and shells made of sandy gravel. The embankment maximum high is of 57 m. The reservoir has an impounding capacity of 42 million of m^3. The embankment subsoil is made of two different rock formations. The one laying towards the left abutment is a clayey-calcareous rock (Corleto Perticara formation), while that close to the right abutment is an arenaceous-clayey rock (Serra Palazzo formation).

Available documents deliver the boundary condition during the construction stages (July 1963–November 1964), the first impounding and operational stages. The monitoring system provides records of internal settlements (cross-arms placed in the core and in the downstream shell) pore water pressures in the core (vibrating wire piezometric cells) and total stresses in the core and in the shells (vibrating wire total stress cells). The dam has been widely described in literature and for more details reference could be made to Sica and Pagano (2009).

The seismic monitoring system installed on the Camastra Dam consists of 5 accelerometer stations (Fig. 1). Two stations were installed on the two rock outcrop formations of Serra Palazzo and Corleto Perticara close to the dam embankment (Fig. 1a). Three were installed along the downstream boundary of the main embankment cross section in correspondence of the crest, intermediate bank and base (Fig. 1b).

Each monitoring station is made of an accelerometer fixed to a concrete base, which was embedded 1 m into the ground. A GPS for each accelerometer allows synchronizing all devices with an accuracy order of 10 μs. Power supply is provided by a photovoltaic panel linked to a battery. A box in plastic material protects each accelerometer from weathering (Fig. 1c).

Each seismic device is a triaxle mass-balance accelerometer sensor with a wide spectrum. The acquisition system is made of three channel data-loggers, which digitalize the analogical signal with a sampling frequency of 200 Hz, a resolution of 18 bit and a dynamic of 108 dB. Output data are stored by an extended-temperature solid-state memory. The device has been levelled to properly measure vertical acceleration. Horizontal sensors have been turned parallel respectively to the dam longitudinal axis (right-left abutment direction) and the transversal (upstream-downstream direction) axis to directly measure longitudinal ad transversal acceleration components.

The trigger threshold was set at 10^{-4} g. In each record, time resolution is set to 0.005 s, corresponding to a maximum sampling frequency of 200 Hz. Pre and post-event durations were set equal to 50 e 120 s, respectively. To eliminate drift phenomena from records and other possible inaccuracies, recorded signals have been preliminary corrected of baseline and other signal errors.

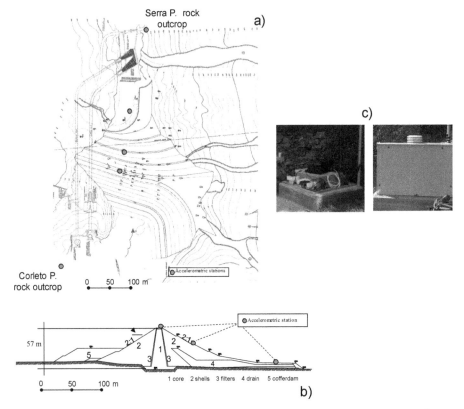

Fig. 1. Camastra Dam: (a) plan view of the dam; (b) main cross section of the embankment; (c) inside and outside view of an accelerometric station

3   Database of Acceleration Signals Recorded at the Dam Site

In five years of continuous monitoring (2002–2007), the accelerometers at the dam site recorded about 50 weak motion events. Among these, 24 events triggered at least 3 monitoring stations and allowed more detailed interpretation. Table 1 provides information on event date, magnitude, seismogenic zone, epicenter and peak acceleration of the recorded longitudinal (L), transversal (T) and vertical (V) components. The sources of the recorded events were variously located with source-site distance spanning from a few kilometers (*near-source events*) to 300 km or more (*far-source events*). In addition, for the three stations placed on the dam embankment the peak transversal (upstream-downstream), longitudinal and vertical accelerations are provided. It is interesting to observe that the peak vertical acceleration often was higher than the two horizontal components, especially for the seismic events originated closer to the dam site (near-source propagation). Most of the recorded events had epicenter in the Apennines, at the border between Campania and Basilicata Regions, where the devastating 1980 Irpinia earthquake took place. The two seismic events that induced higher accelerations at the dam site were the Molise earthquake (31/10/2002) and the Appennino Lucano earthquake (7/9/06). Epicenter of the former event is 300 km far from the dam (*far-source event*), while that of the latter is around 15 km (*near-source*). Neither of these events induced any macroscopic effect on the dam body.

In what follows, some of the seismic events listed in Table 1 were interpreted by means of simplified procedures to derive useful information on dam soil behavior at small strains and on the overall dynamic response of the embankment.

Table 1. Selected recordings of the weak-motion earthquakes recorded on the Camastra Dam (in red the records for which the vertical component was higher than the horizontal ones)

Recording	Date	Magnitudo	Seismogenic zone	source-site distance	Epicenter	dam base a (cm/s²)			dam berm a (cm/s²)			dam crest a (cm/s²)		
						T	L	V	T	L	V	T	L	V
				Km										
1	31/10/2002	5.4	Monti dei Frentani	>300	S. Giuliano di Puglia (CB)	0.51	0.59	0.44	0.71	0.73	0.43	1.35	1.34	0.80
2	01/11/2002	5.3	Monti dei Frentani	>300	Casacalenda (CB)	0.47	0.31	0.21	0.48	0.51	0.29	0.88	0.73	0.42
3	29/03/2003	5.4	Adriatico centro sett.	>300		0.13	0.11	0.07	0.14	0.13	0.07	0.17	0.17	0.10
4	04/05/2003	3.1	Appennino Lucano	30	Ferrandina (MT)	0.13	0.16	0.19	0.13	0.09	0.11	0.20	0.12	0.22
5	21/05/2003	<2	Appennino Lucano	10	zona diga (PZ)	0.14	0.15	0.28	0.09	0.10	0.13	0.15	0.17	0.17
6	27/07/2003	2.9	Appennino Lucano	40	Buccino (SA)	0.13	0.11	0.11	0.16	0.11	0.09	0.19	0.13	0.20
7	14/08/2003	6.2	Grecia	>300		-	-	-	0.17	0.15	0.10	0.21	0.17	0.13
8	23/02/2004	3.5	Appennino Lucano	40	Buccino (SA)	0.17	0.24	0.13	0.17	0.18	0.12	0.39	0.47	0.35
9	24/02/2004	3.7	Appennino Lucano	40	Buccino (SA)	0.37	0.57	0.15	0.48	0.46	0.18	1.05	0.97	0.68
10	03/03/2004	4.2	Costa Calabra	110		0.24	0.25	0.12	0.45	0.33	0.16	0.84	0.53	0.34
11	27/07/2004	2.9	Appennino Lucano	30	Ferrandina (MT)	0.17	0.11	0.20	0.12	0.08	0.14	0.18	0.15	0.21
12	30/07/2004	3	Appennino Lucano	30	Ferrandina (MT)	0.16	0.15	0.33	0.11	0.11	0.22	0.22	0.17	0.33
13	31/07/2004	3.2	Appennino Lucano	30	Ferrandina (MT)	0.26	0.21	0.33	0.23	0.19	0.21	0.46	0.27	0.52
14	02/08/2004	3.3	Appennino Lucano	40	Buccino (SA)	0.17	0.15	0.17	0.08	0.12	0.13	0.14	0.14	0.18
15	15/08/2004	3.2	Appennino Lucano	40	Buccino (SA)	0.18	0.15	0.11	0.13	0.11	0.12	0.23	0.17	0.22
16	03/09/2004	4.1	Appennino Lucano	40	Buccino (SA)	0.57	0.63	0.45	0.80	0.77	0.38	1.16	1.11	0.88
17	08/01/2006	6.7	Grecia	>300		0.44	0.41	0.49	0.45	0.42	0.48	0.73	0.61	0.86
18	17/04/2006	4.3	Golfo di Taranto	60	Rossano Calabro (CS)	0.10	0.06	0.05	0.09	0.10	0.05	0.21	0.13	0.11
19	24/04/2006	2.7	Appennino Lucano	40	Buccino (SA)	0.14	0.14	0.08	0.08	0.12	0.10	0.16	0.14	0.17
20	29/05/2006	4.9	Gargano	50	Carpino (FG)	1.05	0.75	0.48	0.93	0.77	0.50	1.71	1.47	1.22
21	22/06/2006	4.5	Piana di Sibari	60	Corigliano Calabro (CS)	0.28	0.49	0.17	0.25	0.27	0.19	0.60	0.50	0.44
22	06/08/2006	2.9	Appennino Lucano	20	Tito (PZ)	0.21	0.27	0.16	0.21	0.21	0.16	0.37	0.34	0.34
23	07/09/2006	3.9	Appennino Lucano	15	Calciano (MT)	0.72	0.61	0.86	2.37	3.89	4.90	5.24	3.90	8.00
24	26/10/2006	5.7	Tirreno Meridionale	150	Isole Lipari	0.32	0.26	0.36	0.33	0.29	0.33	0.54	0.47	0.58

4 Interpretation of the Recorded Signals

The modification that a seismic signal in its upward propagation in the dam embankment depends on a number of factors, among which the features of the fault mechanism together with the source-site distance, damping and stiffness properties of both construction and foundation soils, dam geometry. Interaction among these factors may increase or decrease the amplitude and frequency content of the seismic signal with respect to the incoming one, generate a phase difference between the motions of points within the dam body.

The interpretative procedures adopted to quantify the above modifications may be distinguished according to the domain of reference, i.e. frequency or time domain. In the following section, the acquired acceleration recordings will be described and interpreted accordingly.

4.1 Frequency Domain

Frequency domain interpretation may be performed by computing the acceleration response spectrum (at 5% of damping) for each recorded accelerogram. Figures 2, 3 and 4 show the response spectra for some events listed in Table 1.

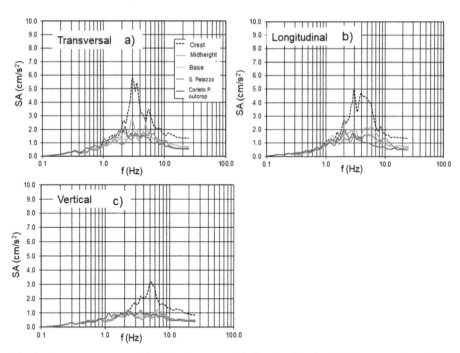

Fig. 2. Response spectra of the signals recorded during the Oct. 31, 2002 event (#1 in Table 1)

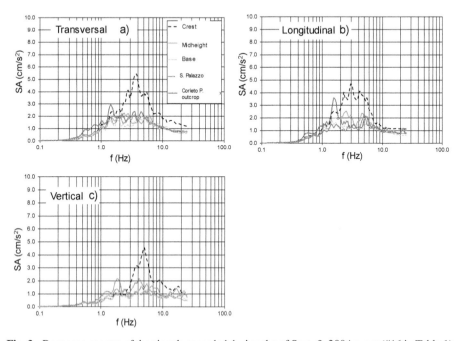

Fig. 3. Response spectra of the signals recorded during the of Sept. 3, 2004 event (#16 in Table 1)

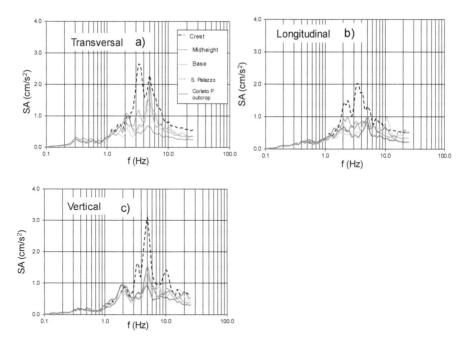

Fig. 4. Response spectra of the signal recorded during the Oct. 26, 2006 event (#24 in Table 1)

The acceleration response spectra of the far source events dated Oct. 31, 2002 (Fig. 2; record n.1 in Table 1), Sept. 3, 2004 (Fig. 3; record n.16 in Table 1) and Oct. 26, 2006 (Fig. 4) show a remarkable filtering effect exerted by the dam body on all components of motion. Signals recorded at the bedrock have a quite flattened spectral shape over the frequency range 2–10 Hz. At the crest, the spectral shapes are narrower with maximum values at frequencies of 3÷4 Hz for both transversal and longitudinal horizontal components, at 5 Hz for the vertical component.

To evaluate the different response of the dam along the three components of motion, Fig. 5 compares the response spectra (event of Sept. 3, 2004) of the signals recorded at single locations. It is possible to notice that in the two horizontal directions the response spectra at the dam crest do not differ significantly each other. Some differences may, instead, be observed on the vertical component, whose maximum spectral ordinate at the crest is shifted towards higher frequency, i.e. 5 Hz.

Similar patterns were found for the other signals listed in Table 1, so the above observations depict a quite general response of the Camastra Dam under weak motion events.

To discriminate the embankment response from the source mechanisms associated to each event, the simple SSR (*Standard Spectral Ratio*) technique has been applied to all recorded signals. In particular, the amplitude of the Fourier spectrum of the accelerogram at a given location of the dam was divided by the Fourier amplitude of the corresponding signal at the bedrock outcrop. For the Camastra dam, this procedure has been applied between the signals (time window containing only S waves) recorded at the dam crest and on the Corleto Perticara outcrop. Figure 6 plots the SSR amplification ratios for all recorded signals. It is evident that the fundamental frequency of the dam embankment is around to 3.3 Hz in both horizontal directions (transversal and longitudinal). In the transversal direction (upstream-downstream) the amplification function shows a number of additional peaks at frequencies of 5 Hz and 14 Hz. As far as the vertical component is concerned, the SSR technique yields peaks at frequencies of 5.9 and 17 Hz.

By comparing the results yielded by the he SSR technique (Fig. 6) and those yielded by the simple representation through response spectra (Figs. 2, 3 and 4), it emerges that the dam fundamental frequency of an earth dam may be conveniently deduced from the response spectra of the far-source events (Figs. 2, 3 and 4). Natural frequencies of higher order require the application of SSR technique only.

From a practical viewpoint, the knowledge of dam natural frequency may be useful to obtain an approximate estimation of a representative shear modulus (G_0) associated to the dam body for calibrating the parameters required by more refined mathematical-numerical models or for selecting suitable input motions from accelerometric databases in order to assess the seismic response of earth dams by means of performance based approaches.

By modelling the embankment as a shear beam (Mononobe et al. 1936; Kramer 1996; Gazetas 1987; Prato e Delmastro 1987), the dam fundamental frequency, f1, in the upstream-downstream direction may be related to the dam height, H, and to the shear wave velocity, Vs, of the whole embankment through the well-known relationship:

$$f_1 = \frac{V_s}{2.59H} \tag{1}$$

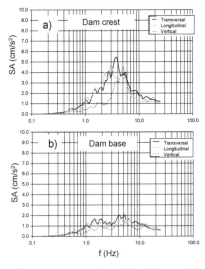

Fig. 5. Comparisons among the response spectra of the three motion components (transversal, longitudinal and vertical) at the dam base and crest during the Sept. 3, 2004 event (#16 in Table 1)

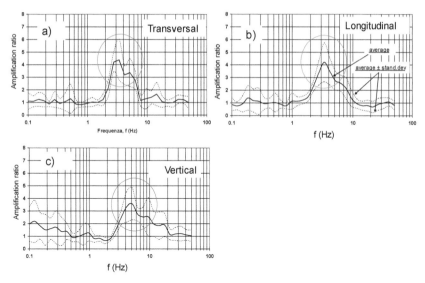

Fig. 6. Amplification ratio computed through the SSR technique (modified from Sica et al., 2008b)

Substituting in Eq. (1) the fundamental frequency estimated from the recorded accelerograms in the transversal direction, ($f_1 = 3.3$ Hz in Fig. 6), a shear wave velocity Vs equal to 490 m/s (shear modulus G_0 of 480 MPa) was obtained. This value, which has the meaning of an equivalent stiffness of the zoned embankment, is close to the values measured on site (Pagano et al. 2008).

4.2 Time Domain

A seismic signal travelling upward throughout the dam body usually amplifies. The monitoring system installed at different elevations of the Conza Dam embankment allows characterizing this phenomenon. In Fig. 7 the maximum accelerations recorded at the crest and bank were normalized with respect to the maximum value recorded on the Corleto Perticara rock outcrop to obtain the Amplification Factors, A_T. The obtained values of A_T point out that often crest peaks are 2–3 times as high as those at rock outcrop or dam basement. At the intermediate bank, A_T is 1–2 times the reference value on Corleto rock outcrop. In most of the analyzed cases, is the transversal component of the motion to denote higher amplification with respect to the incoming signal.

Amplification effects arise due to deformability of the embankment and dam geometry. The trapezoidal shape of the dam, in particular, generates focalization of seismic waves towards the dam crest. At the crest, the vertical component amplifies more than horizontal components (Table 1) because the vertical components of SV and P waves reflected by the dam boundary add up to the direct P and Rayleigh waves (Bouckovalas e Papadimitriou 2005).

In the time domain, it is also possible to observe out-of-phase ground motions between pairs of points of the embankment. These effects, causing asynchronism of the motion, partly compensate the inertial effects within the dam body, thus reducing the destabilizing forces during the earthquake (Bilotta et al. 2010).

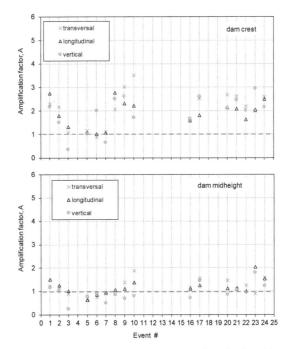

Fig. 7. Amplification ratio at the dam crest and intermediate bank with respect to the rock basement

Figure 8 shows, as an example, the acceleration time histories recorded on the dam body during the *near-source* event of Sept. 7, 2006 record (n. 23). The signals are significantly out-of-phase. This response may be explained considering the high frequencies and therefore the shorter wavelengths associated to this signal, whose epicenter is very close to the dam site.

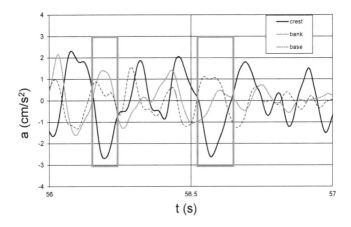

Fig. 8. Out-of-phase motion at different points of the dam body during the Sept. 7, 2006 event.

5 Conclusions

The paper illustrated the important role of weak-motion recordings acquired on earth dams through permanent seismic monitoring stations. These low-intensity signals, often disregarded because not associated to macroscopic effects on the dam body, may be very useful in characterizing important aspects of dam response at relatively low cost.

Once suitably represented in frequency and time domain through basic signal analysis procedures, accelerometric recordings allow the fundamental frequency (and, in some cases, also higher order frequencies) of the embankment to be quantified together with a "representative" shear stiffness of the overall dam body at small strains.

The paper emphasized the high potential of using abundant signals associated to recurrent weak-motion earthquakes, which instead are usually disregarded by the engineers who are entitled to assess the seismic safety of an earth dam. The illustrated interpretive procedures may be extended also to the case of stronger earthquakes in order to characterize dam behavior at higher strain levels.

Acknowledgments. The authors would like to thank Prof. Claudio Mancuso who coordinates the PRIN 2017 project "Risk Assessment of Earth Dams and River Embankments to Earthquakes and Floods" funded by MIUR.

References

AGI: Aspetti geotecnici della progettazione in zona sismica. Linee Guida AGI. Associazione Geotecnica Italiana. Edizione provvisoria. Patron, Bologna (2005)

Bilotta, E., Pagano, L., Sica, S.: Effect of ground-motion asynchronism on the equivalent acceleration of earth dams. Soil Dyn. Earthq. Eng. **30**(7), 561–579 (2010)

Bouckovalas, G.D., Papadimitriou, A.G.: Numerical evaluation of slope topography effects on seismic ground motion. Soil Dyn. Earthq. Eng. **25**, 547–558 (2005)

Elia, G., Amorosi, A., Chan, A.H.C., Kavvadas, M.J.: Fully coupled dynamic analysis of an earth dam. Geotechnique **61**(7), 549–563 (2011)

Gazetas, G.: Seismic response of earth dams; some recent developments. Soil Dyn. Earthq. Eng. **6**(1), 3–47 (1987)

ICOLD: Dam design criteria, the philosophy of their selection. International Commission On Large Dams, Paris, Dam Safety Guidelines, Bull. 61 (1988)

ICOLD: Dam Safety Guidelines. Bull. 72 (1989)

Kramer, S.L.: Geotechnical Earthquake Engineering. Prentice-Hall Inc., New Jersey (1996)

Mononobe, H.A., et al.: Seismic stability of earth dams. In: Proceedings, 2nd Congress on Large Dams, Washington DC, vol. 4 (1936)

NTC2018 - Norme Tecniche per le Costruzioni. DM 17/1/2018, Italian Ministry of Infrastructure and Transportation, G.U. no. 42, 20 February 2018, Rome, Italy (2018). (in Italian)

NTD2014 - Ministero Delle Infrastructure E Dei Trasporti, Decreto N. 26 Giugno 2014 Norme tecniche per la progettazione e la costruzione degli sbarramenti di ritenuta (dighe e traverse). (14A05077) (GU Serie Generale n.156 del 08-07-2014) (2014)

Pagano, L., Mancuso, C., Sica, S.: Prove in sito sulla diga del Camastra: tecniche sperimentali e risultati. Rivista Italiana di Geotecnica **3**, 11–28 (2008)

Pagano, L., Russo, C., Sica, S., Costigliola, R.: Limit states in earth dams during seismic and post-seismic stages. In: Silvestri, F., Moraci, N. (eds.) Proceedings of the VII ICEGE 7th International Conference on Earthquake Geotechnical Engineering (Theme Lecture). Rome 17–20 June 2019. Proceedings in Earth and geosciences, vol. 4, pp. 600–616. CRC press, Balkema (2019). www.crcpress.com, ISBN 978-0-367-14328-2

Prato, C.A., Delamstro, E.: 1-D Seismic analysis of embankment dams. J. Geotech. Eng. ASCE **113**(8), 904–909 (1987)

Sica, S., Pagano, L., Modaressi, A.: Influence of past loading history on the seismic response of earth dams. Comput. Geotech. **35**(1), 61–85 (2008a)

Sica, S., Pagano, L., Rotili, F.: Rapid drawdown on earth dam stability after a strong earthquake. Comput. Geotech. **116**, 103187 (2019)

Sica, S., Pagano, L., Vinale, F.: Interpretazione dei segnali sismici registrati sulla diga di Camastra. Rivista Italiana di Geotecnica **4**(8), 97–111 (2008b)

Sica, S., Pagano, L.: Performance-based analysis of earth dams: procedures and application to a sample case. Soils Found. **49**(6), 921–939 (2009)

Neural Network-Based Monitoring of Rainfall-Induced Pore-Water Pressure in an Unsaturated Embankment

Khonesavanh Vilayvong[1,5(✉)], Duangphachan Khambolisouth[2],
Phanthoudeth Pongpanya[3,5], Phoutthamala Sitthivong[4,5],
and Thipphamala Manivong[3,5]

[1] Kyoto University, Kyoto 606-8502, Japan
sangsinsay@gmail.com
[2] Zeals Co., Ltd, Tokyo, Japan
[3] National University of Laos, Vientiane, Laos
[4] Souphanouvong University, Luang Prabang, Laos
[5] Geotechnical Working Group, Vientiane, Laos

Abstract. Monitoring the behavior of an unsaturated soil embankment under dynamic and complex conditions is still a limit, especially under spatial and temporal variation of soil and foundation properties, boundary conditions, and loading conditions. Pore-water pressure is a major factor for the stability of the embankment. Therefore, this study concentrates on monitoring of pore-water pressure. Finite element analysis (FEA) updated with an artificial neural network (ANN) using an optimally tuned single hidden layer was carried out. The FEA using physical soil properties from literature (soil moisture and soil suction) and actual heavy rainfall data was conducted to generate time-series pore-water pressure. The FEA results were configured to establish input dataset feeding into the ANN model. The ANN was analyzed using open-source libraries for machine learning, parallel and multiprocessing, and python programming to predict the behavior of pore-water pressure. The ANN results showed good performance for monitoring the pore-water pressure.

Keywords: Neural network · Pore-Water pressure · Unsaturated embankment

1 Introduction

During the service life of an embankment dam, monitoring seepage is critical for stability and safety of the dam. Seepage behavior is influenced by a variety of factors such as climate, reservoir level, soil and foundation properties. Pore-water pressure (PWP) variations in an unsaturated soil embankment, whose properties are heterogenous in both temporal and spatial domains, are difficult to solve by traditional methods. In addition, multiphase, multiple physical factors, internal and external loading conditions, initial conditions, and boundary conditions contribute to the degree of complexity in analyzing the pore-water pressures. Besides, durability of monitoring instruments at an

© Springer Nature Switzerland AG 2020
J.-M. Zhang et al. (Eds.): ICED 2020, SSGG, pp. 357–367, 2020.
https://doi.org/10.1007/978-3-030-46351-9_37

embankment site may experience unexpected damages from technical defects, harsh environment, lightning, or low maintenance. Therefore, continuous monitoring of the PWP is essential.

Artificial neural network (ANN) is a computational model using a metaphor to the mimicry of the brain neuron networks. Figure 1 represents a general multiple layer neural network model parallel to the finite element analysis (FEA). The ANN algorithms are widely employed in various fields. For examples: assessing environmental hazard and disaster associated with soil erosion [1], pore-water pressure for slope stability [2, 3], rainfall forecast [4], shear strength of soil [5], dam seepage [6] and flood forecast [7]. However, existing studies of predicting the PWP in an embankment using machine learning is still not widely studied, especially under unsaturated soil conditions. Performance of ANN model in the past also encountered issues associated with availability of computation infrastructure, size of datasets, optimizations, and programming libraries. Therefore, efficient and robust ANN model is required. In this study, a single hidden layer ANN model was configured using a recently developed machine learning libraries, computational infrastructure, optimization and tuned hyperparameters. The ANN was deployed to evaluate the PWP behavior of an earth fill embankment by considering unsaturated soil conditions, pore-water pressure and rainfall.

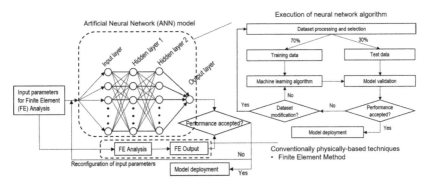

Fig. 1. A schematic diagram of the artificial neural network and the finite element analysis.

2 Methods

2.1 Hydrological Data, Embankment Geometry, and Soil Properties

Dam geometry with a dimension of 17 m height, 80 m width and 22° inclined upstream and downstream (Fig. 3). Toe drain was constructed with filter sand to reduce seepage force in the face of downstream slope. Constant upstream reservoir level was fixed at 13 m above the base. Two properties of the embankment dam, soil-water retention curve [8] and permeability function, were derived from literature and their characteristics are shown in Fig. 4. Actual rainfall of a monitored site in Japan was used in this study (Fig. 2).

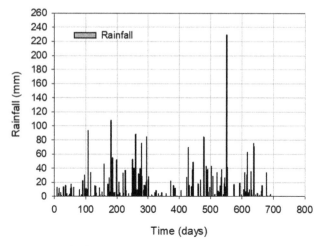

Fig. 2. Hydrological rainfall data of a site.

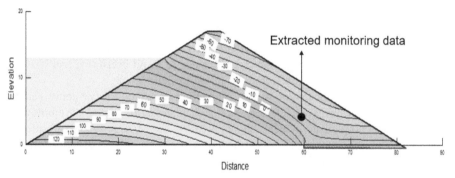

Fig. 3. Geometry and pore-water distribution (in kPa) of the unsaturated soil embankment under steady-stated condition.

Dam embankment was analyzed by finite element (FE) software using SEEP/W from Geoslope. Results of the PWP distribution under steady-stated analysis was initially obtained as shown in Fig. 3 and was set as initial condition before applying rainfall. The PWP is dependent on the rainfall and antecedent rainfall up to 5 days can affect PWP significantly for residual soil slope [9]. Therefore, various combinations of antecedent data were taken into consideration for generating input dataset. Combining same data by different configurations so called "data augmentation" is to increase data dimensionality for improving performance of a machine learning model under data scarcity. A study by [2] used cross-correlation and autocorrelation methods to select appropriate input parameters. However, manually trail method was adopted in this study. Table 2 shows the properties of the input parameters and Table 3 shows the combination of the input parameters used to create the dataset for feeding into the ANN model.

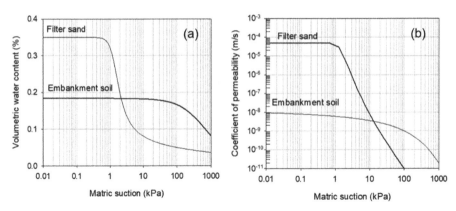

Fig. 4. Soil properties: (a) soil-water characteristic curves and (b) permeability functions.

2.2 Neural Network Algorithm

ANN model with a single hidden layer can represent various functions when feeding with appropriate parameters [10] and was selected in this study. The ANN model was carried out by open-source machine learning libraries using Keras [11], Tensorflow [12], and Scikit-learn [13]. Early-stopping criterion, a form of regulation in machine learning, was opted in order to generalize the model results, to avoid overfitting and to reduce computation time. Detailed configurations are set in Table 1.

Data normalization or feature scaling is an efficient method for fast convergence of gradient descent problems. Standard scaling or standardization, Eq. (1), is employed in this study. Data normalization using Eq. (1) is called "Z-score" or "standardization" to transform Gaussian or non-Gaussian data to have a mean of 0 and unit standard deviation. The Z-score is useful for data with outliers especially climate-change induced rainfall data, which also triggers PWP data to be outliers.

$$Z = \frac{x - \mu}{\sigma} \tag{1}$$

where μ is the mean of the training samples and σ is the standard deviation of the training samples. $\mu = 0$ and $\sigma = 1$ if data is normally distributed.

Activation function is the function used to govern the behavior of neurons for complex problems especially mapping the high degree of nonlinearity. Linear regression problems require no activation function. Hyperbolic tangent, Eq. (2), was selected in this study due to transition range from -1 to $+1$ with zero as center similar to negative and positive pore-water pressures, where the symbol e is the natural number.

$$f(x) = \tan(x) = \frac{\left(e^x - e^{-x}\right)}{\left(e^x + e^{-x}\right)} \tag{2}$$

Table 1. Configuration for operating the ANN model.

Descriptions	Details
Operating System and processor	Ubuntu 18.04 (Linux), 96-core Intel Processor
Machine Learning Infrastructure	Amazon Cloud
Open-source neural network libraries	Keras, Numpy, Pandas, Scikit-learn, Tensorflow
Programming language	Python
Type of operation and batch size	Parallel and multicore processing, 10
ANN architecture and data splitting	Single hidden layer, 70%-30%
Type of machine learning and metrics	Supervise learning, coefficient of determinant
ANN optimization and activation function	Adam optimizer, hyperbolic tangent (tanh)
ANN Learning rates	0.0001, 0.001, 0.01, 0.1
ANN Hyperparameters:	
Activation	None
Use bias	True
Kernel_initializer	Glorot_unitform
Bias inilializer	Zeros
Kernel regularizer	None
Bias regularizer	None
Activity regularizer	None
Kernel constraint	None
Bias constaint	None

Table 2. Considered input parameters and their associated statistics.

Input neurons	Count	Mean	Standard deviation	Min	Max
Daily rainfall parameters (mm):					
Current rainfall, R0	694	4.91	16.053	0	229
1-day antecedent rainfall, R1	694	4.91	16.053	0	229
2-day antecedent rainfall, R2	694	4.91	16.053	0	229
3-day antecedent rainfall, R3	694	4.91	16.053	0	229
4-day antecedent rainfall, R4	694	4.91	16.053	0	229
5-day antecedent rainfall, R5	694	4.91	16.053	0	229
6-day antecedent rainfall, R6	694	4.91	16.053	0	229
7-day antecedent rainfall, R7	694	4.91	16.053	0	229

(continued)

Table 2. (*continued*)

Input neurons	Count	Mean	Standard deviation	Min	Max
Pore-water pressure parameters (kPa):					
1-day antecedent PWP, P1	694	−4.665	1.483	−10.596	0.349
2-day antecedent PWP, P2	694	−4.673	1.499	−10.596	0.349
3-day antecedent PWP, P3	694	−4.680	1.516	−10.596	0.349
4-day antecedent PWP, P4	694	−4.688	1.532	−10.596	0.349
5-day antecedent PWP, P5	694	−4.696	1.549	−10.596	0.349
6-day antecedent PWP, P6	694	−4.703	1.565	−10.596	0.349
7-day antecedent PWP, P7	694	−4.711	1.58	−10.596	0.349
Current PWP, P0, target	694	−4.657	1.466	−10.596	0.349

Note: although the statistical data of the rainfall events R0 to R7 are the same due to shifting the rainfall data, their time-series events are different.

2.3   Optimization, Learning Rates, and Performance Metrics for ANN Model

Adam is one of the commonly used optimization algorithms with growing popularity for neural network and deep learning. Adam optimizer is an adaptive learning rate optimization algorithm [14]. Its attraction and usefulness are the ability to find learning rates for optimization parameters. However, performance of the optimizer is still a case specific. Performing the ANN was conducted following the flow chart in Fig. 1. Before modeling, raw data was pre-processed. Data were split into training dataset (70%) and test dataset (30%). Learning rates used in this study were 0.0001, 0.001, 0.01 and 0.1. Performance metrics are important indexes to gauge performance of machine learning algorithms. Coefficient of determination (R^2), Eq. (3), was used to evaluate performance of the ANN model. The R^2 is a scale-free between the input and output and provides good indexes for predicting future samples. The higher the value of the R^2, with the maximum score of 1.0, the better the model performs.

$$R^2(y, \hat{y}) = 1 - \frac{\sum_{i=1}^{n_{sample}} (y_i - \hat{y_i})^2}{\sum_{i=1}^{n_{sample}} (y_i - \overline{y})^2} \tag{3}$$

$$\overline{y} = \frac{1}{n_{sample}} \sum_{1}^{n_{sample}} y_i \tag{4}$$

where
   $\hat{y}_i$ is the predicted value of the ith sample
   y_i is the corresponding true value
   n_{sample} number of samples.

3 Results and Discussion

3.1 Effect of Neuron Numbers and Learning Rates

Selection of the optimal number both in the input layer and the hidden layer has no specific rule of thumb. Few neurons can result in under-fitting and excessive neurons can result in overfitting. A research by [15] recommends that the neuron number in the hidden layer should be in between the size of the output and input neurons, or 2/3 of the size of the input layer plus the size of the output layer, and twice less than the neuron number of input neurons. However, the neuron number of the hidden layer in this study was preconfigured up to a limit of 50 neurons. This is to compare the results of the ANN model within 50 hidden neurons.

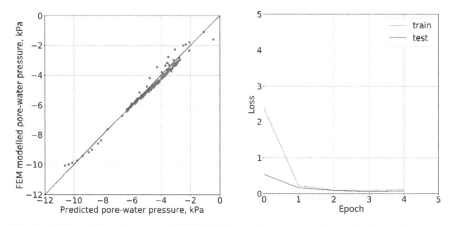

Fig. 5. ANN model performance: RHS (a) fitting performance and LHS (b) loss performance

Table 3 shows the results of the optimal learning rates during the first run of the ANN model by the Amazon Cloud infrastructure. There were different optimal learning rates for any dataset which combines different parameters for the input parameters. The difference in the learning rates was expected due to the random nature of the learning, training, and validating process of the ANN. Besides, the hyperparameters were also the main factors influencing the outcomes. Tuning hyperparameters of the ANN is tedious and case-specific process and subjected to computer framework used. The obtained optimal learning rate indicates that the result can be converged efficiently. However, cautions need to be attended as the optimal learning rate can be fluctuated in every run of the ANN model.

3.2   Performance of the Neural Network Model

The 4-49-1 architecture was identified as the best-performed architecture. The corresponding metric score for the validation was 0.9931 (Table 3). Figure 5a shows the performance result of the neural network application for the test dataset for input dataset No.3 (4-49-1 architecture). All points were closed to the perfect performance straight line (the straight line inclined at 45° to the x-axis and y-axis) (Fig. 5a). Figure 6 showed the result the predicted pore-water pressure using the test dataset. This indicates that using data augmentation in this study under limited dataset can enhance the robustness and performance of the model with respective to the value of R^2 greater than 0.90. In general, it was found that using only the current rainfall is sufficient to obtain the high accuracy of results. However, the ANN model tends to perform better when adding antecedent pore-water pressures than adding the set of antecedent rainfalls. This can be seen in Table 2 that statistics of the antecedent rainfall events (shifted rainfall data with respect to time) are the same.

However, there were different results in each run of the ANN model. Thus, 1000 runs of the model were conducted in order to visualize or observe how the results deviate from each run. Table 4 shows the results of the 1000 runs of the neural network using computer terminal different from that of the Amazon Cloud. The result showed that there was small deviation of the results, indicating that the performance of the neural network was stable, with the mean, standard deviation, minimum value, and maximum value of 0.9829, 0.0042, 0.9337, and 0.9863, respectively.

Figure 5a shows the comparison results for the test data (30% of the total data). It was found that the predicted pore-water pressure was able to mimic the pattern of the FEA modelled pore-water pressure in high accuracy. The Fig. 5b shows the validation results of the trained data and test data with the Early-Stopping criterion. The Early-Stopping criterion was selected in order to stop the training in order to avoid overfitting issue as this can affect the generalization of the result. The maximum epoch was 4 out of 1000 (Fig. 5b), which indicates less execution time to operate the 4-49-1 architecture.

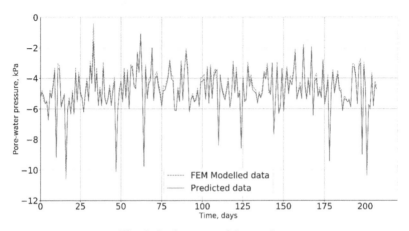

Fig. 6. Performance of the test data.

Table 3. The ANN result performance for the test data.

No.	Input dataset	Neural Network Optimization Properties			
		Learning rate	#Hidden neurons	R^2	Epochs
1	R0, P1	0.1000	17	0.9876	5
2	R0, P1, P2	0.0100	9	0.9908	21
3	R0, P1, P2, P3	0.1000	49	0.9931	6
4	R0, P1, P2, P3, P4	0.0100	49	0.9906	8
5	R0, P1, P2, P3, P4, P5	0.0100	39	0.9876	7
6	R0, P1, P2, P3, P4, P5, P6	0.0100	43	0.9874	8
7	R0, P1, P2, P3, P4, P5, P6, P7	0.0001	41	0.9864	110
8	R0, R1, P1	0.0100	39	0.9828	7
9	R0, R1, P1, P2	0.0100	45	0.9849	8
10	R0, R1, P1, P2, P3	0.0100	4	0.9857	24
11	R0, R1, P1, P2, P3, P4	0.0100	49	0.9855	8
12	R0, R1, P1, P2, P3, P4, P5	0.0100	44	0.9823	5
13	R0, R1, P1, P2, P3, P4, P5, P6	0.0100	42	0.9829	7
14	R0, R1, P1, P2, P3, P4, P5, P6, P7	0.0100	45	0.9830	7
15	R0, R1, R2, P1	0.0010	36	0.9803	45
16	R0, R1, R2, P1, P2	0.0010	50	0.9825	35
17	R0, R1, R2, P1, P2, P3	0.0001	43	0.9877	246
18	R0, R1, R2, P1, P2, P3, P4	0.0010	50	0.9834	23
19	R0, R1, R2, P1, P2, P3, P4, P5	0.0100	50	0.9825	9
20	R0, R1, R2, P1, P2, P3, P4, P5, P6	0.0001	48	0.9832	129
21	R0, R1, R2, P1, P2, P3, P4, P5, P6, P7	0.0010	44	0.9784	19
22	R0, R1, R2, R3, P1	0.0010	50	0.9838	47
23	R0, R1, R2, R3, P1, P2	0.0001	37	0.9880	297
24	R0, R1, R2, R3, P1, P2, P3	0.0001	37	0.9862	262
25	R0, R1, R2, R3, P1, P2, P3, P4	0.0001	43	0.9857	196
26	R0, R1, R2, R3, P1, P2, P3, P4, P5	0.0001	49	0.9834	166
27	R0, R1, R2, R3, P1, P2, P3, P4, P5, P6	0.0010	40	0.9810	22
28	R0, R1, R2, R3, P1, P2, P3, P4, P5, P6, P7	0.0010	48	0.9846	26
29	R0, R1, R2, R3, R4, P1	0.0010	9	0.9822	135
30	R0, R1, R2, R3, R4, P1, P2	0.0001	47	0.9892	290
31	R0, R1, R2, R3, R4, P1, P2, P3	0.0001	48	0.9884	273
32	R0, R1, R2, R3, R4, P1, P2, P3, P4	0.0001	48	0.9865	216
33	R0, R1, R2, R3, R4, P1, P2, P3, P4, P5	0.0001	49	0.9864	165
34	R0, R1, R2, R3, R4, P1, P2, P3, P4, P5, P6	0.0001	46	0.9850	169
35	R0, R1, R2, R3, R4, P1, P2, P3, P4, P5, P6, P7	0.0100	46	0.9802	9
36	R0, R1, R2, R3, R4, R5, P1	0.0010	43	0.9823	42
37	R0, R1, R2, R3, R4, R5, P1, P2	0.0100	48	0.9785	7
38	R0, R1, R2, R3, R4, R5, P1, P2, P3	0.0100	48	0.9833	10
39	R0, R1, R2, R3, R4, R5, P1, P2, P3, P4	0.0100	46	0.9858	8
40	R0, R1, R2, R3, R4, R5, P1, P2, P3, P4, P5	0.0010	49	0.9833	30

(continued)

Table 3. (*continued*)

No.	Input dataset	Neural Network Optimization Properties			
		Learning rate	#Hidden neurons	R^2	Epochs
41	R0, R1, R2, R3, R4, R5, P1, P2, P3, P4, P5, P6	0.0010	50	0.9786	29
42	R0, R1, R2, R3, R4, R5, P1, P2, P3, P4, P5, P6, P7	0.0100	46	0.9843	13
43	R0, R1, R2, R3, R4, R5, R6, P1	0.0010	49	0.9812	52
44	R0, R1, R2, R3, R4, R5, R6, P1, P2	0.0010	43	0.9814	57
45	R0, R1, R2, R3, R4, R5, R6, P1, P2, P3	0.0001	42	0.9846	259
46	R0, R1, R2, R3, R4, R5, R6, P1, P2, P3, P4	0.0001	47	0.9831	243
47	R0, R1, R2, R3, R4, R5, R6, P1, P2, P3, P4, P5	0.0001	50	0.9870	233
48	R0, R1, R2, R3, R4, R5, R6, P1, P2, P3, P4, P5, P6	0.0001	32	0.9791	221
49	R0, R1, R2, R3, R4, R5, R6, P1, P2, P3, P4, P5, P6, P7	0.0100	48	0.9813	9
50	R0, R1, R2, R3, R4, R5, R6, R7, P1	0.0010	36	0.9760	51
51	R0, R1, R2, R3, R4, R5, R6, R7, P1, P2	0.0100	43	0.9814	10
52	R0, R1, R2, R3, R4, R5, R6, R7, P1, P2, P3	0.0001	43	0.9829	264
53	R0, R1, R2, R3, R4, R5, R6, R7, P1, P2, P3, P4	0.0001	50	0.9815	207
54	R0, R1, R2, R3, R4, R5, R6, R7, P1, P2, P3, P4, P5	0.0010	49	0.9816	32
55	R0, R1, R2, R3, R4, R5, R6, R7, P1, P2, P3, P4, P5, P6	0.0001	49	0.9779	161
56	R0, R1, R2, R3, R4, R5, R6, R7, P1, P2, P3, P4, P5, P6, P7	0.0010	48	0.9801	41

Table 4. Metric results of 1000 runs using ANN for input dataset No.3.

Dataset No.	Count	Mean	Standard deviation	Min	Max
#3 (R0, P1, P2, P3)	1000	0.9829	0.0042	0.9337	0.9863

4   Conclusion

Artificial neural network (ANN) was studied with the aid of open-source libraries for neural network, python programming, cloud computing and the finite element analysis (FEA). The ANN model with a single hidden layer architecture was capable of predicting the pore-water pressure variation in an unsaturated soil embankment dam. The optimal ANN architecture was identified using advanced cloud computing infrastructure and multiprocessing and multi-core processing. The established model was viable method for monitoring the pore-water pressure under limited input dataset using the data augmentation for the same data for rainfall and pore-water pressure. The results of the ANN in this study is useful for updating and complementary procedure for the FEA under complex conditions and instrument constraints.

Acknowledgments. Acknowledgements are specially recognized with the support from the Geoslope company in realizing the quality analysis using the finite element method developed by the company. Constructive comments from the reviewers are also noted to make this paper publishable. Lastly, thanks to the institutes where the authors belong to for the enabling and conducive environment and facilities to realize this study.

References

1. Licznar, P., Nearing, M.A.: Artificial neural networks of soil erosion and runoff prediction at the plot scale. Catena **51**(2), 89–114 (2003). Elsevier
2. Mustafa, M.R., Rezaur, R.B., Rahardjo, H., Isa, M.H., Arif, A.: Artificial neural network modeling for spatial and temporal variations of pore-water pressure responses to rainfall. Advances in Meteorology, volume 2015, Article ID273730 (2014)
3. Kumar, S., Basudhar, P.K.: A neural network model for slope stability computations. Geotechnique **8**(2), 149–154 (2018)
4. Ramirez, V., Velho, C.H., Ferrira, N.: Artificial neural network technique for rainfall forecasting applied to the Sao Paulo region. J. Hydrol. **301**(1), 146–162 (2005)
5. Lee, S.J., Lee, S.R., Kim, Y.S.: An approach to estimate unsaturated shear strength using artificial neural network and hyperbolic formulation. Comput. Geotech. **30**(6), 489–503 (2003). Elsevier
6. Tayfur, G., Swiatek, D., Wita, A., Singh, V.P.: Case study: finite element method and artificial neural models for flow through Jeziorsko Earthfill dam in Poland. J. Hydraul. Eng. **131**(6), 431–440 (2005). ASCE
7. Varoonchotikul, P.: Flood Forecasting Using Artificial Neural Networks. CRC Press, Boca Raton (2003)
8. Leong, E.C., Rahardjo, H.: Soil-water characteristic curves of compacted residual soils. In: The 3rd International Conference on Unsaturated Soils (2002)
9. Rahardjo, H., Leong, E.C., Rezaur, R.B.: Effect of antecedent rainfall on pore-water pressure distribution characteristics in residual soil slope under tropic rainfall. Hydrol. Proc. **22**, 506–523 (2008)
10. Wikipedia: Universal approximation theorem. https://en.wikipedia.org/wiki/Universal_approximation_theorem. Accessed 5 Feb 2020
11. Keras. https://keras.io. Accessed 5 Feb 2020
12. Tensorflow. https://www.tensorflow.org. Accessed 11 Dec 2019
13. Scikit-learn. https://scikit-learn.org/. Accessed 5 Feb 2019
14. Kingma, D.P., Ba, J.L.: Adam: a method for stochastic optimization. arXiv (2015)
15. Heaton, J.: The number of hidden layer. https://www.heatonresearch.com/2017/06/01/hidden-layers.html. Accessed 2 May 2020

International Workshop on Prediction of Jinsha River Dam Breaching and Flood Routing

Breaching and Flood Routing Simulation of the 2018 Two Baige Landslide Dams in Jinsha River

Ming Peng[1,2]([✉]), Chenyi Ma[1,2], Danyi Shen[1,2], Jiangtao Yang[1,2], and Yan Zhu[1,2,3]

[1] Key Laboratory of Geotechnical and Underground Engineering of Ministry of Education, Department of Geotechnical Engineering, Tongji University, Shanghai 200092, China
pengming@tongji.edu.cn
[2] Department of Geotechnical Engineering, College of Civil Engineering, Tongji University, Shanghai 200092, China
[3] China Shipbuilding NDRI Engineering Co. Ltd., Shanghai 200063, China

On October 10 and November 3, 2018, the right bank slope of the Jinsha River at Baige village failed twice and formed two huge landslide dams. The dams were 61 m and 96 m in heights and 249 and 757 million m3 in lake volumes, respectively. The breaching floods destroyed several man-made dams and large bridges downstream and caused as much as 4 billion RMB direct loss in Lijiang City, Yunan province, which is 800 km far away from the dam site. Two methods are used to simulate the dam breaching and flood routing process of the Baige landslide dams: (1) Comprised model A, which consists of the DABA model, DTE model and EDDA model, is developed to simulate the sediment propagation and flood evolution during dam breaching. In that model, DABA (Chang and Zhang 2010; Shi et al. 2015) is applied to mimic the erosion of dam materials, EDDA (Chen and Zhang 2015) is applied to mimic the deposition of the sediment downstream of the dam, and the DTE (Peng et al. 2019) is applied to link these two models by mixing the soil and water and obtaining the hydraulic parameters of the mixed fluid; and (2) Comprised model B, which consists of a empirical model based on regression analysis with multi-parameters and dam breaching modular in HEC-RAS model, is developed to simulating the dam breaching. The empirical model (Peng and Zhang 2012) is applied to predict the breaching parameters and HEC-RAS model is applied to mimic dam breaching process. Flood routing after the dam breaching is conducted by using HEC-RAS as shown in Fig. 1.

Three scenarios are considered in the paper: the first dam breaching event (Scenario 1), the second dam breaching event with the excavated spillway (Scenario 2) and the second dam breaching event without spillway (Scenario 3). The simulated results are shown in Table 1 and Fig. 2. The measured values of peak outflow rate during the breaching of the first landslide dam (Scenario 1) and the second landslide dam (Scenario 2) were 10000 m^3/s and 33900 m^3/s, respectively. The predicted peak outflow rates of the first dam were 10052 and 11447 m^3/s for the Comprised models A and B, respectively. The corresponding peak outflow rates of the second dam became 33177 and 32136 m^3/s. The errors of the two scenario calculated by Comprised model A were 0.5% and 2.1%. The corresponding errors calculated by Comprised model B were

J.-M. Zhang et al. (Eds.): ICED 2020, SSGG, pp. 371–373, 2020.
https://doi.org/10.1007/978-3-030-46351-9_38

14.5% and 5.2%. As shown in Fig. 2, the calculated peak discharge in different scenarios show good agreements with the measured data, while the time that the peak discharge occur show little difference with the measured data. The measured values of breach depth for the first landslide dam and the second landslide dam with spillway were 32 m and 61 m, respectively (Zhang et al. 2019). The errors of the two scenario calculated by Comprised model A were 15.3% and 1.0%. The corresponding errors calculated by Comprised model B were 5.6% and 23.1%. The breach depth, top width of breach, and bottom width of breach all increased rapidly after entering the third stage. It is possible to consider the most dangerous scenario that no spillway was excavated for the second landslide dam. The peak flow rate would reach 53060 m³/s calculated by Comprised model A (increased by 57%) and 58789 m³/s calculated by Comprised model B (increased by 73%). Obviously, the excavation of the 15-m depth spillway decreased the risks of the landslide dam significantly.

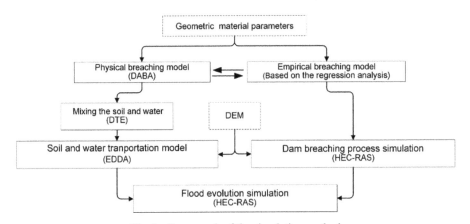

Fig. 1. Framework of the simulation method

Table 1. Predicted breaching parameters of the two Baige landslide dams

Scenario	Peak flow rate (m³/s)		Breach					
			Depth (m)		Bottom width (m)		Top width (m)	
	Model A	Model B	Model A	Model B	Model A	Model B	Model A	Model B
The first landslide dam	10052	11447	36.9	30.2	72.9	90.1	146.7	89.8
The second landslide dam with spillway	33177	32136	60.4	46.9	118.2	145.4	239.0	226.7
The second landslide dam without spillway	53060	58798	65.6	54.7	146.3	178.2	277.5	227.0

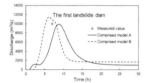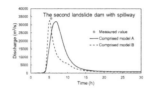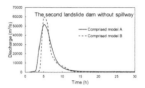

Fig. 2. Discharge rate of breach in Scenarios 1, 2 and 3

Baige landslide dams are so special that the two landslide dams were formed in the same site within one month, which is the first record of such an event in all over the world. Obviously, although the volume of the second landslide is small, the impact is much larger than that of the first dam. At the same time, excavating the spillway has a great effect on reducing peak flow value. Both methods have their advantages and disadvantages. Comprised model A works well for the cases with detailed investigation information. The transportation of both the water and soil near the dam site can be simulated. Comprised model B can be applied to rapidly simulate the breaching and evolution floods for the cases without detailed information.

References

Chang, D.S., Zhang, L.M.: Simulation of the erosion process of landslide dams due to overtopping considering variations in soil erodibility along depth. Nat. Hazards Earth Syst. Sci. **10**(4), 933–946 (2010)

Chen, H.X., Zhang, L.M.: EDDA 1.0: integrated simulation of debris flow erosion, deposition and property changes. Geosci. Model Dev. **8**, 829–844 (2015)

Peng, M., Zhang, L.M.: Breaching parameters of landslide dams. Landslides **9**(1), 13–31 (2012)

Peng, M., Ma, C.Y., Zhang, X.S., Shi, Z.M., Zhu, Y.: Simulation of the sediment propagation during the breaching of the Tangjiashan landslide dam. In: Proceedings of SPIE 9435, 7th International Symposium on Geotechnical Safety and Risk (2019)

Shi, Z.M., Guan, S.G., Peng, M., et al.: Cascading breaching of the Tangjiashan landslide dam and two smaller downstream landslide dam. Eng. Geol. **193**, 445–458 (2015)

Zhang, L.M., Xiao, T., He, J., et al.: Erosion-based analysis of breaching of Baige landslide dams on the Jinsha River, China, in 2018. Landslides (2019). https://doi.org/10.1007/s10346-019-01247-y

Breaching Process of "11·03" Dammed Lake and Flood Routing

Lin Wang[1], Qiming Zhong[2(✉)], Yibo Shan[2], and Simin Cai[1]

[1] College of Water Resources and Hydropower Engineering, Xi'an University of Technology, Xi'an, China
[2] Nanjing Hydraulic Research Institute, Nanjing, China
qmzhong@nhri.cn

Accurate prediction of the breach hydrograph and flood routing is crucial step for emergency response of the breaching of a dammed lake. On October 10 and November 3, 2018, two successive landslides in the same place occurred at Baige village, the border of Sichuan Province and Tibet Autonomous Region, China, which totally dammed the Jinsha River twice. Due to the rapid rising of the dammed lake, engineering measures cannot be taken other than evacuation 20652 people in the risk area. The "10·10" dammed lake burst naturally on October 12, 2018, only two days after the formation. Then, the flow channel formed after the breaching of "10·10" dammed lake was blocked by the "11·03" landslide, resulted in an even larger dammed lake. Since the permission of objective conditions, the measures of excavation of drainage channel were taken to decrease the water level of dammed lake when it burst. On November 12, the dammed lake burst with the peak breach flow of 31000 m^3/s. For "11·03" Baige dammed lake, as detailed hydrological data was well documented, which provided valuable basic data for the study of outburst flood of the dammed lake. In this study, a numerical method was developed to simulate the breaching process of "11·03" dammed lake. The numerical method included two modules, such as module of breaching process of the landslide dam, and module of flood routing after dam breaching. The major highlights of the numerical method are the consideration of the breach mechanism of landslide dam, such as the breach morphology evolution process along the streamwise and transverse directions, as well as the variation of soil erodibility with depth. In addition, Open Source HEC-RAS is used to simulate flood routing along the downstream reach of the Jinsha River after the breaching of "11·03" dammed lake. As for the module of breaching process of the landslide dam, the broad-crested weir equation was adopted the simulate the breach flow discharge, and the variation of water level of the dammed lake was determined according to inflow and breach flow, as well as the relationship curve of water level and surface area of dammed lake. Based on the shear stress of water flow and the critical shear stress of soil materials, the erosion process was simulated by using the erosion formula for wide graded landslide debris. Under the assumption that the breach slope angles remain unchanged during the longitudinal cutting and transverse broadening, the limit equilibrium method is used to analyze the slope instability during the breach evolution process. The coupling process of soil and water during dam breaching was simulated by the algorithm of time step iteration. As for the module of flood routing after dam breaching, the computational domain for the post-"11·03" dammed lake dam breach flood routing simulation covers the reach of the Jinsha River between Baige dammed lake and the Liyuan hydropower

© Springer Nature Switzerland AG 2020
J.-M. Zhang et al. (Eds.): ICED 2020, SSGG, pp. 374–375, 2020.
https://doi.org/10.1007/978-3-030-46351-9_39

station. In the numerical modeling of flood routing, 1D and 2D hydrodynamic coupling model is adopted. The DEM (Digital Elevation Model) data with a resolution of 30 m provided by Geospatial Data Cloud site, Computer Network Information Center, Chinese Academy of Sciences was utilized, further, the terrain data is extracted through Open Source HEC-GeoRAS module and then imported into HEC-RAS to build the channel model. HEC-RAS channel model divides the channel cross-section terrain data into three parts: left floodplain, channel center and right floodplain. The channel model selects different Manning coefficients. The hydrograph calculated for the "11·03" dammed lake breach flood by the above module of breaching process of the landslide dam is applied as the inflow boundary condition for the simulation. The downstream boundary is defined as a free outflow boundary condition, and the other boundaries are defined as closed boundaries. The comparison of measured and calculated results would be conducted to verify the rationality of the numerical method.

Numerical Simulation of the Natural Erosion and Breaching Process of the "10.11" Baige Landslide Dam on the Jinsha River

Chen Xie, Qin Chen, Gang Fan[✉], and Chen Chen

Sichuan University, Chengdu 610065, Sichuan, China
`fangang@scu.edu.cn`

Abstract. Outburst flood of the dammed lake poses a great threat to the downstream residents and properties, in addition, the process of dam erosion and flood routing is very complicated. Thus, the research which using numerical method to simulate and repeat the typical landslide dam breaching, has a great significance for disaster prevention and mitigation in the downstream. The Jinsha River, was dammed twice recently at Baige, Tibet, one on 10 October 2018 and the other on 3 November 2018. Accordingly two large landslide dams were formed in a three-week interval, and breached subsequently, causing a major loss of property and damage in the downstream. This study focuses on the "10.11" Baige landslide-dammed lake. A three-dimensional numerical model of "10.11" Baige landslide dam is established with the terrain data acquired by UAV, and the actual process of dam breaching is simulated and reproduced using the Flow-3D software. The Flow-3D software could be powerful, which has the ability to simulate the phenomenon of turbulence flow and sediment movement under complex terrain conditions. The parameters required for this study, such as hydrological data and particle composition of dam body, are directly referred to the field data. A flow monitoring section is set up near the downstream of the dam, and the simulated peak discharge process is in good agreement with the measured values. Furthermore, the velocity distribution and sediment scour in the natural spillway are analyzed. The results show that: overtopping scour of dam can be divided into four phases, i.e. before the outburst, rapid expansion phase, peak discharge phase and phase with breach develops steadily. In second phase, the sediment erosion rate is large and the terrain changes fast, so the breach expands rapidly. After the flood peak, the upstream water level and discharge decrease gradually, and the development speed of breach becomes slow, finally the terrain tends to be stable. The velocity of flow on the ramp of spillway is large during the releasing period, which leads to the maximum scour depth. On the contrary, the flow velocity in the front and middle of the spillway is small and the scour depth is relatively small. The terrain becomes flat at the downstream of the dam body, so the sediment sinks under the influence of weight and friction with the riverbed, then siltation occurs. There was obvious headward erosion in the process of first dam breaking: with the undercutting of spillway, the bed drop-off moved to the upstream continuously. Headward erosion is an important scour mode of dam breaching and the dominant force of channel expansion, which can produce strong scour effect in a short period of time. This study is significant for understanding the process and mechanism of the "10.11" Baige landslide dam breaching, which could provide technical reference to the management and accommodation of emergency.

© Springer Nature Switzerland AG 2020
J.-M. Zhang et al. (Eds.): ICED 2020, SSGG, pp. 376–377, 2020.
https://doi.org/10.1007/978-3-030-46351-9_40

Keywords: Baige landslide dammed lake · Erosion and breaching · Peak discharge · Breach evolution · Numerical simulation

Simulation of Dam Breaching and Flood Routing on the Jinsha River

Zhengdan Xu[✉], Jian He, and Te Xiao

The Hong Kong University of Science and Technology, Hong Kong, China
zxubv@connect.ust.hk

Abstract. Two landslide dams were formed on the upper reach of the Jinsha River on 10 October 2018 and 3 November 2018, respectively. Considering the great threats to the lives and properties in downstream areas, it is important to rapidly simulate the dam breaching and the flood routing for risk management. This paper focuses on the second landslide dam that has a higher dam height and leads to higher risk. A physically-based model is adopted to simulate the dam breaching process due to overtopping erosion. Subsequently, flood routing analysis along a 640 km-long reach is conducted using a one-dimensional hydrodynamic model. The simulation results of both breaching and flooding analyses are in good agreement with field observations.

Keywords: Landslide dam · Overtopping erosion · Dam breaching · Flood routing

1 Introduction

On 10 October 2018, the upper reach of the Jinsha River was blocked by approximately 25×10^6 m^3 materials from a large landslide at Baige. The 61 m high landslide dam which formed a barrier lake with a capacity of 250×10^6 m^3 was overtopped and breached in 21 h with a peak outflow discharge of 10,000 m^3/s. Three weeks afterwards, another landslide happened at the same location, forming a larger landslide dam. The new dam was as high as 96 m, and can impound at most 750×10^6 m^3 water behind. To reduce the outflow rate, a 15 m-depth channel was excavated on the dam crest, which reduced the barrier lake capacity by one-third. However, even with this risk mitigation measure, the flood triggered by the dam breach could be disastrous to downstream areas. To properly control the risk, it is of great importance to predict the outburst flood and peak discharges along the river. This paper aims to simulate the dam breaching and flood routing processes of the second landslide dam at Baige using efficient analysis models.

2 Dam Breaching Analysis

2.1 Model Description

The physically-based numerical model DABA developed by Chang and Zhang (2010) is adopted for dam breaching simulation. The model considers the dam breaching due

© Springer Nature Switzerland AG 2020
J.-M. Zhang et al. (Eds.): ICED 2020, SSGG, pp. 378–383, 2020.
https://doi.org/10.1007/978-3-030-46351-9_41

to overtopping erosion as a three-stage-process. The evolution of dam geometry during each stage is illustrated by Fig. 1, denoted by C-I to C-III and L-I to L-III for cross section and longitudinal section, respectively.

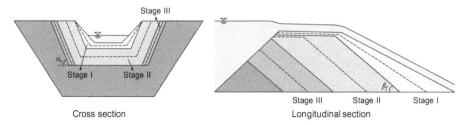

Fig. 1. Three-stage-process of breaching due to overtopping assumed in DABA.

In the model, the erosion rate of soil is evaluated using a shear stress equation (Hanson and Simon 2001) as:

$$E = K_d(\tau - \tau_c) \tag{1}$$

where E is the erosion rate; τ is the shear stress at the soil-water interface; K_d is the coefficient of erodibility; and τ_c is the critical shear stress that initiates soil erosion. The parameters K_d and τ_c can be determined through empirical equations related to some geotechnical parameters that can be easily obtained in the field (Annandale 2006; Chang et al. 2011), such as void ratio, coefficient of uniformity, and median particle size.

Based on the breach geometry evolution, the outflow discharge Q_b is determined using the equation for broad-crested weir flow (Singh and Scarlatos 1988):

$$Q_b = 1.7[B_b + (H - Z)\tan\alpha](H - Z)^{3/2} \tag{2}$$

where B_b is the breach bottom width; α is the angle of side slope; H and Z are elevations of the water surface and the breach bottom, respectively.

2.2 Breaching Parameters

The source of the second landslide dam involves a significant amount of residual landslide materials, which are finer and looser. Therefore, the erodibility of the soil in second dam is considered as "medium-high". Together with information from several samples taken from Baige site (Zhang et al. 2019), the K_d and τ_c profiles along depth are estimated by making adjustment on the basis of erodibility distribution of Tangjiashan landslide dam (Peng et al. 2014), as shown in Fig. 2.

The dam height, originally, was 96 m. After excavating a 15 m-depth diversion channel, its height was reduced to 81 m with a decrease in lake volume from 757×10^6 m^3 to 494×10^6 m^3.

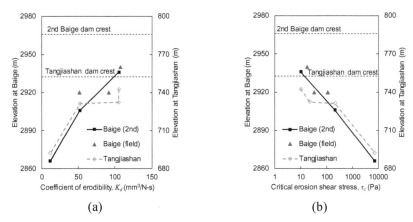

(a)   (b)

Fig. 2. Soil erodibility parameters profile: (a) coefficient of erodibility; (b) critical erosion shear stress.

2.3 Breaching Simulation Results

The entire dam breaching process can be simulated through a designated iterative numerical scheme. The predicted outflow hydrograph during the breaching of the second landslide dam is presented in Fig. 3. It can be seen that both the breaching time and hydrograph patterns from simulation are close to observed values.

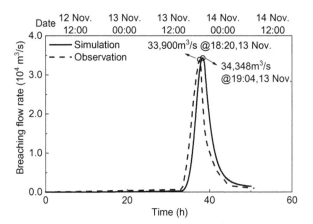

Fig. 3. The simulation results of dam breaching.

3 Flood Routing Analysis

3.1 Model Description

A one-dimensional hydrodynamic model is used for flood routing analysis. The governing equations are:

$$\frac{\partial A}{\partial t} + \frac{\partial Q}{\partial x} = 0 \tag{3}$$

$$\frac{\partial Q}{\partial t} + \frac{\partial \left(Q^2/A\right)}{\partial x} + gA\left(\frac{\partial h}{\partial x} - S_0 + S_f\right) = 0 \tag{4}$$

where t is the elapsed time; x is the distance along the channel bed; A is the cross-sectional flow area; Q is the discharge; h is the flow depth; S_0 is the bed slope; and S_f is the friction slope. S_f is directly related to the Manning's coefficient n, which represents the roughness of the river channel and is estimated to be from 0.04 to 0.06 for the Jinsha River based on Arcement and Schneider (1989). In this study, n is assumed as a constant value along the river.

Flood routing analysis is conducted along the Jinsha River from the landslide dam to Liyuan Reservoir with a total length of 640 km. The outflow hydrograph at the landslide dam predicted by DABA is used for flood routing analysis.

3.2 Results of Flood Routing Analysis

Figure 4 shows the attenuation trend of the peak discharge with distance from the dam. The upper and lower bounds of peak discharges are calculated based on $n = 0.04$ and 0.06, respectively. The predicted peak discharges at downstream hydrological stations are consistent with the observed values, indicating the choice of n is reasonable.

A back analysis is also conducted and results based on $n = 0.055$ are found to be in good agreement with the observations. Figure 5 shows the calculated hydrographs based on $n = 0.055$ and observed hydrographs. Overall, both the peak discharge value and peak time are well consistent with observation records at corresponding hydrologic stations.

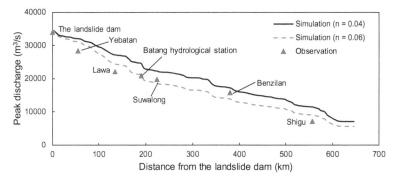

Fig. 4. Simulated peak discharge along the river.

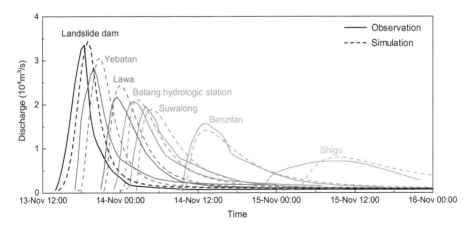

Fig. 5. Comparison of hydrographs between simulation and observation.

4 Summary

The simulation of landslide dam breaching and flooding routing on the Jinsha River in November 2018 are performed by a physically-based model DABA and a one-dimensional hydrodynamic model, respectively. The breaching analysis gives a peak outflow discharge at 34,348 m^3/s, which is fairly close to the observation 33,900 m^3/s. In the flooding simulation, the roughness of river channel is assumed to be constant along the river and $n = 0.055$ is indicated according to back analysis for the concerned reach of the Jinsha River. When compared with recorded data, the computed flooding hydrographs show good consistency with respect to both peak discharge value and time to peak at each station. It is noteworthy that both the dam breaching simulation and the flood routing analysis can be implemented within several minutes. Because of the high efficiency and accuracy of the models adopted in this study, the predicted results can serve as a basis for timely landslide dam risk control.

References

Annandale, G.W.: Scour Technology - Mechanics and Engineering Practice. McGraw-Hill, New York (2006)

Arcement, G.J., Schneider, V.R.: Guide for selecting Manning's roughness coefficients for natural channels and flood plains. U.S. Geological Survey Water-Supply Paper, vol. 2339, U.S. Geological Survey, Washington (1989)

Chang, D.S., Zhang, L.M.: Simulation of the erosion process of landslide dams due to overtopping considering variations in soil erodibility along depth. Nat. Hazards Earth Syst. Sci. **10**(4), 933–946 (2010)

Chang, D.S., Zhang, L.M., Xu, Y., Huang, R.Q.: Field testing of erodibility of two landslide dams triggered by the 12 May Wenchuan earthquake. Landslides **8**(3), 321–332 (2011)

Hanson, G.J., Simon, A.: Erodibility of cohesive streambeds in the loess area of the midwestern USA. Hydrol. Process. **15**(1), 23–38 (2001)

Singh, V.P., Scarlatos, P.D.: Analysis of gradual earth-dam failure. J. Hydraul. Eng. **114**(1), 21–42 (1988)

Peng, M., Zhang, L.M., Chang, D.S., Shi, Z.M.: Engineering risk mitigation measures for the landslide dams induced by the 2008 Wenchuan earthquake. Eng. Geol. **180**, 68–84 (2014)

Zhang, L.M., Xiao, T., He, J., Chen, C.: Erosion-based analysis of breaching of Baige landslide dams on the Jinsha River, China, in 2018. Landslides **16**, 1965–1979 (2019)

Prediction of a Multi-hazard Chain by an Integrated Numerical Simulation Approach: The Baige Landslide Along the Jinsha River, China

Fan Yang, Zetao Feng, Maomao Liu, and Xuanmei Fan[✉]

State Key Laboratory of Geohazard Prevention and Geoenvironment Protection,
Chengdu University of Technology, Chengdu 610059, China
fxm_cdut@qq.com

Abstract. Successive landslides during October and November 2018 in Baige village, eastern Tibet, China dammed the Jinsha River twice and the dam-breach had flooded many towns downstream, instigating a catastrophic disaster chain. In order to evaluate and understand the disaster chain effect that may be caused by the potentially unstable rock mass, we systematically studied the multi-hazard scenarios through an integrated numerical modelling approach. The model starts from the landslide failure probability to runout and river damming, and then to dam breach and dam-breach induced flood, hence predicting and visualizing an entire disaster chain. The parameters required for the modelling were calibrated using measured data from the two Baige landslides. Then, we predict the future cascading hazards based on seven scenarios according to all possible combinations of the potentially unstable rock mass failures. For each scenario, the landslide runouts, dam-breach process and flooding are numerically simulated with consideration of uncertainties of model input parameters. The maximum dam-breach flood extent, depth, velocity and peak arrival time at different places downstream are predicted. As a first attempt to simulate the whole process of a landslide induced multi-hazard chain, this study provides some insights and substantiates the necessity of landslide induced disaster chain modelling. The integrated approach proposed by this study can be applied for simulating similar types of landslide induced chains of hazards in other regions.

Keywords: Disaster chain · Landslide runout · Landslide dam breach · Outburst flood

1 Introduction

Successive landslides on 11 October and 3 November 2018 dammed the Jinsha River twice and caused catastrophic floods (Fan et al. 2019b). Although the first landslide dam breached naturally, the second one formed a barrier lake with a volume of ~524 million m^3, seriously threatening the lives of people living upstream and downstream. Despite immediate and prolonged disaster mitigation measures, natural and manual breaching of

© Springer Nature Switzerland AG 2020
J.-M. Zhang et al. (Eds.): ICED 2020, SSGG, pp. 384–392, 2020.
https://doi.org/10.1007/978-3-030-46351-9_42

the dam caused floods in the lower reaches of the Jinsha River, inundating >3400 houses and destroying a 3.5×10^3 km^2 area of croplands (Fig. 1). More than 102,000 people were evacuated (Liang et al. 2019; Ouyang et al. 2019; Zhang et al. 2019a, 2019b). According to field and remote-sensing investigations, there still exists three large and potentially unstable rock masses at the trailing edge of the landslide and some major discontinuities (i.e. cracks and joints) have clearly propagated and enlarged, threatening to trigger a failure that is likely to block the Jinsha River again, causing future mayhem (Fan et al. 2019b).

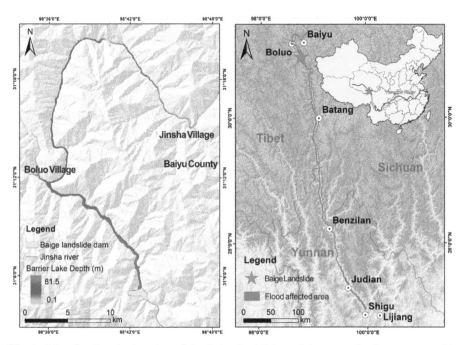

Fig. 1. Map showing the location of the Baige landslides. a lake water-depth upstream and b towns affected by flooding downstream (light blue line in the inset marks the Jinsha river)

Previous studies on landslides, landslide dams, and dam breach floods have mostly focused on a single type of geomorphic hazard based on the perspective that different types of disasters are quasi-independent and the interactions between them can hence be ignored (Yutao and Shengxie 2009). However, the successive landslides, sequential damming, and aftermath flooding in Jinsha River provide a valuable opportunity to study the causes and effects of interconnected multi-hazard events. To this end, our study presents analyses of landsliding, damming, natural dam-breaching, and flooding, as well as the interactions between each of these processes by constructing realistic scenarios based on our observations at Baige.

2 Data and Methods

2.1 Numerical Modelling Approach

In this study, we integrate different numerical models to simulate and predict the entire disaster chain of the future possible failures of the Baige landslides. For the landslide runout simulation, the MassFlow program developed by Ouyang et al. (2014). has been adopted. The dam-breach is simulated by the DABA program (Chang and Zhang 2010) and the dam-breach induced flood is simulated by the HEC-RAS program (Brunner 1995, Brunner 2002). A four-step methodology has been adopted as shown in Fig. 2. Data preparation and model calibration by previous failures is done in Step 1. Anticipated disaster scenarios are designed, and landslide initiation and runout modelling are done in Step 2. Dam-breach and consequent flooding are simulated respectively in Step 3 and Step 4. Initially, the geotechnical parameters needed for the landslide runout model were calibrated by back calculating the two events of Baige landslides using the MassFlow program. The parameters of the dam-breach model and the flood model are calibrated according to the measured peak discharge, peak arrival time and final flood area recorded by the hydrological station located downstream.

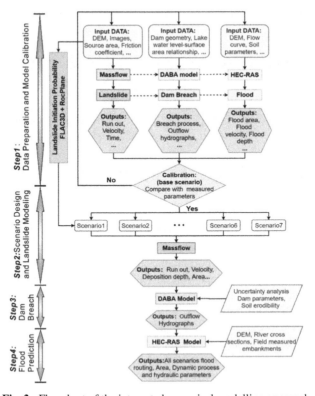

Fig. 2. Flowchart of the integrated numerical modelling approach.

2.2 Dynamic Modelling of Landslides by MassFlow

Through comparing the DEM obtained before and after the landslides, the difference in topographic elevation, the initial sliding position, thickness of the sliding body and the accumulation area of both landslides were determined respectively. The back calculations were done separately for both the first and second landslide events by Massflow. Using the topographic data after the second landslide, the landslide runout and accumulation characteristics of the unstable rock mass are studied based on Scenario 1 to 7(Fig. 3, Fig. 4). For these future scenarios simulation, the runout was modelled using the same mechanical parameters calibrated by the second Baige landslide.

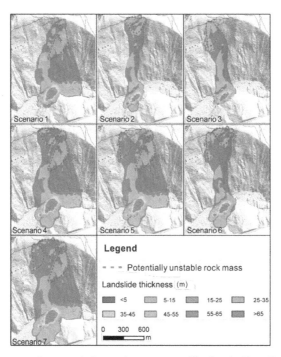

Fig. 3. Runout and accumulation at time step, t = 80 s for the 7 studied scenarios

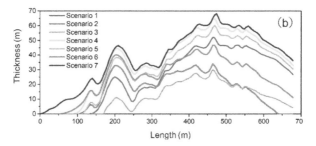

Fig. 4. Simulated landslide accumulation thickness for scenarios 1 to 7

2.3   Landslide Dam Breach Process Modelling by the DABA Model

The DABA model is used to calculate the hydrograph of dam-breach under different anticipated scenarios. The landslide dam dimensions were determined by the MassFlow simulations (Table 1).

Table 1. Landslide dam geometries obtained from simulations using MassFlow

Scenarios	Dam height (m)	Dam width (m)	Downstream slope length (m)	Downstream slope angle (°)	Upstream slope angle (°)
Scenario 1	60.38	269	207	8.25	6.34
Scenario 2	50.77	200	164	5.92	8.85
Scenario 3	46.38	200	156	6.58	6.38
Scenario 4	70.42	300	210	9.73	7.52
Scenario 5	66.38	300	168	10.12	6.36
Scenario 6	59.91	276	120	13.59	12.26
Scenario 7	73.82	355	300	7.03	7.85

The second flood discharge from the artificially breached landslide dam in November 2018 is recorded (Ouyang et al. 2019), and we use that to calibrate the DABA model. After calibrating the model parameters, the same properties were used to predict the hydrograph of the dam breach under scenarios 1 to 7 (Fig. 5).

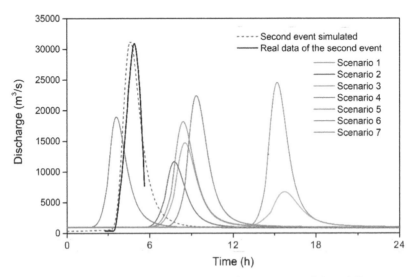

Fig. 5. Flow discharge curve during the dam breach for the seven anticipated disaster scenarios and second landslide dam event

Table 2. Breaching parameters for scenarios 1 to 7 and the second landslide event using DABA

Simulation scenarios	Breach channel depth (m)	Breach channel top width (m)	Breach channel bottom width (m)	Breach time (h)	Peak outflow rate (m³/s)
2nd Event	52.37	254.95	167.06	66.38	31227.36
Scenario 1	36.52	411.93	350.65	193.88	14731.43
Scenario 2	30.31	246.38	195.51	125.32	11661.06
Scenario 3	33.25	384.47	328.67	180.23	6725.30
Scenario 4	43.32	431.47	358.77	307.73	22424.10
Scenario 5	40.38	460.58	392.81	196.38	18175.89
Scenario 6	37.00	244.31	182.21	106.92	18942.27
Scenario 7	44.63	428.49	353.59	176.67	24549.02

2.4 Dam-Breach Flood Modelling

The peak discharge values obtained from the DABA model were input to the HEC-RAS model to simulate the impacts of flooding in the downstream areas. This has been done first for the second landslide dam breach flooding and then for scenarios 1 to 7. (Table 2) The parameters of the HEC-RAS model are calibrated with the hydrological monitoring data. The flood simulation results correspond well with the measured data to 300 km downstream.

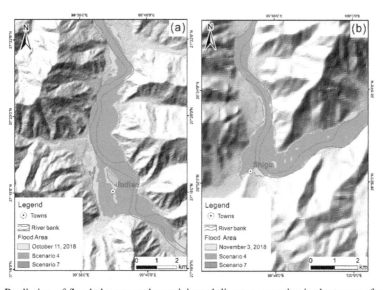

Fig. 6. Prediction of flooded areas under anticipated disaster scenarios in the towns of a Judian and b Shigu

The simulation results show that scenario 7 has the largest flood area. The areal extents of the flooded regions are shown for the second landslide dam breach event, scenario 4 and scenario 7 in Fig. 6. The peak flow of scenario 4 is smaller compared with other scenarios, and the flood area is the smallest. The flood area of the second event is between scenarios 4 and 7. The flood simulation results of all scenarios are summarized in Table 3. The peak arrival time (Table 3) can be used to predict flood overtopping time and flood arrival time in downstream towns under specified inflow rates.

Table 3. Peak discharge and peak arrival time of scenarios 1 to 7 at two sites downstream along the Jinsha River

Simulation scenarios	Yebatan hydropower station (54 km)		Shigu town (557 km)	
	Peak discharge (m^3/s)	Peak arrival time (h)	Peak discharge (m^3/s)	Peak arrival time (h)
Scenario 1	16828	2018-10-13 15:00 $\pm$ 0:30	5077	2018-10-15 9:30 $\pm$ 0:30
Scenario 2	12189	2018-10-13 13:30 $\pm$ 0:30	3943	2018-10-15 10:30 $\pm$ 0:30
Scenario 3	9638	2018-10-13 10:00 $\pm$ 0:30	3351	2018-10-15 9:00 $\pm$ 0:30
Scenario 4	24905	2018-10-13 16:30 $\pm$ 0 : 30	6864	2018-10-15 8:00 $\pm$ 0:30
Scenario 5	20873	2018-10-13 15:00 $\pm$ 0:30	5978	2018-10-15 7:30 $\pm$ 0:30
Scenario 6	18983	2018-10-13 5:30 $\pm$ 0:30	5042	2018-10-15 0:30 $\pm$ 0:30
Scenario 7	26924	2018-10-14 4:00 $\pm$ 0:30	7767	2018-10-15 18:00 $\pm$ 0:30

3   Conclusions

The Baige landslides provide an excellent example of a landslide-induced disaster chain that propagated to far-field areas hundreds of km downstream of the trigger point. Based upon the numerical analyses performed in this study, we provide insights to the processes and interactions between landslide runout, dam-breaching, and flood propagation under a set of hypothetical scenarios representative of steep, landslide-dominated terrain. The integration of different numerical models (i.e. MassFlow, DABA, and HEC-RAS) simulated successfully the events triggered by the two landslides at Baige. The simulation results are in good agreement with the observations. Using the well-calibrated integrated model, we attempted to predict the flood disaster caused by future possible landslides under different anticipated scenarios. The prediction of dam-breaching and the consequent flooding downstream can provide invaluable guidance to assist with the evacuation of local people during emergency scenarios. An improved fundamental understanding of disaster chains coupled with refinement of such integrated numerical simulations promises to alleviate future catastrophic losses of human and non-human life.

References

Brunner, G.W.: HEC-RAS (river analysis system), pp. 3782–3787. ASCE (2002)

Brunner, G.W.: HEC-RAS River Analysis System. Hydraulic Reference Manual. Version 1.0. Hydrologic Engineering Center, Davis, CA (1995)

Chai, H.-J., Liu, H.-C., Zhang, Z.-Y.: The temporal-spatial distribution of damming landslides in China. J. Mt. Res. S 1 (2000)

Chang, D.S., Zhang, L.M.: Simulation of the erosion process of landslide dams due to overtopping considering variations in soil erodibility along depth. Nat. Hazards Earth Syst. Sci. **10**, 933–946 (2010)

Fan, X., Xu, Q., Alonso-Rodriguez, A., Siva Subramanian, S., Li, W., Zheng, G., Dong, X., Huang, R.: Successive landsliding and damming of the Jinsha River in eastern Tibet, China: prime investigation, early warning, and emergency response. Landslides **16**, 1003–1020 (2019b)

Iverson, R.M., Ouyang, C.: Entrainment of bed material by Earth-surface mass flows: review and reformulation of depth-integrated theory. Rev. Geophys. **53**, 27–58 (2015)

Liang, G., Wang, Z., Zhang, G., Wu, L.: Two huge landslides that took place in quick succession within a month at the same location of Jinsha River. Landslides **16**, 1059–1062 (2019)

Ouyang, C., An, H., Zhou, S., Wang, Z., Su, P., Wang, D., Cheng, D., She, J.: Insights from the failure and dynamic characteristics of two sequential landslides at Baige village along the Jinsha River, China. Landslides **16**(7), 1397–1414 (2019)

Ouyang, C., He, S., Tang, C.: Numerical analysis of dynamics of debris flow over erodible beds in Wenchuan earthquake-induced area. Eng. Geol. **194**, 62–72 (2015)

Ouyang, C., He, S., Xu, Q., Luo, Y., Zhang, W.: A MacCormack-TVD finite difference method to simulate the mass flow in mountainous terrain with variable computational domain. Comput. Geosci. **52**, 1–10 (2013)

Ouyang, C., He, S., Xu, Q.: MacCormack-TVD finite difference solution for dam break hydraulics over erodible sediment beds. J. Hydraul. Eng. **141**, 06014026 (2014)

Ouyang, C., Zhou, K., Xu, Q., Yin, J., Peng, D., Wang, D., Li, W.: Dynamic analysis and numerical modeling of the 2015 catastrophic landslide of the construction waste landfill at Guangming, Shenzhen, China. Landslides **14**, 705–718 (2017)

Peng, M., Zhang, L.M.: Analysis of human risks due to dam break floods—part 2: application to Tangjiashan landslide dam failure. Nat. Hazards **64**, 1899–1923 (2012b)

Peng, M., Zhang, L.M.: Breaching parameters of landslide dams. Landslides **9**, 13–31 (2012a)

Peng, M., Zhang, L.M., Chang, D.S., Shi, Z.M.: Engineering risk mitigation measures for the landslide dams induced by the 2008 Wenchuan earthquake. Eng. Geol. **180**, 68–84 (2014)

Peng, S.-H.: 1D and 2D numerical modeling for solving dam-break flow problems using finite volume method. J. Appl. Math. **2012**, 14 (2012)

Scaringi, G., Fan, X., Xu, Q., Liu, C., Ouyang, C., Domènech, G., Yang, F., Dai, L.: Some considerations on the use of numerical methods to simulate past landslides and possible new failures: the case of the recent Xinmo landslide (Sichuan, China). Landslides **15**, 1359–1375 (2018)

Shi, Z.M., Guan, S.G., Peng, M., Zhang, L.M., Zhu, Y., Cai, Q.P.: Cascading breaching of the Tangjiashan landslide dam and two smaller downstream landslide dams. Eng. Geol. **193**, 445–458 (2015)

Shi, Z.-M., Zheng, H.-C., Yu, S.-B., Peng, M., Jiang, T.: Application of cfd-dem to investigate seepage characteristics of landslide dam materials. Comput. Geotech. **101**, 23–33 (2018)

Yutao, F., Shengxie, X.: Chain mechanism and optimized control of collapses, landslides and debris flows. J. Catastrophol. 3 (2009)

Zhang, L., Xiao, T., He, J., Chen, C.: Erosion-based analysis of breaching of Baige landslide dams on the Jinsha River, China, in 2018. Landslides **16**, 1965–1979 (2019a)

Zhang, Z., He, S., Liu, W., Liang, H., Yan, S., Deng, Y., Bai, X., Chen, Z.: Source characteristics and dynamics of the October 2018 Baige landslide revealed by broadband seismograms. Landslides **16**, 777–785 (2019b)

Author Index

© Springer Nature Switzerland AG 2020
J.-M. Zhang et al. (Eds.): ICED 2020, SSGG, pp. 393–394, 2020.
https://doi.org/10.1007/978-3-030-46351-9

CPSIA information can be obtained
at www.ICGtesting.com
Printed in the USA
LVHW082022260420
654475LV00002B/95